# WEAK CONVERGENCE OF MEASURES

This is a volume in
PROBABILITY AND MATHEMATICAL STATISTICS

A Series of Monographs and Textbooks

Editors: Z. W. Birnbaum and E. Lukacs

A complete list of titles in this series appears at the end of this volume.

# WEAK CONVERGENCE OF MEASURES

HARALD BERGSTRÖM

Department of Mathematics
Chalmers University of Technology
and University of Göteborg
Göteborg, Sweden

1982

ACADEMIC PRESS

A Subsidiary of Harcourt Brace Jovanovich, Publishers

New York London
Paris San Diego San Francisco São Paulo Sydney Tokyo Toronto

ACADEMIC PRESS, INC.
111 Fifth Avenue, New York, New York 10003

*United Kingdom Edition published by*
ACADEMIC PRESS, INC. (LONDON) LTD.
24/28 Oval Road, London NW1 7DX

**Library of Congress Cataloging in Publication Data**

Bergstrom, Harald.
Weak convergence of measures.

(Probability and mathematical statistics)
Bibliography: p.
Includes index.
1. Probabilities. 2. Measure theory.
3. Convergence. I. Title. II. Series.
QA273.43.B47 519.2 82-6789
ISBN 0-12-091080-2 AACR2



PRINTED IN THE UNITED STATES OF AMERICA

82 83 84 85 9 8 7 6 5 4 3 2 1

# CONTENTS

**Chapter V**

## WEAK CONVERGENCE ON THE *C*- AND *D*-SPACES



**Chapter VI**

## WEAK CONVERGENCE IN SEPARABLE HILBERT SPACES



**Appendix**

# PREFACE

Weak convergence is a fundamental concept in probability theory. There we deal with random variables, i.e., measurable mappings from a probability space $(\Omega,\mathscr{B},P)$ into a topological space $S$. A random variable $X$ from $(\Omega,\mathscr{B},P)$ into $S$ determines the measure $P(X^{-1}\cdot)$ on $S$ called the distribution of $X$. A sequence $\{X_n\}$ of random variables from $(\Omega,\mathscr{B},P)$ into $S$ then determines a sequence $\{P(X_n^{-1}\cdot)\}$ of measures on $S$. We are interested in the limit of such sequences. Then, however, we do not require that $P(X_n^{-1}E)$ converges for all measurable sets $E$ in $S$ but only for sets that matter when we deal with integrals over real-valued, bounded, continuous functions. Integrals are functionals. Weak convergence of measures means convergence of functionals and thus reduces to the convergence of sequences of real numbers. Clearly, we may consider any sequences of measures on $S$ and weak convergence of such sequences without any connection with random variables.

The short presentation of weak convergence given above already tells us that it concerns elementary and fundamental concepts in mathematics, such as sets, topology, mappings, measurability, continuity, integrals, functionals, measures, etc. These concepts are closely related to each other and are to a great extent unified in the concept of weak convergence. Indeed, a topology determines a class of continuous functions. Besides the usual topology, we deal with a sigma-topology, which is determined by a certain class of functions, which then are continuous in this $\sigma$-topology. Measures determine integrals, which are functionals, and functionals determine measures. I have arranged the material, aiming at a unification as indicated above.

In the first chapter elementary fundamental notions are treated, such as sets, different classes of sets, different topological and $\sigma$-topological spaces, and different classes of functions and measures. They are presented as far as our need for them, though to a great extent they may be apparent. I have done so because some of these concepts are used here in a general sense, which is not common in ordinary textbooks. For instance, our measures need not be $\sigma$-additive. If they are, they are called $\sigma$-smooth. It may be appropriate for most readers only to survey the content in this chapter at a first reading and then return to the different parts when needed in the following.

The main purpose of Chapter II is to give the connection between functionals and measures. For this reason, a rather detailed introduction of the abstract integral as a bounded, linear functional is presented.

The main part of the book is Chapter III, which deals with weak convergence of sequences of measures, i.e., convergence of sequences of bounded, linear functionals. Such convergence is considered on general normal spaces, where the conditions for convergence may be difficult to verify. It is shown how they simplify in more special spaces, particularly metric spaces. Two essential methods are presented. One method makes use of seminorms of measures and is suitable for dealing with convolution of measures. A second method, called a reduction method, reduces the weak limit problem in a space to a less difficult limit problem in another space. Of particular interest in this chapter is a deep theorem of A. D. Alexandrov concerning the $\sigma$-smoothness of the weak limit of a sequence of $\sigma$-smooth measures (Section 4).

Chapter IV deals with weak convergence of sequences of $\sigma$-smooth measures on finite-dimensional vector spaces. The seminorm method is here applied. This chapter contains the main part of my earlier book [4] but here as application of the general theory, presented in Chapter III.

In Chapter V weak convergence in the $C$- and $D$-spaces is considered and by the help of the reduction reduced to limit problems, which have been solved in Chapter IV. In the same way limit problems in Hilbert spaces, particularly the $l^2$-spaces are reduced in Chapter IV. Infinitely divisible $\sigma$-smooth measures are described and expressed in suitable representations in the finite-dimensional vector spaces (Chapter IV) and in the $l^2$-space (Chapter VI).

In the notes at the end of the book, I give some information about papers and books that have had an influence on my presentation of weak convergence. However, I would like to point out here that the main source for Chapter III and connected parts in the preceding chapters has been A. D. Alexandrov's important framework [2].

# WEAK CONVERGENCE OF MEASURES

CHAPTER I

# SPACES, MAPPINGS, AND MEASURES

## 1. CLASSES OF SETS

In the following, $R$ is the set of all real numbers and $N$ the sequence of positive integers. Sequences of elements (numbers, sets, etc.) are denoted by $\{x_i\}_{i} \in N$ or $\{x_i\}_{i=1}^{\infty}$ if they are infinite or simply by $\{x_i\}$ when there can be no misunderstanding, and by $\{x_i\}_{i=1}^{m}$ if they are finite. In a set $S$ of elements $x$, $y$, etc., we consider subsets and then call $S$ a space. Subsets are denoted by $E$, $E_i$, $E^{(i)}$, and sometimes by other capital letters. We also let the empty set be a subset of $S$ and denote it by $\varnothing$. For these subsets we deal with the common set operations: union $E_1 \cup E_2$; intersection $E_1 \cap E_2$, also written $E_1 E_2$; countable union $\bigcup_{n \in N} E_n$; countable intersection $\bigcap_{n \in N} E_n$; arbitrary union $\bigcup_{t \in \Upsilon} E_t$; and arbitrary intersection $\bigcap_{t \in \Upsilon} E_t$, where $\Upsilon$ is a countable or noncountable set. A difference is denoted by $E_1 \backslash E_2$, and the particular difference $S \backslash E$, called the complement of $E$, also written $\complement E$ or $E^c$. By de Morgan's rules we have

(i) $$\complement \bigcup_{t \in \Upsilon} E_t = \bigcap_{t \in \Upsilon} \complement E_t,$$

(ii) $$\complement \bigcap_{t \in \Upsilon} E_t = \bigcup_{t \in \Upsilon} \complement E_t.$$

We verify these relations in a way that is common for proving set relations. If $x$ belongs to the set on the left-hand side of (i), then it does not belong to any $E_t$, $t \in \Upsilon$. Hence it belongs to the set $\complement E_t$ for all $t \in \Upsilon$ and thus to the set on the right-hand side. The converse follows in the same way, and in the same way (ii) can be proved.

We shall use the signs $\Rightarrow$ and $\Leftrightarrow$ for implications. Hence, if $\gamma_1$ is one statement and $\gamma_2$ another statement, we shall read "$\gamma_1$ implies $\gamma_2$" if we write $\gamma_1 \Rightarrow \gamma_2$, and read "$\gamma_1$ implies $\gamma_2$ and $\gamma_2$ implies $\gamma_1$" if we write $\gamma_1 \Leftrightarrow \gamma_2$.

For a set we put $x \in E$ if $x$ belongs to $E$ and $x \notin E$ if it does not belong to $E$. For any two sets $E_1$ and $E_2$, we put $E_1 \subset E_2$ ($E_2 \supset E_1$) if $E_1$ is a subset of $E_2$. We also call $E$ a subset of $E$.

Different classes of subsets of $S$ are characterized by different set operations. We say that a class of sets is closed under a certain set operation if the performance of the set operation on members in the class gives a member in the class as result.

A class $\mathscr{G}$ of subsets of $S$ is called a $\sigma$-topological class of open sets if it is closed under countable unions and finite intersections and if $S \in \mathscr{G}$, $\emptyset \in \mathscr{G}$. The members of $\mathscr{G}$ are thus called open sets. They will be denoted by $G$, $G^{(i)}$, $G_i$, etc.

A class $\mathscr{F}$ of subsets of $S$ is called a $\sigma$-topological class of closed sets if it is closed under countable intersections and finite unions and if $S \in \mathscr{F}$, $\emptyset \in \mathscr{F}$. Thus the members of the class, which will be denoted by $F$, $F_i$, $F^{(i)}$, etc., are called closed sets; note that $F = \complement G$ forms a $\sigma$-topological class of closed sets for $G \in \mathscr{G}$, where $\mathscr{G}$ is a $\sigma$-topological class of open sets. In the same way, $G = \complement F$ is open if $F$ is closed. This follows from de Morgan's rules.

If a $\sigma$-topological class of open sets is closed under arbitrary unions, it is called a topological class of open sets, and if a $\sigma$-topological class of closed sets is closed under arbitrary intersections, it is called a topological class of closed sets.

A class $S$ of subsets of a set $S$ is called a Boolean algebra, or just algebra, of sets if $S \in \mathscr{S}$, $\emptyset \in \mathscr{S}$, and if it is closed under finite unions and the compliment $E^c$ of any $E \in \mathscr{S}$ also belongs to $S$. Note then that $S$ is also closed for finite intersections. If the algebra is closed for countable unions, hence also for countable intersections, it is called a $\sigma$-algebra. We shall often denote a $\sigma$-algebra by $\mathscr{S}_\sigma$. Clearly a $\sigma$-algebra is also an algebra. Note that a $\sigma$-algebra $\mathscr{S}_\sigma$ contains the sets

$$\limsup_{n\to\infty} E_n = \bigcap_{n=1}^{\infty} \bigcup_{k\geq n} E_k, \qquad \liminf_{n\to\infty} E_n = \bigcup_{n=1}^{\infty} \bigcap_{k\geq n} E_k$$

for any sequence

$$\{E_n\}_{n=1}^{\infty}, \qquad E_n \in \mathscr{S}_\sigma.$$

The notations lim sup and lim inf are used because the first set contains any element that belongs to infinitely many $E_n$ and the second set contains any element that belongs to all $E_n$ except at most finitely many.

We may consider a $\sigma$-algebra as a $\sigma$-topological class of open sets and thus of closed sets as well. Then any $E \in \mathscr{S}_\sigma$ is both open and closed.

Let $\mathscr{M}$ be a given class of subsets of a set $S$, and C a given collection of conditions such that the class of all subsets of $S$ satisfy these conditions.

Then there exists a smallest class $\mathcal{M}_C$ of subsets of $S$ such that $\mathcal{M} \subset \mathcal{M}_C$ and $\mathcal{M}_C$ satisfies the conditions C. Indeed, the intersection $\mathcal{M}_C$ of all classes that contain $\mathcal{M}$ and satisfy C contains $\mathcal{M}$ and satisfies C. Clearly $\mathcal{M}_C$ is the smallest class having the required properties. We say that $\mathcal{M}_C$ is generated by $\mathcal{M}$ under C.

Hence, if $\mathcal{M}$ is any class of subsets of $S$, it generates a $\sigma$-topological class of open sets (closed sets), a topological class of open sets (closed sets), an algebra, and a $\sigma$-algebra. If $\mathcal{M}$ is the class of open intervals $(a,b)$ with rational end points $a$ and $b$, it thus generates the just-mentioned classes of subsets on the real line $R$. We call these classes Borel classes: the Borel $\sigma$-topological class, the Borel topological class, etc.

A class $\mathcal{D}$ of subsets of a set $S$ is called a Dynkin class if it has the following properties: (i) $S \in \mathcal{D}$, (ii) $E_1 \in \mathcal{D}$, $E_2 \in \mathcal{D}$, $E_1 \subset E_2 \Rightarrow E_2 \backslash E_1 \in \mathcal{D}$, (iii) for any sequence $\{E_i\}$, $E_i \cap E_j = \varnothing$ for $i \neq j$, $E_i \in \mathcal{D}$, and $\bigcup_{j=1}^{\infty} E_j \in \mathcal{D}$.

Example: A $\sigma$-algebra is a Dynkin class.

A class $\mathcal{M}$ of subsets of a set $S$ is called a monotone class if $E_i \in \mathcal{M}$, $i = 1, 2, \ldots$ implies

$$\bigcap_{i=1}^{\infty} E_i \in \mathcal{M}, \qquad \bigcup_{i=1}^{\infty} E_i \in \mathcal{M}.$$

**Theorem 1.** *A Dynkin class is a $\sigma$-algebra if and only if $E_1 \in \mathcal{D}$, $E_2 \in \mathcal{D} \Rightarrow E_1 \cap E_2 \in \mathcal{D}$.*

*Proof:* We have only to prove that if it has the mentioned property $\mathcal{D}$ is a $\sigma$-algebra. Now if it has this property, then for $E_1, E_2 \in \mathcal{D}$, $E_1 \cup E_2 = E_1 \cup (E_2 \backslash E_1) \in \mathcal{D}$.

Hence $\mathcal{D}$ is an algebra. Further, if $E_i \in \mathcal{D}$ where $i = 1, 2, \ldots$, putting

$$E_1' = E_1, \qquad E_j' = E_j \Big\backslash \bigcup_{i<j} E_i',$$

we get

$$\bigcup_{j=1}^{\infty} E_j = \bigcup_{j=1}^{\infty} E_j' \in \mathcal{D}$$

since the $E_j'$ are disjoint and belong to $\mathcal{D}$. □

**Theorem 2.** *Let $\mathcal{E}$ be a class of subsets of a set $S$ such that $E_1, E_2 \in \mathcal{E} \Rightarrow E_1 \cap E_2 \in \mathcal{E}$. Then the Dynkin class generated by $\mathcal{E}$ is identical with the $\sigma$-algebra generated by $\mathcal{E}$.*

**Theorem 3.** *The monotone class generated by an algebra $\mathcal{S}$ of subsets of a set $S$ is identical with the $\sigma$-algebra generated by $\mathcal{S}$.*

*Proof of Theorem 2:* Since a $\sigma$-algebra is a Dynkin class, it remains to be proven that the Dynkin class generated by $\mathscr{E}$ is a $\sigma$-algebra. Let $\mathscr{D}$ be the Dynkin class generated by $\mathscr{E}$. For $D \in \mathscr{D}$, put

$$\mathscr{D}_D = \{E : E \cap D \in \mathscr{D}\}, \qquad E \text{ subset of } S.$$

It easily follows that $\mathscr{D}_D$ is a Dynkin class. If $E \in \mathscr{E}$, then $E \in \mathscr{D}_D$ by the assumption made. Hence $\mathscr{E} \subset \mathscr{D}_D$ and $\mathscr{D} = \mathscr{D}_D$ for any $D \in \mathscr{D}$. That means that $E, D \in \mathscr{D} \Rightarrow E \cap D \in \mathscr{D}$, and thus $\mathscr{D}$ is a $\sigma$-algebra, according to Theorem 1. □

*Proof of Theorem 3:* Clearly, a Dynkin class is a monotone class, and thus the Dynkin class $\mathscr{D}$ generated by the algebra $\mathscr{S}$ is contained in the monotone class generated by $\mathscr{S}$. By Theorem 2, $\mathscr{D}$ is a $\sigma$-algebra. The monotone class generated by $\mathscr{D}$ is identical with $\mathscr{D}$ and contains the monotone class generated by $\mathscr{S}$. □

## 2. ALEXANDROV SPACES, TOPOLOGICAL SPACES, AND MEASURABLE SPACES

A set $S$ together with a $\sigma$-topological class of closed (open) subsets of $S$ is said to form a $\sigma$-topological space. We shall call this space an Alexandrov space, A-space for short. A set $S$ together with a topological class of closed (open) subsets of $S$ is said to form a topological space. Note that the closed sets determine the open sets and the open sets determine the closed sets. We say that these sets determine a $\sigma$-topology in the A-space and a topology in the topological space.

A neighborhood of an element $x \in S$ in an A-space or topological space is any open set containing $x$. Such a space is called a Hausdorff space if for any distinct points there exist disjoint neighborhoods. We shall only deal with Hausdorff A-spaces and Hausdorff topological spaces and then assume that the space contains at least two points (=elements). A basis of an A-space (topological space) is a class $\mathscr{B}$ of open sets such that for any $x \in S$ and any neighborhood $U$ of $x$ there is a set $B \in \mathscr{B}$ such that $x \in B \subset U$. An A-space is said to be second-order countable if it has a countable basis. Clearly, any topological space is an A-space. The converse is not true, but we have

**Theorem 1.** *If an A-space is second-order countable, it is a topological space.*

*Proof:* Let $\{G_i'\}_{i \in N}$ be a basis of the A-space and $G$ any open set. Then $G$ is a neighborhood of any $x \in G$ and thus $x \in B_x \in G$, where $B_x$ is a set in the basis. Hence

$$G = \bigcup_{x \in G} B_x.$$

However, the right-hand side only consists of countably many distinct basis elements. Hence any open set is a countable union of basis elements. Then any union of open sets, countable or noncountable, is a countable union of basis elements, and thus it belongs to the $\sigma$-topological class of open sets. Hence the A-space is a topological space. □

If $E$ is a subset of a topological space with the classes $\mathscr{G}$ of open and $\mathscr{F}$ of closed sets, we put

$$E^\circ = \bigcup_{G \subset E,\, G \in \mathscr{G}} G, \qquad \bar{E} = \bigcap_{E \subset F,\, F \in \mathscr{F}} F,$$

and call $E^\circ$ the interior of $E$ and $\bar{E}$ the closure of $E$. Note that $E^\circ$ is an open set and $\bar{E}$ a closed set.

In an A-space (topological space) $S$, a point $y \in S$ is called a point of accumulation (or limit point) of a set $E$ if any open set containing $y$ contains some element in $E$ that is different from $y$. Note that $y$ need not belong to $E$.

**Theorem 2.** *A set in a topological space is closed if and only if it contains all its points of accumulation.*

*Proof:* Let the set $E$ contain all its points of accumulation. Then if $y \in S \backslash E$, it certainly belongs to some open set, since it belongs to an open set $G_y$, not containing any point of $E$, since $y$ is not a point of accumulation. Hence

$$S \backslash E = \bigcup_{y \in S \backslash E} G_y$$

is open and $E$ closed. Let $E$ be closed and suppose that there is a point of accumulation $y$ of $E$ not contained in $E$ and hence in $S \backslash E$. Then any open set $G$ containing $y$ should contain points in $E$, but $G \cap (S \backslash E)$ is an open set containing $y$ but no point in $E$. This contradiction proves that any point of accumulation of $E$ belongs to $E$. □

An open covering of a set $E$ in an A-space is any union $\bigcup_{t \in \Upsilon} G_t$ of open sets ($\Upsilon$ countable or noncountable) such that $E \subset \bigcup_{t \in \Upsilon} G_t$. A set $E$ is called compact if any open covering of $E$ contains a finite open covering $\bigcup_{i=1}^{m} G_{t_i}$ of $E$, and $E$ is called $\sigma$-compact if any countable open covering contains a finite open covering.

**Theorem 3.** *A compact set in a Hausdorff topological space is closed.*

*Proof:* Let $A$ be a compact set in the Hausdorff topological space $S$, and suppose that $y$ is a point of accumulation of $A$ not contained in $A$. Then since

$S$ is a Hausdorff space for any $x \in A$, there exist disjoint open sets $G_x^{(1)}$ and $G_x^{(2)}$ such that $x \in G_x^{(1)}$, $y \in G_x^{(2)}$, where $G_x^{(2)} \cap A = \varnothing$. The open sets $G_x^{(1)}$ cover $A$: $A \subset \bigcup_{x \in A} G_x^{(1)}$. According to the compactness of $A$, there then exists an open covering $\bigcup_{i=1}^{m} G_{x_i}^{(1)}$ of $A$, and the open set $\bigcap_{i=1}^{m} G_{x_i}^{(2)}$ contains $y$ but no point in $A$. This is a contradiction, $y$ being a point of accumulation of $A$. Hence $y \in A$ and $A$ is closed according to Theorem 2. □

If a set $S$ and an algebra $\mathscr{S}$ of subsets of $S$ are given, we say that $S$ together with $\mathscr{S}$ forms a measurable space with respect to the algebra $\mathscr{S}$, and call the members of $\mathscr{S}$ measurable ($\mathscr{S}$). If, furthermore, $\mathscr{S}$ is a $\sigma$-algebra, we call $(S, \mathscr{S})$ a measurable space and the members of $\mathscr{S}$ measurable sets. When we consider A-spaces (topological spaces) and talk about measurable spaces and sets, we require that the algebra, respectively $\sigma$-algebra, is generated by the closed sets.

If $\mathscr{S}$ is a $\sigma$-algebra, we may deal with $\mathscr{S}$ as a $\sigma$-topological class of closed sets. Any closed set is then also open, and the measurable space is also an A-space in the $\sigma$-topology determined by $\mathscr{S}$.

## 3. MAPPINGS

The $\sigma$-topology and the topology are of importance for the description of mappings. Let $S$ and $S'$ be A-spaces (topological spaces). We call $f$ a mapping or a function from $S$ into $S'$ if for any $x \in S$ there exists an element $f(x)$ in $S'$ determined by $x$. If for any $x' \in S'$ there exists an element $x$ in $S$ such that $f(x) = x'$, we say that $f$ is a mapping of $S$ onto $S'$. By definition, $f$ is continuous $S/S'$ if for any open set $G' \in S'$, the inverse set

$$f^{-1}(G') = \{x \in S, f(x) \in G'\}$$

is an open set, or equivalently, $f^{-1}(F')$ is closed for any closed set $F'$ in $S'$. If $\mathscr{S}$ and $\mathscr{S}'$ are algebras ($\sigma$-algebras) on $S$ and $S'$, respectively, a function $f$ from $S$ into $S'$ is said to be measurable $\mathscr{S}/\mathscr{S}'$. If $\mathscr{S}'$ is the Borel $\sigma$-algebra on $R$ and $\mathscr{S}$ a $\sigma$-algebra, we simply call $f$ measurable. If $\mathscr{S}$ and $\mathscr{S}'$ are $\sigma$-algebras, they may be chosen as $\sigma$-topological classes of closed sets, and then, of course, the concepts continuity and measurability coincide. Thus any theorem about continuity that holds true for an A-space is also a theorem about measurability. When we deal with an algebra or $\sigma$-algebra on an A-space (topological space), we assume that it is generated by the closed sets. Let $S$, $S'$, and $S''$ be sets. If $f$ is a function from $S$ into $S$ and $g$ a function from $S'$ into $S''$, then $x \to g(f(x))$ is a function from $S$ into $S''$. We call this function a composite function and denote it by $g \circ f$.

**Theorem 1.** *Let $S$, $S'$, and $S''$ be A-spaces or topological spaces (measurable spaces). If $f$ is continuous (measurable) $S/S'$ and $g$ is continuous (measurable) $S'/S''$, then $g \circ f$ is continuous (measurable) $S/S''$.*

*Proof:* For any set $E''$ in $S''$, we have

$$(g \circ f)^{-1}(E'') = f^{-1}(g^{-1}E''),$$

and $g^{-1}E''$ is open (measurable) in $S'$ if $E''$ is open (measurable) in $S''$, and $g$ is continuous (measurable) $S'/S''$. Repeating this argument for the inverse mapping $f^{-1}$, we get the theorem. □

**Theorem 2.** *Let $S$ and $S'$ be A-spaces or topological spaces with σ-algebras generated by the closed sets. If $f$ is continuous $S/S'$, then it is measurable $S/S'$.*

*Proof:* It is easy to verify that

$$f^{-1}\left(\bigcup_{t \in \Upsilon} E'_t\right) = \bigcup_{t \in \Upsilon} (f^{-1}E'_t), \tag{1}$$

$$f^{-1}\left(\bigcap_{t \in \Upsilon} E'_t\right) = \bigcap_{t \in \Upsilon} (f^{-1}E'_t) \tag{2}$$

for any function $f$ from $S$ into $S'$, any subsets $E'_t$ in $S'$, and any (countable or noncountable) index set $\Upsilon$. If $f$ is a continuous function, then $f^{-1}(E'_t)$ is an open set in $S$ if $E'_t$ is an open set in $S'$. The open sets in $S'$ generate the σ-algebra $\mathscr{S}'_\sigma$. It follows by (1) and (2) that the inverse mappings of the sets in $\mathscr{S}'_\sigma$ by $f$ form a σ-algebra that must be contained in the σ-algebra $\mathscr{S}_\sigma$ generated by the open sets in $S$. □

**Lemma 1.** *Let $S$ and $S'$ be A-spaces (measurable spaces) and $\mathscr{G}$ and $\mathscr{G}'$ the σ-topological classes of open sets (the σ-algebras) in $S$ and $S'$, respectively, and $f$ a mapping of $S$ into $S'$. Suppose that $\mathscr{G}'$ is generated by a subclass $\mathscr{C}' \in \mathscr{G}'$. If $f^{-1}(G') \in \mathscr{G}$ for all $G' \in \mathscr{C}'$, then $f$ is continuous (measurable).*

*Proof:* It follows from (1) and (2) that those $G' \in \mathscr{G}'$ for which $f^{-1}(G') \in \mathscr{G}$ form a σ-topological class of open sets (a σ-algebra). This class contains those $G'$ that belong to $\mathscr{C}'$, and $\mathscr{C}'$ generates $\mathscr{G}'$. Hence $f^{-1}(G') \in \mathscr{G}$ for any $G' \in \mathscr{G}'$. □

The following lemma is a corollary of the lemma above.

**Lemma 2.** *A function $f$ from an A-space $S$ (measurable space) into $R$ is continuous (measurable) if and only if the sets $\{x: f(x) < a\}$ and $\{x: f(x) > a\}$ are open (belong to the σ-algebra in $S$) for any rational number $a$.*

*Proof:* Indeed, these sets are open (belong to the $\sigma$-algebra in $S$), and by definition the open intervals $(-\infty, a)$ and $(a, \infty)$ generate the $\sigma$-topological class of open sets in $R$ (the $\sigma$-algebra on $R$). □

Consider functions from a set $S$ into $R$. Put

$$\|f\| = \sup_{x \in S} |f(x)|,$$

$$f^+(x) = \begin{cases} f(x) & \text{for } f(x) \geq 0, \\ 0 & \text{for } f(x) < 0, \end{cases} \tag{3}$$

$$f^-(x) = \begin{cases} -f(x) & \text{for } f(x) \leq 0, \\ 0 & \text{for } f(x) > 0. \end{cases}$$

Denote the function $x \to \max[f_1(x), f_2(x)]$ by $\max(f_1, f_2)$, and define $\min(f_1, f_2)$ correspondingly. We say that $f$ is bounded if $\|f\| < \infty$.

**Lemma 3.** *Let $f$, $f_1$, and $f_2$ be continuous functions (measurable functions) from an A-space (measurable space) into $R$ and $a$ be any real number. Then $af, f_1 + f_2, |f|, \max(f_1, f_2)$ are continuous (measurable).*

*Proof:* We consider continuity. The statements about measurability follow in the same way. Applying Lemma 2, we find directly that $af$ is continuous. The fact that $f_1 + f_2$ is continuous follows by this lemma and the relations for open sets:

$$[x: f_1(x) + f_2(x) < a] = \bigcup_{r_1 + r_2 < a} \{[x: f_1(x) < r_1] \cap [x: f_2(x) < r_2]\},$$

$$[x: f_1(x) + f_2(x) > a] = \bigcup_{r_1 + r_2 > a} \{[x: f_1(x) > r_1] \cap [x: f_2(x) > r_2]\},$$

where $r_1, r_2$ are any rational numbers satisfying the inequalities. The continuity of $\max(f_1, f_2)$ follows by the same lemma and the relations ($r$ rational)

$$\{x: \max[f_1(x), f_2(x) > a]\} = \bigcup_{r > a} \{[x: f_1(x) > r] \cup [x: f_2(x) > r]\},$$

$$\{x: \max[f_1(x), f_2(x) < a]\} = \bigcup_{r < a} \{[x: f_1(x) < r] \cap [x: f_2(x) < r]\}.$$

Since $\min(f_1, f_2) = -\max(-f_1, -f_2)$, we also conclude that $\min(f_1, f_2)$ is continuous. Further, $f^+ = \max(f, 0)$ and $f^- = \max(-f, 0)$, and $f = f^+ + f^-$ is therefore continuous. In order to prove that $f_1 f_2$ is continuous it is sufficient to show that this is true for functions $f_1$, $f_2$ with $f_1(x) \geq 1$, $f_2(x) \geq 1$ for all $x$, since for any $f_1$ and $f_2$

$$f_1 f_2 = (f_1^+ - f_1^-)(f_2^+ - f_2^-) = f_1^+ f_2^+ + f_1^- f_2^- - f_1^- f_2^+,$$

and $f_i^+ + 1 \geq 1$, $f_i^- + 1 \geq 1$. Since all open Borel sets on $[0, +\infty)$ belong to $(0, +\infty)$, we need only prove that $f_1 f_2$ is continuous when $f_1$ and $f_2$ are larger than unity and continuous. Using the relation

$$\ln f_1 f_2 = \ln f_1 + \ln f_2,$$

and the fact that $t \to \ln t$ and its inverse are continuous for $t \geq 1$, and that the sum of two continuous functions are continuous, we conclude that $f_1 f_2$ is continuous. □

As we have remarked above, theorems about continuity in A-spaces remain true about measurability in measurable spaces. The converse, however, is not true. This depends on the fact that a $\sigma$-algebra may be considered as a $\sigma$-topological class of closed sets, but the converse is not true.

For a sequence of real-valued functions $f_n$ from a measurable space, we use the notations $\sup_{n>m} f_n$ and $\inf_{n>m} f_n$ for the functions $x \to \sup_{n>m} f_n(x)$ and $x \to \inf_{n>m} f_n(x)$. These functions always exist, but they may be infinite for some $x$. Further we put

$$\overline{\lim_n} f_n = \limsup_n f_n = \inf_{n>0} \sup_{m>n} f_m,$$

$$\underline{\lim_n} f_n = \liminf_n f_n = \sup_{n>0} \inf_{m>n} f_m.$$

These functions exist. Indeed, putting

$$g_m = \sup_{m>n} f_n,$$

we find that $\{g_m(x)\}$ is a nondecreasing sequence. Hence $\limsup_n f_n$ exists. In the same way, we find that $\liminf_n f_n$ exists. If $\lim_n f_n(x)$ exists for all $x$, we use the notation $\lim_n f_n$ for the function

$$x \to \lim_n f_n(x).$$

**Theorem 3.** *If $\{f_n\}$ is a sequence of real-valued measurable functions from a measurable space into R, then $\sup_{n>m} f_n$, $\inf_{n>m} f_n$, $\limsup_n f_n$, $\liminf_n f_n$, and $\lim_n f_n$ (if it exists) are measurable.*

*Proof:*

$$\left\{x : \sup_{n>m} f_n(x) < a\right\} = \bigcap_{n>m} \{x : f_n(x) < a\},$$

$$\left\{x : \sup_{n>m} f_n(x) > a\right\} = \bigcup_{n>m} \{x : f_n(x) > a\}.$$

Hence, by Lemma 2, $\sup_{n>m} f_n$ is measurable. Since

$$\inf_{n>m} f_n = -\sup_{n>m} \{-f_n\},$$

$\inf_{n>m} f_n$ is also measurable. It then follows that $\limsup_n f_n$, $\liminf_n f_n$, and (since $\lim_n f_n = \limsup_n f_n = \liminf_n f_n$) also $\lim_n f_n$ are measurable. □

Let $(S, \mathscr{S}_\sigma)$ be a measurable space and $E \in \mathscr{S}_\sigma$. The function $1_E$ from $S$ into $R$, defined by

$$1_E(x) = \begin{cases} 1 & \text{for} \quad x \in E, \\ 0 & \text{for} \quad x \notin E, \end{cases}$$

is called the indicator function of $E$. If $E_i$, $i = 1, \ldots, m$, are pairwise disjoint sets and $\bigcup_{i=1}^m E_i = S$, the function

$$f = \sum_{i=1}^{m} a_i 1_{E_i},$$

where $a_i$ is a real number, is called simple, and if the sequence $\{E_i\}$ is infinite it is called elementary.

**Theorem 4.** *A function $f$ from a measurable space $(S, \mathscr{S}_\sigma)$ into $R$ is measurable if and only if $f$ is the limit of a sequence of simple functions. Any measurable function $f$ is also the limit of a monotone sequence of simple functions.*

*Proof:* Since an indicator function is obviously measurable and a sum of measurable functions is measurable, a simple function is measurable. By Theorem 3, a limit of a sequence of simple functions is measurable.

Let $f$ be measurable, and define

$$E_i^{(n)} = \{x : (i-1)2^{-n} < x \leq i2^{-n}\}$$

for

$$i = -n2^n + 1, -n2^n + 2, \ldots, n2^n,$$
$$E_n^- = \{x : f(x) \leq -n\}, \qquad E_n^+ = \{x : f(x) > n\},$$

and

$$f_n = \sum_{i=-n2^n+1}^{n2^n} (i-1)2^{-n} 1_{E_i^{(n)}} - n1_{E_n^-} + n1_{E_n^+}.$$

Then $f_n$ is a measurable simple function for all $n$ and $\lim_n f_n = f$. Note that $f_n \uparrow f$, i.e., $f_n(x) \uparrow f(x)$ for all $x$ as $n \uparrow \infty$. We then say that $f_n$ tends nondecreasingly to $f$. □

## 4. CLASSES OF BOUNDED, REAL-VALUED, CONTINUOUS FUNCTIONS AND MEASURABLE FUNCTIONS

Different classes of continuous functions and measurable functions are characterized by different properties.

For a class $\psi$ of functions from a set $S$ into $R$ we introduce the distance

$$\|f - g\| = \sup_{x \in S} |f(x) - g(x)|$$

between $f, g \in \psi$. It is called the uniform metric. General metrics are considered in Section 7. It is easy to prove that $\|f - g\|$ satisfies the following relations required for a metric:

$$\|f - g\| \geq 0, \quad \|f - g\| = \|g - f\|, \quad \|f - g\| = 0 \qquad \Rightarrow f = g,$$
$$\|f - g\| \leq \|f - h\| + \|h - g\|$$

for functions $f$, $g$, $h$ from $S$ into $R$.

A sequence $\{f_n\}$, $f_n \in \psi$ is called Cauchy convergent in the uniform metric if

$$\|f_n - f_m\| \to 0 \qquad (n \to \infty, m \to \infty). \tag{1}$$

Clearly this convergence implies the pointwise convergence $\{f_n(x)\}$ to $f(x)$. However, $f$ may not belong to $\psi$. If this limit of any Cauchy-convergent sequence $\{f_n\}$, $f_n \in \psi$ has its limit in $\psi$, then $\psi$ is said to be complete in the uniform metric. Note that $\|f_n - f\| \to 0$ $(n \to +\infty)$ implies (1) since

$$\|f_n - f_m\| \leq \|f_n - f\| + \|f_m - f\|.$$

If $\psi$ is not complete in the uniform metric, we add the limits of the Cauchy-convergent sequences in the metric $\|\cdot\|$ to $\psi$ and so get a larger class $\bar{\psi}$. We call $\bar{\psi}$ the closure of $\psi$ with respect to the metric $\|\cdot\|$. We say that a class $\psi$ is closed in the metric $\|\cdot\|$ if the limit $f$ in this metric of any sequence $\{f_n\}_{n \in N}$, $f_n \in \psi$ belongs to $\psi$.

**Theorem 1.** *The closure in the uniform metric of a class of functions from a set $S$ into $R$ is closed and hence complete.*

*Proof:* We have to show that the limit of any Cauchy-convergent sequence $\{f_n\}_{n \in N}$ in the closure $\bar{\psi}$ of $\psi$ belongs to $\bar{\psi}$. Now if $f_n \in \bar{\psi}$, we have

$$\lim_{m \to +\infty} \|f_{n,m} - f_n\| = 0 \tag{2}$$

for any $n$, where $f_{n,m} \in \psi$. Further, according to the Cauchy convergence of $\{f_n\}_{n \in N}$, we get

$$\lim_{n \to +\infty} \|f_n - f\| = 0$$

with a function $f$ from $S$ to $R$. We may choose $m = m_n$ such that $\|f_{n,m_n} - f_n\| < 1/n$, and then

$$\|f_{n,m_n} - f\| \le \|f_{n,m_n} - f_n\| + \|f_n - f\| \to 0 \qquad (n \to +\infty),$$

and thus $f \in \bar{\psi}$ since $f_{n,m_n} \in \psi$. □

A class $\psi$ of bounded, real-valued functions is called a Stone vector lattice if the following conditions hold:

$$\text{(i)} \qquad f \in \psi \Rightarrow af \in \psi \qquad \text{for any real number } a,$$

$$\text{(ii)} \qquad f_1, f_2 \in \psi \Rightarrow f_1 + f_2 \in \psi, \qquad \max(f_1, f_2) \in \psi.$$

Note that (i) and (ii) imply $\min(f_1, f_2) \in \psi$ since $\min(f_1, f_2) = -\max(-f_1, -f_2)$. If the Stone vector lattice contains $fg$, and contains $f$ and $g$, we call it a Stone vector lattice ring.

**Lemma 1.** *The closure $\bar{\psi}_0$ of a Stone vector lattice $\psi_0$ is a complete Stone vector lattice, and if, furthermore, $\psi_0$ is a Stone vector lattice ring, then $\bar{\psi}_0$ is a complete such ring.*

*Proof:* The fact that $\bar{\psi}_0$ is a complete class follows from Theorem 1. Let $f, g \in \psi_0$. Then there exist sequences $\{f_n\}, \{g_n\}, f_n, g_n \in \psi_0$ such that

$$\|f - f_n\| \to 0, \qquad \|g - g_n\| \to 0 \qquad (n \to +\infty).$$

This implies $\|af_n - af\| \to 0$ for any real number $a$, and

$$\|f_n + g_n - f + g\| \le \|f_n - f\| + \|g_n - g\| \to 0 \qquad (n \to +\infty).$$

For sufficiently large $n$ and $\varepsilon > 0$, we further have

$$f - \varepsilon \le f_n \le f + \varepsilon, \qquad g - \varepsilon \le g_n \le g + \varepsilon,$$

and hence

$$f_n(x) \le \max[f(x), g(x)] + \varepsilon.$$

Here we may interchange $f$ and $g$. Thus

$$\max[f_n(x), g_n(x)] \le \max[f(x), g(x)] + \varepsilon.$$

In the same way we obtain

$$\max[f(x), g(x)] \le \max[f_n(x), g_n(x)] + \varepsilon.$$

Since $\varepsilon$ is arbitrarily small for sufficiently large $n$, these inequalities imply that

$$\|\max(f_n, g_n) - \max(f, g)\| \to 0 \qquad (n \to +\infty).$$

Also

$$\|f_n g_n - fg\| \leq \|f_n\| \cdot \|g_n - g\| + \|g\| \cdot \|f_n - f\| \to 0 \qquad (n \to +\infty). \quad \square$$

Consider an algebra $\mathscr{S}$ of subsets of a set $S$. We call a finite disjoint sequence $\{E_i\}_{i=1}^n$, $E_i \in \mathscr{S}$ such that $\bigcup_{i=1}^n E_i = S$ a partition of $\mathscr{S}$. To such a partition and real numbers $a_i$ there belongs the simple function

$$f = \sum_{i=1}^{n} a_i 1_{E_i}, \tag{3}$$

where $1_{E_i}$ is the indicator function.

**Theorem 2.**

(i) *The class $\psi_0$ of all simple functions belonging to $\mathscr{S}$ form a Stone vector lattice ring and so does its closure $\bar{\psi}_0$ with respect to the uniform metric.*

(ii) *If $S$ is an A-space and $\mathscr{S}$ the algebra generated by the closed sets, then $\bar{\psi}_0$ contains the class $\psi$ of all real-valued, bounded, continuous functions, and $\psi$ is a complete Stone vector lattice ring. Furthermore, any $f \in \psi$ is the limit in the uniform metric of a sequence $\{f_n\}$, $f_n \in \Psi_0$:*

$$f_n = \sum_{i=1}^{m_n} a_i^{(n)} 1_{F_{i-1}^{(n)} \setminus F_i^{(n)}},$$

*where the $a_i^{(n)}$ may be chosen such that $f_n \uparrow f$ or $f_n \downarrow f$, and the $F_i$ are closed sets:*

$$S = F_0^{(n)} \supset F_1^{(n)} \supset F_2^{(n)} \supset \cdots \supset F_{m_n}^{(n)} = \varnothing.$$

(iii) *If $\mathscr{S}$ is a $\sigma$-algebra, then $\bar{\psi}_0$ is the class of all real-valued, bounded, measurable functions.*

*Proof:* Let $f$ be given by (3), and

$$g = \sum_{i=1}^{m} b_j 1_{E'_j}, \quad E'_j \cap E'_i = \varnothing \quad \text{for} \quad i \neq j, \quad \bigcup_{j=1}^{m} E'_j = S \quad (b_j \text{ real numbers}).$$

Then $f$ and $g$ have also the representations

$$f = \sum_{i=1}^{n} \sum_{j=1}^{m} a_i 1_{E_j \cap E'_j}, \qquad g = \sum_{i=1}^{n} \sum_{j=1}^{m} b_j 1_{E_i \cap E'_j}, \tag{4}$$

where some sets $E_j \cap E'_j$ may be empty. By these representations it is seen that $f + g$ and $fg$ belong to $\psi_0$.

By Lemma 1, the closure $\bar{\psi}_0$ of $\psi_0$ is a complete Stone vector lattice ring. Now let $S$ be an A-space, and $\mathscr{S}$ the algebra generated by the closed sets in $S$ and $\psi$ the class of real-valued, bounded, continuous functions from $S$.

For $f \in \psi$, there exist real numbers $a$ and $b$ such that

$$a < \inf_{x \in S} f(x), \qquad b > \sup_{x \in S} f(x).$$

Choose real numbers $a_i^{(n)}$, $a = a_0^{(n)} < a_1^{(n)} < \cdots < a_{m_n}^{(n)} = b$, $a_i^{(n)} - a_{i-1}^{(n)} < n^{-1}(b - a)$ for $n = 1, 2, \ldots$, and put $F_i^{(n)} = \{x : f(x) \geq a_i^{(n)}\}$. Then $F_i^{(n)}$ is a closed set, $F_0^{(n)} = S$, $F_{m_n}^{(n)} = 0$, and $F_i^{(n)} \subset F_{i-1}^{(n)}$. Put

$$f_n = \sum_{i=1}^{m_n} a_i^{(n)} 1_{F_{i-1}^{(n)} \setminus F_i^{(n)}}. \tag{5}$$

Clearly, $f_n \in \psi_0$, and $\|f - f_n\| \to 0$ $(n \to +\infty)$. Thus $\psi \subset \bar{\psi}_0$. By Lemma 3.2, $\psi$ is a Stone vector lattice ring. To show that $\psi$ is complete, we have to prove that $f \in \psi$ if $\|f - f_n\| \to 0$ $(n \to +\infty)$ with $f_n \in \psi$. This statement follows by Lemma 3.1 and the obvious relations

$$\{x : f(x) < a\} = \bigcup_{n \in N} \{x : f_n(x) < a - \|f_n - f\|\},$$
$$\{x : f(x) > a\} = \bigcup_{n \in N} \{x : f_n(x) > a + \|f_n - f\|\}$$

for any real number $a$.

If we choose $a_i^{(n)} - a_{i-1}^{(n)} = 2^{-n}$ in (5), it is easily seen that $f_n \downarrow f$, and if we then change $a_i^{(n)}$ to $a_{i-1}^{(n)}$ in (5) for all $i$, then $f_n \uparrow f$. The statement (iii) follows by (ii) if we consider the $\sigma$-algebra as a $\sigma$-topological class of closed sets. Note that the approximation (5) then holds with measurable sets. □

If $f$ is a real-valued, bounded, continuous function, the set $\{x : f(x) \leq a\}$ is closed for any real number $a$. The $\sigma$-topological class $\mathscr{F}$ of closed sets determines the $\sigma$-topology in the A-space. Hence there is a connection between a certain class of real-valued, continuous functions and a $\sigma$-topological class of closed sets. Such a connection is expressed by the following theorem. (Compare further Section 6, where this connection is given more explicitly.)

**Theorem 3.** *Let $S$ be a set and $\bar{\psi}$ a complete Stone vector lattice of bounded, real-valued functions from $S$, where $\bar{\psi}$ contains the function $x \to 1$ (hence any constant). Then the class of all sets $F = \{x : f(x) \leq a\}$, $f \in \bar{\psi}$, $a$ real number, is a $\sigma$-topological class of closed sets. If $F$ is given as above, then it has also the representation $[x : h(x) = 0]$ with $h \in \bar{\psi}$, $0 \leq h(x) \leq 1$ for $x \in S$.*

*Proof:* Let $F = [x : f(x) \leq a]$. Define the functions $g$ and $h$ by

$$g(x) = a - \min[f(x), a], \qquad h(x) = \min[1, g(x)].$$

Since $x \to 1$ belongs to $\bar{\psi}$, $g$ and $h$ also belong to $\bar{\psi}$. Clearly, $f(x) \leq a$ if and only if $h(x) = 0$. Further $0 \leq h(x) \leq 1$ for $x \in S$. Thus $F$ has the representation $[x:h(x) = 0]$. In order to prove the theorem, we may consider the representations of the last form. By definition, a $\sigma$-topological class of closed sets on a set $S$ is a class of subsets that is closed for finite unions and countable intersections and that contains the set $S$ and the empty set. Consider sets $F$ with representations $F = [x:h(x) = 0]$, $h \in \bar{\psi}$, and $0 \leq h(x) \leq 1$ for $x \in s$, and let $\mathscr{F}$ be the class of all such sets. Since $\bar{\psi}$ contains the functions $x \to 0$ and $x \to 1$, the sets $S$ and the empty set are contained in $\mathscr{F}$. Let $F_i = [x:h_i(x) = 0]$, $0 \leq h_i(x) \leq 1$ for $x \in S$, $g(x) = \sum_{i=1}^{+\infty} 2^{-i} h_i(x)$. Then we have

$$F_1 \cup F_2 = \{x:\min[h_1(x), h_2(x)] = 0\},$$
$$F_1 \cap F_2 = \{x:\tfrac{1}{2}[h_1(x) + h_2(x)] = 0\},$$
$$\bigcap_{i=1}^{+\infty} F_i = [x:g(x) = 0],$$

where the functions $\min[h_1, h_2]$, $\frac{1}{2}[h_1 + h_2]$, and $g$ belong to $\psi$ and take values between 0 and 1 or 0 and $x$. Hence $\mathscr{F}$ is a $\sigma$-topological class of closed sets. □

## 5. NORMAL SPACES AND COMPLETELY NORMAL SPACES

If $F_1$ and $F_2$ are disjoint closed sets in an A-space $S$, we say that $F_1$ is connected with $F_2$ if there exists a real-valued, continuous function $f$ from $S$ such that $f(x) = 0$ for $x \in F_1$, $f(x) = 1$ for $x \in F_2$, and $0 \leq f(x) \leq 1$ for $x \in S$. If any two disjoint sets in an A-space are connected, the A-space is called normal. The sets $[x:f(x) = 0]$, $f$ continuous, $0 \leq f(x) \leq 1$ are called totally closed. The concepts introduced here for A-spaces also make sense, of course, for topological spaces, since they are A-spaces.

**Theorem 1.** *The totally closed sets in an A-space S form a $\sigma$-topological class of closed sets, i.e., the class of totally closed sets is closed for finite unions and countable intersections, and it contains S and the empty set. Any set $[x:f(x) \leq a]$, $f$ bounded and continuous and a real number, is totally closed.*

*Proof:* The class of bounded, continuous functions is a complete Stone vector lattice by Theorem 4.2 and contains the function $x \to 1$. Hence our statement here follows by Theorem 4.3. □

If all closed sets in an A-space are totally closed, we call the A-space a completely normal space. This denomination is motivated by the following lemma, according to which such a space is normal.

**Lemma 1.** *If $F_1$ and $F_2$ are totally closed sets in an A-space S, then $F_1$ and $F_2$ are connected.*

*Proof:* Let $F_i = \{x : f_i(x) = 0\}$, $0 \le f_i(x) \le 1$ for $x \in S$, $f_i$ a real-valued, continuous function rom $S$, $i = 1, 2$. Put $f = f_{1/(f_1 + f_2)}$. Observing that $f_1(x) + f_2(x) > 0$ for all $x$, we find by Lemma 3.2 that $f$ is continuous, and clearly $0 \le f(x) \le 1$ for $x \in S$, $f_1(x) = 0$ on $F_1$, and $f_1(x) = 1$ on $F_2$. □

If $F$ is totally closed, we call $G = S \backslash F$ totally open. A class $\psi$ of continuous functions in a completely normal space $S$ is called a representative class if any closed set in $S$ has a representation $\{x : f(x) \le a\}$ with $f \in \psi$.

**Theorem 2.** *Let $F_1$ and $F_2$ be totally closed disjoint sets in an A-space S. Then there exist totally open disjoint sets $G_1$ and $G_2$ such that $F_1 \subset G_1 \backslash F_2 \subset G_2$.*

**Corollary.** *To any disjoint closed sets $F_1$ and $F_2$ in a normal A-space, there exist disjoint open sets* (even totally open) *$G_1$ and $G_2$ such that $F_1 \subset G_1$, $F_2 \subset G_2$.*

*Proof:* We have $F_i = \{x : f_i(x) = 0\}$, $i = 1, 2$, where $0 \le f_i(x) \le 1$ and $f_i$ is continuous. The sets

$$G_i^{(n)} = \bigcap_{k=1}^{n} \left\{x : f_i(x) < \frac{1}{k}\right\}$$

are totally open for any positive integer $n$ and the sets $G_1^{(n)}$ and $G_2^{(n)}$ must be disjoint for some $n$. Otherwise they have a point in common for all $n$ and this point must be contained in $F_1$ and $F_2$, which is impossible. □

The corollary follows since, in a normal space, the disjoint sets $F_1$ and $F_2$ are connected, i.e., there exists a continuous function $f$ such that $f(x) = 0$ on $F_1$, $f(x) = 1$ on $F_2$, and $0 \le f(x) \le 1$ on $S$. The sets

$$F_1' = [x : f(x) = 0], \qquad F_2' = [x : f(x) = 1]$$

are totally closed and $F_1 \subset F_1'$ and $F_2 \subset F_2'$. □

*Remark:* The corollary of Theorem 2 is often used as a definition of a normal space. Then the famous Urysohn's lemma states that any two disjoint closed sets are connected. However, we shall not use this lemma.

We have seen in Section 3 that a $\sigma$-topological class of closed subsets of a set $S$ determines the class of real-valued, bounded, continuous functions and this class is a complete Stone vector lattice that contains the function $x \to 1$. We find the converse by Theorem 4.3 and the proof of Lemma 1 above.

**Theorem 3.** *A complete Stone vector lattice $\psi$ of bounded, real-valued functions from a set S, where $\psi$ contains the function $x \to 1$, determines a $\sigma$-topological class $\mathscr{F}$ of totally closed sets on S. In the $\sigma$-topology, determined by $\mathscr{F}$, any $f \in \psi$ is continuous. The A-space S, determined by $\mathscr{F}$, is completely normal.*

## 6. SEQUENCES OF SETS

In this section we shall give some properties of sequences of sets, which will be used mainly in Chapter III. But we present them here, since we have reason to apply methods that will now be used.

A sequence $\{E_n\}$ of sets is called nondecreasing (noted $E_n\uparrow$) if $E_n \subset E_{n+1}$ and nonincreasing (noted $E_n\downarrow$) if $E_{n+1} \subset E_n$ for all $n$. If $E_n\downarrow$, then $\bigcap_{n\in N} E_n$ is the limit of the sequence $\{E_n\}$, and if $E_n\uparrow$, then $\bigcup_{n\in N} E_n$ is the limit of $\{E_n\}$. Of course, an element $x$ belongs to the first-mentioned limit if it belongs to all $E_n$, and to the second limit if it belongs to some $E_n$.

A sequence $\{E_n\}$ is called disjoint if the sets $E_n$ are pairwise disjoint and $E_i \cap E_j = 0$ for $i \neq j$. A sequence $\{F_n\}$ of closed sets in an A-space is called divergent if it is disjoint and if any union $\bigcup_{i\in N} F_{n_i}$, $n_1 < n_2 < \cdots$ is closed. Of course, any finite union of closed sets is a closed set.

**Lemma 1.** *If $\{F_n\}$ is a sequence of closed sets, $\{G_n\}$ a sequence of open sets in an A-space, and $F_n \subset G_n$, $F_n \downarrow \varnothing$, and $G_n \downarrow \varnothing$, then $\{F_n\backslash G_{n+1}\}$ is divergent.*

*Proof:* Let the conditions in the lemma be satisfied. Then $F_n - G_{n+1} = F_n \cap (S\backslash G_{n+1})$ is closed, and $[F_k\backslash G_{k+1}] \cap [F_n\backslash G_{n+1}] = \varnothing$ for $n > k$ since $F_n \subset G_{k+1}$ for $n > k$. Our statement then follows from the identity

$$\bigcup_{i=1}^{\infty} [F_{n_i}\backslash G_{n_i+1}] = F_{n_1} \cap \bigcap_{i=1}^{\infty} [S\backslash G_{n_i+1} \cup F_{n_{i+1}}] \qquad (n_i < n_{i+1}). \tag{1}$$

We verify this identity. If $x$ belongs to the set on the left-hand side of (1), then for some index $j \geq 1$, $x \in F_{n_j}$ and $x \neq G_{n_j+1}$; therefore $x \in F_{n_i}$ for $i \leq j$ and $x \subset S\backslash G_{n_i+1}$ for $i \geq j$. Hence $x$ belongs to the set on the right-hand side of (1). Conversely, if $x$ belongs to the set on the right-hand side, then $x$ cannot belong to all $F_{n_i}$ since $F_n\downarrow\varnothing$. Thus there exists a first index $j$ such that $x \in F_{n_j}$ and $x \notin F_{n_{j+1}}$. Then, however, $x \notin G_{n_j+1}$ since $x \in (S\backslash G_{n_j+1}) \cup F_{n_{j+1}}$. Consequently, $x$ belongs to the left-hand side of (1). □

**Lemma 2.** *If $\{F_n\}$ is a divergent sequence of closed sets in an A-space and $F^{(n)}$ is a closed subset of $F_n$ for $n = 1, 2, \ldots,$ then $\{F^{(n)}\}$ is divergent.*

*Proof:* For an infinite sequence of positive integers $\{n_i\}$, $n_{i+1} > n_i$, put

$$F_{n_j}^{(i)} = \begin{cases} F_{n_j} & \text{for } \ j \neq i, \\ F^{(n_i)} & \text{for } \ j = i, \end{cases} \qquad \bar{F}_i = \bigcup_{j=1}^{\infty} F_{n_j}^{(i)}.$$

Since $\{F_n\}$ is divergent, $\bar{F}_i$ is closed for all $i$. Further we easily verify the identity

$$\bigcup_{i=1}^{\infty} F^{(n_i)} = \bigcap_{i=1}^{\infty} \bar{F}_i. \quad \square$$

**Lemma 3.** *If $\{F_n\}$ is a divergent sequence in a normal A-space, then there exists a disjoint sequence $\{G_n\}$ of open sets such that $F_n \subset G_n$ for all n.*

*Proof:* Since $F_1$ and $\bigcup_{i\geq 2} F_i$ are disjoint closed sets in a normal space, there exist disjoint open sets $G_1$ and $G^{(1)}$ such that

$$F_1 \subset G_1, \qquad \bigcup_{i\geq 2} F_i \subset G^{(1)}$$

(corollary of Theorem 2). Using induction, we assume that there exist disjoint sets $G_j$, $j = 1, 2, \ldots, n$ and $G^{(n)}$ such that

$$F_i \subset G_i, \qquad i = 1, 2, \ldots, n,$$
$$\bigcup_{i\geq n+1} F_i \subset G^{(n)} \qquad \text{for} \quad n \leq n_0.$$

Since $F_{n+1}$ and $\bigcup_{i\geq n+2} F_i$ are disjoint, we find as above that there exist disjoint open sets $G'$ and $G''$ such that $F_{n+1} \subset G'$ and $\bigcup_{i\geq n+2} F_i \subset G''$. Putting

$$G_{n+1} = G' \cap G^{(n)}, \qquad G^{(n+1)} = G'' \cap G^{(n)},$$

we conclude that $F_i \subset G_i$, $i = 1, \ldots, n+1$ and $\bigcup_{i\geq n+2} F_i \subset G^{(n+1)}$. □

**Lemma 4.** *Let $\{F_n\}$ be a divergent sequence and $\{G_n\}$ a disjoint sequence of open sets in a normal A-space, and suppose that $F_n \subset G_n$, $n = 1, 2, \ldots$. Then there exist functions $f_n$ connecting $S\backslash G_n$ with $F_n$ for any n such that any sum $\sum_i f_{n_i}$, $n_1 < n_2 < \cdots$* (finite or infinite sum) *is continuous.*

*Proof:* The sets $\bigcup F_n$ and $S\backslash\bigcup G_n$ are disjoint, and both are closed sets (the former set is closed since $\{F_n\}$ is divergent). Hence, in the normal A-space there exists a continuous function $f$ connecting $S\backslash\bigcup G_n$ with $\bigcup F_n$, i.e.,

$$f(x) = 0 \quad \text{for} \quad x \in S\backslash\bigcup G_n, \qquad f(x) = 1 \quad \text{for} \quad x \in \bigcup F_n,$$
$$0 \leq f(x) \leq 1, \quad x \in S, \quad f \text{ continuous.}$$

Define

$$f_n(x) = \begin{cases} f(x) & \text{for} \quad x \in G_n, \\ 0 & \text{for} \quad x \in S\backslash G_n. \end{cases}$$

Then

$$0 \leq f_n(x) \leq 1, \qquad f_n(x) = 0 \quad \text{for} \quad x \in S\backslash G_n, \qquad f_n(x) = 1 \quad \text{for} \quad x \in F_n.$$

Hence, $f_n$ connects $S\backslash G_n$ with $F_n$ if it is continuous. We will prove that

$$g = \sum_i f_{n_i}, \qquad n_1 < n_2 < \cdots \tag{2}$$

is continuous for any finite or infinite sequence $\{n_i\}$ of positive integers. Indeed the sum is convergent since

$$g(x) = \begin{cases} f_{n_i}(x) = f(x) & \text{for} \quad x \subset G_{n_i}, \\ 0 & \text{for} \quad x \subset S \backslash \bigcup G_{n_i}. \end{cases}$$

We have to prove by Theorem 3.1 that the sets $[x:g(x) > a]$ and $[x:g(x) < a]$ are open for any real number $a$. If $a < 0$, then $[x:g(x) < a]$ is the empty set and $[x:g(x) > a]$ is the set $S$. Hence, it is sufficient to deal with $a \geq 0$. We show then that

$$[x:g(x) > a] = [x:f(x) > a] \cap \left[ \bigcup_i G_{n_i} \right], \tag{3}$$

$$[x:g(x) < a] = [x:f(x) < a] \cup \left[ \bigcup_{n \neq n_i} G_n \right]. \tag{4}$$

[In the last union, $n$ should run through all positive integers that do not belong to the set $(n_1, n_2, \ldots)$.] Now if $x \in G_{n_i}$ for some $i$, then $g(x) = f_{n_i}(x) = f(x)$, and $x$ belongs to the sets on both sides of the equality (3) if it belongs to one of these sets. If $x \notin G_{n_i}$, $i = 1, 2, \ldots$, then $g(x) = 0$ and the sets on both sides of (3) are empty (since $a \geq 0$). Hence (3) holds for all $x$. Now consider (4). If $x \notin G_{n_i}$ for some $i$, we again have $g(x) = f(x)$, and $x$ belongs to the sets on both sides of the equality (4) if it belongs to one of these sets. If $x \notin G_{n_i}$, $i = 1, 2, \ldots$, then $g(x) = 0$, and $x$ belongs to both sides of the equality (4). Hence (4) holds for all $x$. □

**Lemma 5.** *If $\{F_n\}$ is a nonincreasing sequence of closed sets in a completely normal space, then there exists a nonincreasing sequence $\{G_n\}$ of open sets such that*

$$F_n \subset G_n, \quad \text{where } n = 1, 2, \ldots, \qquad \bigcap_{n \in N} F_n = \bigcap_{n \in N} G_n.$$

*Observe that we require here that the space is completely normal.*

*Proof:* Let $F_i = [x:f^{(i)}(x) = 0]$ be the representation of $F_i$, where $f^{(i)}$ is continuous and $0 \leq f^{(i)}(x) \leq 1$ for all $x$. Put

$$f_n = \sum_{i=1}^{n} f^{(i)}, \qquad G_n = \left[ x:f_n(x) < \frac{1}{n} \right].$$

Then $G_{n+1} \subset G_n$ since $f_n \leq f_{n+1}$. Further $F_n \subset G_n$ since $f^{(i)}(x) = 0$ for $i \leq n$ and $x \in F_n$. If $x \in G_n$ for all $n$, then

$$f_m(x) \leq f_n(x) < 1/n$$

for $m < n$. Letting $n \to +\infty$, we get $f_m(x) = 0$ for all $m$. Hence $x \in \bigcap_{n \in N} G_n$ implies $x \in \bigcap_{n \in N} F_n$. □

## 7. METRIC SPACES

An important subclass of the completely normal spaces is formed by the metric spaces. Let $S$ be a set and $\rho$ a function from the product set $S \times S$ such that for any elements (= points) $x$, $y$, $z$ in $S$:

$$\text{(i)} \qquad \rho(x, y) < \infty, \qquad \rho(x, y) \geq 0,$$
$$\text{(ii)} \qquad \rho(x, y) = \rho(y, x),$$
$$\text{(iii)} \qquad \rho(x, y) = 0 \Leftrightarrow x = y,$$
$$\text{(iv)} \qquad \rho(x, y) \leq \rho(x, z) + \rho(z, y).$$

Then $\rho$ is called a metric on $S$. We call (ii) the symmetry relation and (iv) the triangle inequality. The metric is a generalization of the concept "absolute value" for numbers.

The subset

$$G_\varepsilon(y) = [x : \rho(x, y) < \varepsilon]$$

is called an open sphere. The open spheres generate a topological class $\mathscr{G}$ of open sets. We introduce the topology determined by $\mathscr{G}$ on $S$ and so get a topological space, which is called a metric space. It is easily seen that a metric space is a Hausdorff space. A sphere is also called a ball.

A real-valued function $f$ from a metric space $S$ is called uniformly continuous if for any number $\varepsilon > 0$ there exists a number $\delta > 0$ such that

$$|f(x) - f(y)| < \varepsilon \qquad \text{for} \quad \rho(x, y) < \delta. \tag{1}$$

The notion uniformly continuous is justified by

**Theorem 1.** *A uniformly continuous function from a metric space is continuous.*

*Proof:* Let $f$ be uniformly continuous and put $E = [x : f(x) > a]$, $a$ real number. For $y \in E$ and sufficiently small $\delta_y$, the open sphere $G_{\delta_y}(y)$ belongs to $E$ according to (1) and clearly

$$E = \bigcup_y G_{\delta_y}(y).$$

Hence $E$ is open. Since $-f$ is uniformly continuous if $f$ is uniformly continuous, and $[x : f(x) < a] = [x : -f(x) > -a]$, we find that $[x : f(x) < a]$ is also open. Thus it follows by Lemma 3.2 that $f$ is continuous. □

For any given set $E$ in the metric space $S$, we define the function $\rho(\cdot, E)$ and the distance $\rho(E_1, E_2)$ between two sets by

$$\rho(x, E) = \inf_{y \in E} \rho(x, y), \qquad \rho(E_1, E_2) = \inf_{x \in E_1,\, y \in E_2} \rho(x, y). \tag{2}$$

**Lemma 1.** *$\rho(\cdot, E)$ is uniformly continuous on any set $E'$, where it is bounded.*

*Proof:* Let $x_1$ and $x_2$ be points on $E'$ such that $\rho(x_1, x_2) < \delta$ and let $y \in E'$. By the triangle inequality we get for $y \in E'$:

$$\rho(x_1, E) \leq \rho(x_1, y) \leq \rho(x_1, x_2) + \rho(x_2, y) \leq \rho(x_2, y) + \delta$$
$$\Rightarrow \rho(x_1, E) \leq \rho(x_2, E) + \delta.$$

Here $x_1$ and $x_2$ may be interchanged. Thus

$$|\rho(x_1, E) - \rho(x_2, E)| \leq \delta \qquad \text{for} \quad \rho(x_1, x_2) < \delta. \quad \square$$

*Remark:* According to this lemma, for any given $x$ the function $\rho(x, \cdot)$ is continuous, and by Lemma 3.1 the sets $\{y : \rho(x, y) \leq a\}$ and $\{y : \rho(x, y) \geq a\}$ are closed for any positive number $a$. Clearly $\{y : \rho(x, y) \leq a\}$ is the closure of $\{y : \rho(x, y) < a\}$ and this set is open. The fact that $\rho(x, \cdot)$ is continuous follows in the same way as Theorem 1. Compare also the more general statement in Lemma 2 in Section 8.

Many properties of a metric space $S$ can be described by the help of sequences of elements. An infinite sequence $\{x_n\}$ in a subset $A$ of $S$ is called Cauchy convergent if $\lim_{m \to \infty, n \to \infty} \rho(x_m, x_n) = 0$, and convergent if $\lim_{n \to \infty} \rho(x_n, x) = 0$ for an element $x \in S$ but not necessarily belonging to $A$. If any convergent sequence in $A$ has its limit in $A$, then $A$ is said to be complete. Note that we permit elements $x_i$ to be identical for different $i$. A limit $x$ of $\{x_n\}$ is uniquely determined by the sequence since

$$\rho(x, y) \leq \rho(x, x_n) + \rho(x_n, y).$$

An $\varepsilon$-net of a set $A$ is a set of points such that the distance between any point in $A$ and some point in the net is smaller than $\varepsilon$. If $A$ has a finite $\varepsilon$-net, then it is said to be totally bounded.

**Lemma 2.** *The following properties imply each other:* (i) *The subset $A$ of the metric space is compact,* (ii) *any sequence $\{x_n\}$ in $A$ has a convergent subsequence with its limit in $A$,* (iii) *$A$ is totally bounded and complete.*

*Proof:* (i) $\Rightarrow$ (ii). The compact set $A$ is covered by finitely many spheres of radius $1/m$ for any integer $m$ since it is covered by the spheres $\{x : \rho(x, y) < 1/m\}$, $y \in S$. Consider an infinite sequence $\{x_n\}$, $x_n \in A$. We may assume that all $x_n$ are distinct since either $\{x_n\}$ contains an infinite subsequence of distinct $x_i$ or all $x_i$ are equal, say equal to $x$ for $i > n_0$ (then $x$ is the limit of the sequence). If $\{x_n\}$ is an infinite sequence of distinct points, then there exists an open sphere $B_1$ of radius 1 that contains infinitely many distinct $x_i$. Assume that we have found open spheres $B_1, B_2, \ldots, B_m$ such that $A \supset B_1 \supset B_2 \supset \cdots \supset B_m$ and the distance between any two points in $B_i$ for $i \leq$ m is smaller than $2/m$, and all $B_i$ contain infinitely many distinct points

$x_j$. Since $A$ is covered by finitely many open spheres of radius $1/m+1$, $B_m$ is also covered by these spheres and thus contains an open sphere $B_{m+1}$ of radius $1/m+1$ that has infinitely many points $x_i$ in the given sequence. Let $\bar{B}_m$ be the closure of $B_m$. The set $\bigcap_{m=1}^{\infty} \bar{B}_m$ is not empty. Indeed, if it were, $A$ would be covered by $\bigcup_{m=1}^{\infty} B_n^c$ and then by finitely many $\bar{B}_j^c$, say by $\bar{B}_{j_1}^c \cup \bar{B}_{j_2}^c \cup \cdots \cup \bar{B}_{j_n}^c$, which is impossible since the complement of this set contains infinitely many points $x_i$ of the sequence. Having proved that $\bigcap_{m=1}^{\infty} \bar{B}_m$ contains a point $x$ which necessarily belongs to $A$ since $A$ is closed, we choose a point $x_{i_m}$ from the sequence in $\bar{B}_m$ for all $m$ such that the points $x_{i_m}$ are distinct. Clearly $\rho(x_{i_k}, x) < 2/m$ for $k \geq m$. Hence $\{x_{i_n}\}$ converges to $x$.

(ii) $\Rightarrow$ (iii). By (ii) $A$ is complete. Suppose that $A$ is not totally bounded. Then there exists an $\varepsilon > 0$ and an infinite sequence $\{x_n\}$ such that $\rho(x_n, x_{n-1}) \geq \varepsilon$ for all $n$. But then $\{x_n\}$ cannot contain a convergent subsequence.

(iii) $\Rightarrow$ (i). By (iii), $A$ is covered by a finite sequence $\{B_{mj}\}$, $j = 1, \ldots, k_m$ of open spheres with radius $2^{-m}$. Suppose that there exists an open covering $\{G_t\}$ of $A$ such that $A$ is not covered by a finite subset of $\{G_t\}$. Then at least one $B_{mi} \cap A$ is not covered by a finite subset of $\{G_t\}$. We can then successively choose open spheres $B^{(1)}, B^{(2)}, \ldots, B^{(m)}, \ldots$, such that $B^{(m)}$ has radius $2^{-m}$, $B^{(m-1)} \cap B^{(m)}$ is not empty for $m > 1$, and $B^{(m)}$ is not covered by any finite subset of $\{G_t\}$. Indeed, we may choose $B^{(1)}$ in this way. Having chosen $B^{(2)}$ for $k \leq m-1$, we observe that the $B_{mj}$ cover $B^{(m-1)} \cap A$. Hence we may choose $B^{(m)}$ such that $B^{(m)} \cap B^{(m-1)} \cap A$ is not covered by a finite subset of $\{G_t\}$. Clearly, then $B^{(m)} \cap A$ is not covered by a finite subset of $\{G_t\}$.

The distance between two points in $B^{(m)}$ is at most equal to $2 \times 2^{-m}$. Hence the distance between two points in $B^{(m-1)} \cap B^{(m)}$ is at most $6 \times 2^{-m}$. We choose a point $x_m \in B^{(m)} \cap A$ for all $m$. Then $\rho(x_{m-1}, x_m) < 6 \times 2^{-m}$ and $\rho(x_m, x_{m+n}) < 6 \times 2^{-m}$. Hence $\{x_m\}$ is Cauchy convergent, and since $A$ is complete, $\{x_m\}$ converges to a limit $x$ in $A$, and we have $\rho(x, x_m) < 6 \times 2^{-m}$ for $x \subset B^{(m)}$. Now $x$ belongs to some $G_t$, and since $G_t$ is open, there exists an open sphere $\{y : \rho(x, y) < \varepsilon\} \subset G_t$ for some $\varepsilon > 0$. But then $B^{(m)} \subset G_t$ for sufficiently large $m$, which contradicts the fact that $B^{(m)}$ is not covered by any finite subset of $\{G_t\}$. □

We call a metric space $S$ pseudocompact if any two disjoint closed sets have a positive distance. Note that any compact space is pseudocompact. Indeed, if $F_1$ and $F_2$ are any two disjoint closed sets in a compact space, the distance $\rho(F_1, F_2)$ between $F_1$ and $F_2$ is attained for points $x_1 \in F_1$ and $x_2 \in F_2$.

**Theorem 2.** *A metric space is completely normal. Any closed set $F$ in such a space $S$ has a representation $F = \{x : f(x) = 0\}$ with a uniformly continuous function $f$, $0 \leq f(x) \leq 1$ for $x \in S$. If $F_1$ and $F_2$ are disjoint closed sets, $F_1 = \{x : f_1(x) = 0\}$ and $F_2 = \{x : f_2(x) = 0\}$, $0 \leq f_i(x) \leq 1$ for $x \in S$ with such* *representations, then* $F_1$ *and* $F_2$ *are connected by* $f_1/(f_1 + f_2)$, *and this function is uniformly continuous if* $S$ *is pseudocompact.*

*Remark 1:* We may choose the function $f$ above as $f = g_\varepsilon \circ (1/\varepsilon)\rho(\cdot, F)$, where $\rho(\cdot, F)$ is the function considered in Lemma 1 and $g_\varepsilon(t) = 0$ for $t < 0$, $g_\varepsilon(t) = t$ for $\circ \le t \le \varepsilon$, and $g_\varepsilon(t) = \varepsilon$ for $t \ge \varepsilon$.

*Proof:* It is easily seen that $F = \{x : f(x) = 0\}$ for the function given in Remark 1. Consider the second statement in Theorem 2.

Let $\rho(F_1, F_2) \ge 2\varepsilon$ for two closed sets $F_1$ and $F_2$, and let $f_1$ and $f_2$ belong to $\varepsilon$ and $F_1$ and $F_2$, respectively, as in Remark 1. For any $z$ in $S$ we have

$$2\varepsilon \le \rho(F_1, F_2) \le \rho(F_1, z) + \rho(F_2, z).$$

Hence either $\rho(F_2, z) \ge \varepsilon$ [and then $f_2(z) \ge 1$] or $\rho(F_1, z) \ge \varepsilon$ [and then $f_1(z) \ge 1$]. Hence $f_1(z) + f_2(z) \ge 1$ for all $z$. Since $f_1$ and $f_2$ are uniformly continuous, we then find that $f_1/(f_1 + f_2)$ is uniformly continuous and certainly it connects $F_1$ and $F_2$. □

Note that two different metrics may determine the same topology and hence the same class of real-valued, bounded, continuous functions. The subclass of uniformly continuous functions, however, may not be the same.

**Theorem 3.** *A metric space is second-order countable if and only if it is separable.*

*Proof:* Let the metric space $S$ with metric $\rho$ be separable. This means that $S$ contains a countable dense set $E$ such that $\bar{E} = S$. The open spheres $\{x : \rho(x, y) < (1/r)\}$, where $y \in E$ and $r = 1, 2, \ldots$, form countable bases of $S$. Let $S$ have countable bases $\{B_i\}_{i=1}^{\infty}$ and choose a point $x_i$ in $B_i$ for all $i$. Let $E$ be the set of all these points $x_i$. We claim that $\bar{E} = S$. Indeed, if $x \in S$, then for any positive integer $r$ there exists $B_{j_r}$ such that $x \in B_{j_r} \subset \{y : \rho(x, y) < (1/r)\}$ and thus $\rho(x, x_{j_r}) < (1/r)$ with $x_{j_r} \in E$. Then $x \in \bar{E}$. □

## 8. MAPPINGS INTO METRIC SPACES

If $S$ and $S'$ are metric spaces with metrics $\rho$ and $\rho'$, respectively, we call a point $x \in S$ a continuity point of the mapping $T$ of $S$ into $S'$ if for any $\varepsilon > 0$ there exists $\delta > 0$ such that $\rho(x, y) < \delta$, $y \in S$ implies $\rho(Tx, Ty) < \varepsilon$. Otherwise $x$ is called a discontinuity point of $T$. This is a generalization of the elementary continuity concept.

**Lemma 1.** *A mapping* $T$ *from a metric space* $S$ *with metric* $\rho$ *into a metric space* $S'$ *with metric* $\rho'$ *is continuous at* $x \in S$ *if and only if for any sequence* $\{x_n\}$ *with* $\rho(x_n, x) \to 0$ $(n \to \infty)$ *we have* $\rho'(Tx_n, Tx) \to 0$ $(n \to \infty)$.

*Proof:* If $T$ is continuous at $x$ then by the definition above the relation $\rho(x_n, x) \to 0$ $(n \to \infty)$ implies $\rho'(Tx_n, Tx) \to 0$ $(n \to \infty)$. Conversely, suppose that the second relation follows from the first one and that $T$ is not continuous. Then to some $\varepsilon > 0$ there exists a sequence $\{y_n\}$ such that $\rho'(Tx, Ty_n) \geq \varepsilon$ and $\rho(x, y_n) < \delta_n$ with arbitrarily $\delta_n > 0$ for sufficiently large $n$. This is a contradiction. □

If the mapping $T$ from $S$ into $S'$ is continuous at all points $x \in S$, we say that $T$ is continuous everywhere.

**Lemma 2.** *Let $D_T$ be the set of discontinuity points of a measurable mapping $T$ from a metric space $S$ into a metric space $S'$. Then $D_T$ is measurable $S/S'$. For any closed set $F'$ in $S'$ we have*

$$\overline{T^{-1}F'} \subset T^{-1}F' \cup D_T. \tag{1}$$

*Hence if $T$ is continuous everywhere it is continuous.*

*Proof:* Put

$$A_{\varepsilon,\delta} = \bigcup_{\rho'(Ty,Tz) \geq \varepsilon} \{x : \rho(x, y) < \delta, \rho(y, z) < \delta\}.$$

This set is open for any $\varepsilon > 0$, $\sigma > 0$. We have

$$D_T = \bigcup_{\varepsilon > 0} \bigcap_{\delta > 0} A_{\varepsilon,\delta},$$

where $\varepsilon$, $\delta$ run through all positive rational numbers. Thus $D_T$ is measurable.

Suppose that $x \notin D_T$, $x \in \overline{T^{-1}F'} = E$. Thus $x$ is a continuity point of $T$. Then either $x$ belongs to the interior $E^\circ$ of $E$ and $Tx \subset F'$ according to the definition of continuity, or $x$ is a limit point of $T^{-1}F'$. That is, $\rho(x, x_n) \to 0$ $(n \to \infty)$ with $x_n \in E_0$, and then by Lemma 1 $\rho'(Tx_n, Tx) \to 0$ $(n \to \infty)$, which means that $Tx$ is a limit point of $F'$ and belongs to $F'$ since $F'$ is closed. □

**Lemma 3.** *If $\omega \to \xi_n(\omega)$ are measurable mappings from a measurable space $(\Omega, \mathscr{B})$ into a metric space $S$ with metric $\rho$ and $\rho[\xi_n(\omega), \xi(\omega)] \to 0$ $(n \to \infty)$, then $\omega \to \xi(\omega)$ is measurable.*

*Proof:* For an open sphere $G_\varepsilon(y)$ in $S$ we have

$$\{\omega : \xi(\omega) \in G_\varepsilon(y)\} = \bigcup_{m=1}^{\infty} \bigcap_{n=m}^{\infty} \{\omega : \xi_n(\omega) \in G_\varepsilon(y)\},$$

and this set belongs to $\mathscr{B}$ since $\omega \to \xi_n(\omega)$ is measurable for all $n$. Hence, $\xi^{-1}G_\varepsilon(y) \in \mathscr{B}$. Since the $\sigma$-algebra in $S$ is generated by the open spheres, it follows that $\omega \to \xi(\omega)$ is measurable. □

**Lemma 4.** *If $T$ is a continuous mapping of a metric space $S$ with metric $\rho$ into $S$, then the mapping $x \to \rho(x, Tx)$ is continuous. If $T$ is only measurable but $S$ is separable, then $x \to \rho(x, Tx)$ is measurable.*

*Proof:* Let $T$ be continuous. Using Lemma 1, we have to prove that $\rho(x, x_n) \to 0$ $(n \to \infty)$ implies

$$\rho(x, Tx) - \rho(x_n, Tx_n) \to 0 \qquad (n \to \infty).$$

Now

$$\rho(x, Tx) \leq \rho(x, x_n) + \rho(x_n, Tx_n) + \rho(Tx_n, Tx),$$

where $x$ and $x_n$ may be interchanged. Hence

$$|\rho(x, Tx) - \rho(x_n, Tx_n)| \leq \rho(x, x_n) + \rho(Tx_n, Tx).$$

Since $T$ is continuous, the right-hand side tends to 0 as $\rho(x_n, x) \to 0$. This proves the first part of the lemma. Let $T$ be only measurable but $S$ separable. Then there exists a countable set $\{x_i\}$ of elements in $S$ such that for any $x$ and any $\varepsilon > 0$, $\rho(x, x_i) < \varepsilon$ for some $i$. Hence, observing that

$$\rho(x, Tx) \leq \rho(x, x_i) + \rho(x_i, x_j) + \rho(x_j, Tx),$$

we get

$$\rho(x, Tx) < a = \bigcup_{ij} \bigcap_{r=1}^{\infty} \left\{x : \rho(x_i, x_j) < a,\ \rho(x, x_i) < \frac{1}{r},\ \rho(x_j, Tx) < \frac{1}{r}\right\}$$

and a corresponding inequality with $> a$ instead of $< a$. Hence, $x \to \rho(x, Tx)$ is measurable. □

## 9. PRODUCT SPACES

The set of all pairs $(x', x'')$, $x' \in S'$, and $x'' \in S''$ for sets $S'$ and $S''$ is called the Cartesian product of $S'$ and $S''$. We denote it by $S' \otimes S''$. Open sets $G'$ and $G''$ in $\sigma$-topological spaces $S'$ and $S''$, respectively, determine the Cartesian product $G' \otimes G''$, which is called an open rectangle in $S' \otimes S''$. Correspondingly, $F' \otimes F''$ for closed sets in $S'$ and $S''$ is called a closed rectangle. The open rectangles generate a $\sigma$-topological class of open sets in $S' \otimes S''$, and this class determines a $\sigma$-topology in $S' \otimes S''$. We call $S' \otimes S''$ a product A-space. If the $S'$ and $S''$ are topological spaces, we obtain a topological product space in this way.

Note that the $\sigma$-topological (topological) class of open sets is also generated by the open rectangles $G' \otimes S''$ and $S' \otimes G''$, and the $\sigma$-topological (topological) class of closed sets is also generated by the rectangles $F' \otimes S''$ and $S' \otimes F''$.

**Theorem 1.** *The product space $S' \otimes S''$ of two spaces $S', S''$ is a topological* ($\sigma$-topological) *space if $S'$ and $S''$ are topological ($\sigma$-topological) spaces, a completely normal A-space if $S'$ and $S''$ are completely normal A-spaces, and a metric space if $S'$ and $S''$ are metric spaces.*

*Remark:* Note that the product space has a $\sigma$-topology when we consider A-spaces, and a topology when we consider topological spaces.

*Proof:* It only remains to prove the statements for normal and completely normal A-spaces, and particularly for metric spaces. Thus let $S'$ and $S''$ be normal A-spaces and $\mathscr{S}'$ and $\mathscr{S}''$ be the algebras generated by the closed sets in $S'$ and $S''$, respectively. Form the simple functions:

$$f = \sum_{i=1}^{n} \alpha_i 1_{E_i' \otimes E_i''}, \qquad E_i' \in \mathscr{S}', \quad E_i'' \in \mathscr{S}'', \quad \alpha_i \in R,$$
$$E_i' \otimes E_i'' \cap E_j' \otimes E_j'' = \varnothing \qquad \text{for} \quad i \neq j, \quad \bigcup_{i=1}^{n} E_i' = S', \quad \bigcup_{j=1}^{n} E_j'' = S''. \tag{1}$$

Denote the class of all these simple functions by $\Psi_0$. In the same way as in Section 4, we give representations of any two functions $f$ and $g$ in $\Psi_0$ in the form (1) with the same sets $E_i' \otimes E_j''$ and we find that $\Psi_0$ is a Stone vector lattice ring. By Theorem 4.1, its closure $\bar{\Psi}_0$ with respect to the uniform metric is a complete Stone vector lattice ring of bounded functions. Clearly the class $\bar{\Psi}$ of continuous functions contained in $\bar{\Psi}_0$ is also a Stone vector lattice ring.

If $f'$ is a real-valued, bounded, continuous function from $S'$ into $R$, then the function $f$ defined by

$$f(x) = f'(x') \qquad \text{for} \quad x = (x', x''), \quad x' \in S', \quad x'' \in S'' \tag{2}$$

belongs to $\bar{\Psi}$. Indeed, we easily find by Lemma 3.2 that $f$ is continuous. Moreover, $f \in \bar{\Psi}_0$ since $f'$ is the limit in the uniform metric of a sequence $f_n'$ of elementary functions from $S'$ into $R$, and hence $f = f' \cdot 1_{S'}$ is the limit in the uniform metric of the sequence $\{f_n' \cdot 1_{S''}\}$ of elementary functions from $S' \otimes S''$ into $R$. In the same way, we conclude that $f'' \cdot 1_{S'} \in \bar{\Psi}_0$ if $f''$ is a real-valued, bounded, continuous function from $S''$. Since $\bar{\Psi}$ is a complete Stone vector lattice ring, according to Theorem 5.3, it determines a $\sigma$-topological class $\bar{\mathscr{F}}$ of closed sets consisting of sets $\bar{F}$ with a representation $\bar{F} = \{x : f(x) = 0\}$ with $f \in \bar{\Psi}$, $0 \leq f(x) \leq 1$ for $x \in S$. Now suppose that $S'$ and $S''$ are completely normal. Then any closed sets $F'$ in $S'$ and $F''$ in $S''$ have representations

$$F' = \{x : f'(x') = 0\}, \qquad F'' = \{x'' : f''(x'') = 0\},$$

where

$$0 \leq f'(x') \leq 1 \quad \text{for} \quad x' \in S', \qquad 0 \leq f''(x'') \leq 1 \quad \text{for} \quad x'' \in S''.$$

If $f$ belongs to $f'$ as in (2), then $f \in \bar{\Psi}$ and $F' \otimes S''$ has the representation

$$F' \otimes S'' = \{x : f(x) = 0\}. \tag{3}$$

Thus $F' \otimes S''$ belongs to $\bar{\mathscr{F}}$. But the $\sigma$-topological class $\mathscr{F}$ of the product $\sigma$-topological space $S' \otimes S''$ is generated by the closed sets $F' \otimes S''$ and $S' \otimes F''$. Hence $\mathscr{F} \subset \bar{\mathscr{F}}$. This means that $S' \otimes S''$ is completely normal.

If $S'$ and $S''$ are metric spaces with metrics $\rho'$ and $\rho''$, respectively, then for $x = (x', x'')$, $y = (y', y'')$,

$$\rho(x, y) = \{[\rho'(x', y')]^2 + [\rho''(x'', y'')]^2\}^{1/2} \tag{4}$$

is a metric of the product space $S' \otimes S''$. The fact that $\rho$ is a metric is easy to show. We prove that the metric space determined by the metric (4) and the product space with the topology given above have the same topology. Indeed, let $\bar{\mathscr{G}}$ be the class of open sets generated by the open balls $B_\varepsilon(y) = \{x : \rho(x, y) < \varepsilon\}$, and $\mathscr{G}$ the class of open sets generated by the open rectangles $G' \otimes S''$ and $S' \otimes G''$, where $G'$ is the open rectangle in $S'$ and $G''$ is the open rectangle in $S''$. For sufficiently small $\varepsilon$ and $y' \subset G'$, $y'' \in S''$, the open ball $B_\varepsilon(y)$, where $y = (y', y'')$, belongs to $G' \otimes S''$; and this set is the union of such balls belonging to all points $(y', y'')$ in $G' \otimes S''$. The union of open balls is in the same $S' \otimes G'$. Let $B_\varepsilon(y_0)$ be a given ball and $y = (y', y'') \in B_\varepsilon(y)$. The set

$$\{x' : \rho'(x', y') < \eta\} \cap \{x'' : \rho''(x'', y'') < \eta\}$$

belongs to $B_\varepsilon(y)$ if $\eta\sqrt{2} < \varepsilon$ and $B_\varepsilon(y)$ is the union of such intersections belonging to all points $y \in B_\varepsilon(y_0)$. □

Example: $R^{(k)}$ is the product space of $k$ factors $R$, all factors being metric spaces.

If $f$ is a real-valued function from a product space $S' \times S''$, the function $f'_{x''}$ from $S'$, defined by

$$f'_{x''}(x') = f(x) \qquad \text{for} \quad x = (x', x''),$$

is called the section of $f$ from $S'$ at $x''$. In the same way, the section $f''_{x'}$ of $f$ from $S''$ at $x'$ is defined. For a set $E$ on $S' \times S''$, the set $\{x' : (x', x'') \in E\}$ in $S'$ is called the section of $E$ on $S'$ at $x''$. In the same way, the section of $E$ on $S''$ at $x'$ is defined.

**Theorem 2.** *The sections of a closed (measurable) set of a product A-space are closed. The sections of a continuous function from such a space into R are continuous.*

*Proof:* Let $\mathscr{C}$ be the class of all closed sets $F$ in $S' \times S''$ such that the sections of $F$ are closed. It is easy to verify that $\mathscr{C}$ is a $\sigma$-topological class of closed sets. Further it contains the sets $F' \times S''$ and $S' \times F''$, where $F'$ and $F''$ are closed sets on $S'$ and $S''$, respectively. But the sets $F' \times S''$ and $S' \times F''$ generate the $\sigma$-topological class of all closed sets. Hence $\mathscr{C}$ is identical with this class. The statements about continuity follow from Lemma 3.2, since $\{x:f'_{x''}x') \leq a\}$ and $\{x':f'_{x''}(x')\} \geq a$ are the sections of $\{x:f(x) \leq a\}$ and $\{x:f(x) \geq a\}$ on $S'$ at $x''$, respectively; correspondingly for the sections on $S''$. The class $\Psi_0$, considered above, does not always contain all continuous functions. However, let $\bar{\Psi}_0$ be the class of all pointwise bounded limits of sequences $\{f_n\}$, $f_n \in \Psi_0$. Then $\bar{\bar{\Psi}}_0$ is closed with respect to pointwise bounded limits and clearly $\bar{\Psi}_0 \subset \bar{\bar{\Psi}}_0$. If $F$ is a closed set and $F = \{x:f(x) = 1\}$, $0 \leq f(x) \leq 1$ for $x \in S' \times S''$, $f \in S$, then $1_F$ is the bounded (moreover monotone) pointwise limit of $\{f_n\}$ as $n \to \infty$, and thus $1_F \in \bar{\bar{\Psi}}_0$. It then follows by Theorem 4.2 that $\bar{\bar{\Psi}}_0$ contains all continuous functions.

If $S'$ and $S''$ are measurable spaces with $\sigma$-algebras $\mathscr{S}'_\sigma$ and $\mathscr{S}''_\sigma$, we define the product-measurable space $S' \times S''$ as the product space whose $\sigma$-algebra $\mathscr{S}_\sigma$ is generated by the rectangles $E' \times E''$, $E' \in \mathscr{S}'_\sigma$, $E'' \in \mathscr{S}''_\sigma$. Consider then $S' \times S''$ as the product A-space of the A-spaces $S'$ and $S''$ with the $\sigma$-topological classes $\mathscr{S}'_\sigma$ and $\mathscr{S}''_\sigma$, respectively. Clearly $S'$ and $S''$ are completely normal spaces according to this $\sigma$-topology. We apply the theory above. Then $\Psi_0$ is the class of all simple functions (Theorem 1) with $E' \in \mathscr{S}'_\sigma$ and $E'' \in \mathscr{S}''_\sigma$. We find that any "closed" set $E$ in $S = S' \times S''$ has a representation $E = \{x:f(x) = 0\}$ with $f \in \bar{\Psi}_0$, $0 \leq f(x) \leq 1$ for $x \in S$. But $E$ is now any measurable set. Further, by the remark above, the class $\bar{\bar{\Psi}}_0$ contains any indicator function $1_E$, $E$ a measurable set in $\mathscr{S}_\sigma$ and any bounded, measurable function. □

**Theorem 3.** *The sections of any measurable space in a product-measurable space are measurable. The sections of any measurable function from a product-measurable space are measurable.*

*Proof:* Theorem 3 follows from Theorem 2 if we deal with $\sigma$-algebras as $\sigma$-topological classes of closed sets. □

## 10. PRODUCT SPACES OF INFINITELY MANY FACTORS

Let $\Upsilon$ be a countable set of distinct points on the line and $S_t$ an A-space for any $t \in \Upsilon$. To the space $S_t$ we form a space $S_\Upsilon$ such that any point in $S_\Upsilon$ is the collection $x = \{x\}_{t \in \Upsilon}$ where $x^{(t)} \in S_t$. We say that $x^{(t)}$ is the coordinate of $x$ at the place $t$ on the line. $\Upsilon$ is called an index set, any subsequence $\{t_i\}$ of $\Upsilon$ an index sequence, and any finite vector $(t_1, t_2, \ldots, t_r)$ with distinct numbers $t_i$ an index vector. For any two index sequences (vectors) $\mathbf{t} = (t_1, t_2, \ldots)$ and $\mathbf{s} = (s_1, s_2, \ldots)$, we put $\mathbf{s} \prec \mathbf{t}$ if for any $i$ we have $s_i \leq t_i$. Note that for any index sequence $\mathbf{s}$ and $\mathbf{t}$ there exists an index sequence $\boldsymbol{\omega}$ such that $\mathbf{s} \prec \boldsymbol{\omega}$, $\mathbf{t} \prec \boldsymbol{\omega}$. According to this property, the index sequences (vectors) are said to be partially ordered.

We now consider index vectors. To the index vector $\mathbf{t} = (t_1, t_2, \ldots, t_r)$ there belongs the mapping $\pi_\mathbf{t}$ of $S$ onto the finite-dimensional product space $S_\mathbf{t} = S_{t_1} \otimes S_{t_2} \otimes \cdots \otimes S_{t_r}$. It is called a projection of $S$ onto $S_\mathbf{t}$. Two index vectors $\mathbf{s}$ and $\mathbf{t}$, $\mathbf{s} \prec \mathbf{t}$ determine the product spaces $S_\mathbf{s}$ and $S_\mathbf{t}$ and the projection $\pi_{\mathbf{t},\mathbf{s}}$ of $S_\mathbf{t}$ onto $S_\mathbf{s}$. This mapping was considered in the preceding section and it was shown that it is continuous.

We introduce a $\sigma$-topology on $S$ as follows. If $F_\mathbf{s}$ is a closed set on $S_\mathbf{s}$, we call $\pi_\mathbf{s}^{-1} F_\mathbf{s}$ a closed set on $S_\Upsilon$. All such sets generate a $\sigma$-topological class of closed sets, which we let be the $\sigma$-topological class of closed sets on $S_\Upsilon$. If the $S_\mathbf{t}$ are topological spaces, we may form a topological space in the same way. In this $\sigma$-topology (topology) $\pi_\mathbf{t}$ is continuous. Indeed, for any closed set $F_\mathbf{t}$ in $S_\mathbf{t}$, $\pi_\mathbf{t}^{-1} F_\mathbf{t}$ is closed by definition.

If $\Upsilon = (1, 2, 3, \ldots)$ and the $S_i$, $i = 1, 2, \ldots$ are metric spaces with metrics $\rho_i$, respectively, then $\rho$, defined by

$$\rho(x, y) = \sum_{i=1}^{\infty} 2^{-i} \frac{\rho_i[x^{(i)}, y^{(i)}]}{1 + \rho_i[x^{(i)}, y^{(i)}]}, \tag{1}$$

is a metric on $S_\Upsilon$. To show this, it is sufficient to show that $\rho$ satisfies the triangle inequality. Other properties of a metric can be verified at once. Clearly, the sum in (1) satisfies the triangle inequality if the terms satisfy this inequality, or $\rho/1 + \rho$ satisfies this inequality if $\rho$ satisfies it. Now $1/(1 + t)$ is increasing on $(0, \infty)$, and hence

$$\frac{\rho(x, y)}{1 + \rho(x, y)} \leq \frac{\rho(x, z) + \rho(z, y)}{1 + \rho(x, z) + \rho(z, y)}.$$

Thus, we have only to verify the simple inequality

$$\frac{\alpha + \beta}{1 + \alpha + \beta} \leq \frac{\alpha}{1 + \alpha} + \frac{\beta}{1 + \beta}$$

for nonnegative numbers $\alpha$ and $\beta$. The fact that the metric (1) determines the same topology as the topology, determined as for general product spaces, can easily be shown, as for product spaces of finitely many factors.

By the definition of the metric on the metric product space $S$ of metric spaces, a sequence $\{x_n\}$ in $S$ converges if and only if the sequences of coordinates $\{x_n^{(i)}\}$ converge for all $i$. Hence we get, applying Lemma 7.2,

**Theorem 1.** *The metric product space of countably many compact metric spaces is compact.*

Product-measurable spaces of countably many measurable spaces are defined in the same way as product A-spaces. In fact, as $\sigma$-topological classes of closed sets they are special A-spaces for the $\sigma$-algebras in the factor spaces.

## 11. SOME PARTICULAR METRIC SPACES

The real line $R$, already considered in Section 1, is the metric space with the absolute value $|x - y|$ as metric. Any closed interval $[a, b]$ is compact. Indeed, any sequence of numbers in $[a, b]$ contains a convergent subsequence. This is a simple consequence of the definition of real numbers. Applying Lemma 7.2, we conclude that $[a, b]$ is compact.

The space $R^{(k)}$ is the product space of $k$ factors $R$ and, as such, it has the metric

$$\rho(x, y) = \left\{\sum_{i=1}^{k} (x^{(i)} - y^{(i)})^2\right\}^{1/2}, \tag{1}$$

where both here and in what follows $x = [x^{(1)}, \ldots, x^{(k)}]$ for the elements, also called vectors, in $R^{(k)}$. We write $x < y$ or $y > x$ if $x^{(i)} < y^{(i)}$ for all $i$, and $x \leq y$ or $y \geq x$ if $x^{(i)} \leq y^{(i)}$ for all $i$. The set $(a, b) = \{x : a < x < b\}$ is called an open and $[a, b] = \{x : a \leq x \leq b\}$ a closed interval. Correspondingly, we use notations $(a, b]$ and $[a, b)$.

By the definition of the metric, it follows that a sequence $\{x_n\}$ of vectors in the closed interval $[a, b]$ is convergent if and only if the sequences $\{x_n^{(i)}\}$ for the coordinates of the vectors $x_n$ converge. Thus, using the same arguments as for $R$, we conclude that any closed set $[a, b]$ is compact.

We use the vector notations

$$x \cdot y = \sum_{i=1}^{k} x^{(i)} y^{(i)}, \qquad x \cdot x = \|x\|^2,$$

where $x \cdot y$ is called the scalar product and $\|x\|$ the norm of $x$. Note that by (1) we have

$$\rho(x, y) = \|x - y\|. \tag{2}$$

A mapping $\pi(t)$ of $R^{(k)}$ into $R$, defined by $\pi(t)x = x \cdot t$ for a given vector $\mathbf{t} \in R^{(k)}$, is called a projection. It is continuous, according to Lemma 8.1. The space $R^{(\infty)}$ is a product space of countably many factors $R$, which may be numbered $R_1$, $R_2$, etc. Hence, any element $x$ in $R^{(\infty)}$ is an infinite sequence

$x = (x^{(1)}, x^{(2)}, \ldots)$. As product space of factors $R$, it has the metric

$$\rho(x, y) = \sum_{i=1}^{\infty} \frac{|(x^{(i)} - y^{(i)}|}{1 + |x^{(i)} - y^{(i)}|} \cdot 2^{-i}. \tag{3}$$

A sequence $\{x_n\}$ in $R^{(\infty)}$ is convergent if and only if $\{x_n^{(i)}\}$ is convergent for the sequence $\{x_n^{(i)}\}$ of coordinates for $i = 1, 2, \ldots$. Using the same arguments as for $R$ and $R^{(k)}$, we conclude that $R^{(\infty)}$ is compact on the set

$$\{x : |x^{(i)} - y^{(i)}| \leq a^{(i)},\ i = 1, 2, \ldots\}.$$

The real Hilbert space $l^2$ consists of all sequences $\{x^{(i)}\}_{i=1}^{\infty}$ with real "coordinates" $x^{(i)}$ and

$$\sum_{i=1}^{\infty} (x^{(i)})^2 < \infty.$$

We denote the sum on the left-hand side by $\|x\|^2$, where we choose $\|x\| \geq 0$. Note that we add sequences by adding the coordinates in the usual way and multiply a sequence by a real number by multiplying its coordinates by this number. Then $l^2$ is a linear space over the real number field, i.e., it has the properties

$$x, y \in l^2 \Rightarrow \alpha x + \beta y \in l^2, \qquad x \in l^2 \Rightarrow \alpha x \in l^2 \tag{4}$$

for any real numbers $\alpha$ and $\beta$. Any space $S$ such that $\alpha x$ is defined and belongs to $S$, where $x \in S$ and $\alpha$ is a real number, and that has property (4) ($l^2$ changed to $S$) is called a linear space over the real number field. [We use the notation $(-1)x = -x$.]

A linear space $S$ over the real number field has a scalar product $x \cdot y$ over $S$ if $x \cdot y$ is defined as a real number for $x, y \in S$ and if this "product" has the properties

$$\begin{gathered} x \cdot y = y \cdot x, \qquad (\alpha x, y) = \alpha(x \cdot y), \qquad (x + y) \cdot z = x \cdot z + y \cdot z, \\ |x \cdot y| \leq [x \cdot x]^{1/2} \cdot [y \cdot y]^{1/2}, \end{gathered} \tag{5}$$

where $x$, $y$, $z \in S$ and $\alpha$ is a real number. The square root $\sqrt{x \cdot x}$ is called the norm and is denoted by $\|x\|$. On $l^2$, the scalar product $x \cdot y$ is defined as the sum

$$x \cdot y = \sum_{i=1}^{\infty} x^{(i)} y^{(i)}. \tag{6}$$

By Cauchy's inequality, we get from (6)

$$|x \cdot y| \leq \left\{\sum_{i=1}^{\infty} [x^{(i)}]^2\right\}^{1/2} \left\{\sum_{i=1}^{\infty} [y^{(i)}]^2\right\}^{1/2} = \|x\| \cdot \|y\|. \tag{7}$$

Using the properties of the scalar product, we obtain for $x, y \in l^2$

$$\|x + y\|^2 = \|x\|^2 + \|y\|^2 + 2x \cdot y,$$

and regarding (7),

$$\|x + y\| \leq \|x\| + \|y\|. \tag{8}$$

Further it follows from the properties of the scalar product that

$$\|\alpha x\| = |\alpha| \, \|x\|. \tag{9}$$

A real-valued function $\|x\|$ from a linear space $S$ is called a norm if it has the properties (8) and (9). The norm determines the metric

$$\rho(x, y) = \|x - y\|. \tag{10}$$

If a linear space has a scalar product and a norm and is complete in the metric (10), it is called a Hilbert space. [The inequality $|x \cdot y| \leq \|x\| \cdot \|y\|$ in (5) is in fact a consequence of the other properties in (5) for any linear space over $R$ with a scalar product.]

We show that $l^2$ is complete and thus is a Hilbert space. Indeed, if

$$\|x_n - x_m\| \to 0 \qquad (n \to \infty, m \to \infty) \tag{11}$$

for a sequence $x_m$ in $l^2$, then

$$\|x_n\| \leq \|x_m\| + \|x_n - x_m\|.$$

Hence, $\sup_n |x_n| < \infty$, which means that the coordinates $x_n^{(i)}$ are uniformly bounded with respect to $i$ and $n$:

$$\sup_{i,n} |x_n^{(i)}| < \infty. \tag{12}$$

Using (12), we first show that there exists a sequence $\bar{N}$ of positive integers and an element $x$ in $l^2$ such that

$$\|x_m - x\| \to 0 \qquad (n \to \infty, n \in \bar{N}).$$

We determine $\bar{N}$ and $x$ by the selection principle, also called the diagonal method, as follows. First choose an infinite sequence $\bar{N}_1$ of positive integers such that $x_m^{(1)}$ converges to a number $x^{(1)}$ as $m \to \infty, m \in \bar{N}_1$. Then from $\bar{N}_1$, choose an infinite subsequence $\bar{N}_2$ such that $x_m^{(2)}$ converges to a number $x^{(2)}$ as $m \to \infty$, $m \in \bar{N}_2$, and so on. From $\bar{N}_{k-1}$, choose a subsequence $\bar{N}_k$ such that $x_m^{(k)}$ converges to a number $x^{(k)}$ as $m \to \infty$, $m \in \bar{N}_k$, etc. Let $\bar{N}$ be the sequence of positive integers such that the first term in $\bar{N}$ is the first term in $\bar{N}_1$, the second term in $\bar{N}$ is the second term in $\bar{N}_2$, etc. Then clearly

$$x_m^{(k)} \to x^{(k)} \qquad (m \to \infty, m \in \bar{N}).$$

Having found the sequence $x = \{x^{(k)}\}_{k=1}^{\infty}$, we get by (11)

$$\|x_n - x\| \leq \|x_n - x_m\| + \|x_m - x\| \to 0$$

as $n \to \infty$, $m \to \infty$, and $m \in \bar{N}$. Hence, $\{x_n\}$ converges to $x$ in the metric. □

**Lemma 1.** *A sequence $\{x_n\}$ of elements in a Hilbert space $H$ converges to an element $x$ in $H$ if and only if $x_n \cdot t \to x \cdot t \quad (n \to \infty)$ for any $t \in H$ and $\|x_n\| \to x$.*

*Proof:* If $x_n \cdot t \to x \cdot t$ for any $t \in H$ and $\|x_n\| \to \|x\|$, then

$$\|x_n - x\|^2 = \|x_n\|^2 - 2x \cdot x_n + \|x\|^2 \to \|x\|^2 - 2\|x\|^2 + \|x\|^2 = 0.$$

Conversely, if $\|x_n - x\| \to 0$, $n \to \infty$, then

$$\|x_n \cdot t - x \cdot t\| = |t \cdot (x_n - x)| \leq \|t\|\,\|x_n - x\| \to 0.$$

Moreover,

$$\|x_n\| \leq \|x\| + \|x_n - x\|, \qquad \|x\| \leq \|x_n\| + \|x - x_n\|,$$

and thus $\lim\|x_n\| = \|x\|$. □

## 12. MEASURES ON AN ALGEBRA OF SUBSETS

Let $S$ be a set and $\mathscr{S}$ an algebra of subsets of $S$. We call a set function $\mu$ on $\mathscr{S}$ a signed measure if it has the following properties:

(i) additivity:

$$\mu(E_1 \cup E_2) = \mu(E_1) + \mu(E_2) \qquad \text{for} \quad E_1, E_2 \in \mathscr{S}, \quad E_1 \cap E_2 = \varnothing,$$

(ii) $\mu(\varnothing) = 0, \qquad \sup_{E \in S} |\mu(E)| < \infty.$

If $\mu(E) \geq 0$ for $E \in \mathscr{S}$, the signed measure is called a measure. The notation "signed measure" is justified by

**Theorem 1.** *If $\mu$ is a signed measure on the algebra $\mathscr{S}$ and the set functions $\bar{\mu}$, $\mu$ and $|\mu|$ are defined on $\mathscr{S}$ by*

$$\bar{\mu}(E_0) = \sup_{E \subset E_0,\, E \in \mathscr{S}} \mu(E), \qquad \mu(E_0) = \sup_{E \subset E_0,\, E \in \mathscr{S}} -\mu(E), \qquad |\mu| = \bar{\mu} + \mu,$$

*then $\mu$, $\bar{\mu}$, $|\mu|$, and $\tilde{\mu} = |\mu| - \mu$ are measures on $\mathscr{S}$.*

Then note that $\mu = |\mu| - \tilde{\mu}$, so that $\mu$ is a difference between two measures.

*Proof:* Let $\mu$ be a signed measure on $\mathscr{S}$. Clearly $\bar{\mu}(E_0) \geq 0$, since $\mu(\varnothing) = 0$. For $E_1$, $E_2 \in \mathscr{S}$, $E_1 \cap E_2 = \varnothing$, the set $E_i \cap E$ runs through all subsets of

$E_i$ $(i = 1, 2)$ when $E$ runs through all subsets of $E_1 \cap E_2$. Hence

$$\begin{aligned}\bar{\mu}(E_1 \cup E_2) &= \sup_{E \subset E_1 \cup E_2} \mu(E) \\ &= \sup_{E \in E_1 \cup E_2} [\mu(E_1 \cap E) + \mu(E_2 \cap E)] \leq \sup_{E_1' \subset E_1} \mu(E_1') + \sup_{E_2' \subset E_2} \mu(E_2') \\ &= \bar{\mu}(E_1) + \bar{\mu}(E_2).\end{aligned}$$

However, equality must hold in these relations, since $E_1' \cup E_2' \subset E_1 \cup E_2$. Thus $\bar{\mu}$ is additive and hence a measure on $\mathscr{S}$. In the same way we find that $\underline{\mu}$ is a measure on $\mathscr{S}$, and then clearly $|\mu|$ and $|\mu| - \mu$ are measures on $\mathscr{S}$. □

A set function $\mu$ on $\mathscr{S}$ is said to be a $\sigma$-smooth measure on $\mathscr{S}$ if it is a measure and if

$$\mu\left(\bigcup_{i \geq 1} E_i\right) = \sum_{i \geq 1} \mu(E_i) \tag{1}$$

for pairwise disjoint sets, such that $E_i \in \mathscr{S}$, $\bigcup_{i \geq 1} E_i \in \mathscr{S}$.

**Theorem 2.** *A measure $\mu$ on $\mathscr{S}$ is $\sigma$-smooth if and only if $\lim_{n \to +\infty} \mu(E_n) = 0$ for any sequence $\{E_n\}$, $E_n \in \mathscr{S}$ such that $E_n \downarrow \varnothing$.*

*Proof:* Let $\{E_n\}$ be a sequence in $\mathscr{S}$ such that $E_n \downarrow \varnothing$. Then $\{E_{n-1} \backslash E_n\}_{n \in N}$ is a disjoint sequence and

$$E_1 = \bigcup_{n \geq 2} (E_{n-1} \backslash E_n).$$

Hence, if $\mu$ is $\sigma$-smooth, we get

$$\mu(E_1) = \sum_{n \geq 1} \mu(E_{n-1} \backslash E_n) = \lim_{n \to +\infty} \sum_{j=2}^{n} [\mu(E_{j-1}) - \mu(E_j)] = \mu(E_1) - \lim_{n \to +\infty} \mu(E_n),$$

and thus $\lim_{n \to \infty} \mu(E_n) = 0$. Let $\{E^{(n)}\}_{n \in N}$ be a disjoint sequence and put $E = \bigcup_{j \geq 1} E^{(j)}$, $E_n = \bigcup_{j=1}^{n} E^{(j)}$. Then

$$\mu(E) = \mu(E_n) + \mu(E \backslash E_n) = \sum_{j=1}^{n} \mu[E^{(j)}] + \mu(E \backslash E_n),$$

and if $\mu(E \backslash E_n) \downarrow 0$, we get

$$\mu(E) = \sum_{j \geq 1} \mu(E^{(j)}). \quad \square$$

Besides measures, also called finite measures, on a measurable space $(S, \mathscr{S}_\sigma)$, we consider additive set functions $\mu$, having all the properties of a $\sigma$-smooth measure except finiteness, i.e., we have $\mu(S) = +\infty$, and hence $\mu(E)$ may be infinite for some $E \in \mathscr{S}_\sigma$. Such an additive set function is called a $\sigma$-finite measure if there exist a disjoint sequence $\{S_i\}_{i\in N}$ of sets $S_i \in \mathscr{S}_\sigma$ such that $\bigcup_{i\geq 1} S_i = S$ and $\mu(S_i) < \infty$ for all $i$. Putting

$$\mu_i(E) = \mu(E \cap S_i), \qquad i = 1, 2, \ldots,$$

it is easily seen that $\mu_i$ is a finite measure on $(S, \mathscr{S}_\sigma)$ and that

$$\mu(E) = \sum_{i\geq 1} \mu_i(E)$$

according to the $\sigma$-smoothness of $\mu$, which means that $\mu(E) = \sum_{i\geq 1} \mu(E_i)$ for a disjoint sequence $\{E_i\}$ with $\bigcup_{i\geq 1} E_i = E$. □

A measurable space $(S, \mathscr{S}_\sigma)$, together with a measure $\mu$ on $\mathscr{S}_\sigma$, is called a measure space and is denoted by $(S, \mathscr{S}_\sigma, \mu)$. If $\mu$ is a probability measure, i.e., a $\sigma$-smooth measure with $\mu(S) = 1$, we call $(S, \mathscr{S}_\sigma, \mu)$ a probability space.

**Theorem 3.** *Let $\mathscr{E}$ be a class of subsets of a set $S$ such that $E_1, E_2 \in \mathscr{E} \Rightarrow E_1 \cap E_2 \in \mathscr{E}$, and let $\mathscr{S}_\sigma$ be the $\sigma$-algebra generated by $\mathscr{E}$. A $\sigma$-smooth measure $\mu$ on $\mathscr{S}_\sigma$ is then uniquely determined by its values on $\mathscr{E}$.*

*Proof:* Let $\mu_1$ and $\mu_2$ be any two $\sigma$-smooth measures on $\mathscr{S}_\sigma$, and denote the class of subsets $E$ in $\mathscr{S}_\sigma$, for which

$$\mu_1(E) = \mu_2(E), \tag{2}$$

by $\mathscr{D}$. It is easy to show that $\mathscr{D}$ is a Dynkin class and contains $\mathscr{E}$ if (2) holds for $E \in \mathscr{E}$. By Theorem 1.1, the class $\mathscr{D}$ is identical with the $\sigma$-algebra $\mathscr{S}_\sigma$. Hence (2) holds for any $E \in \mathscr{S}_\sigma$. □

## 13. MEASURES ON A-SPACES

Consider an A-space $S$, and let $\mathscr{S}$ be the algebra generated by the closed sets. If $\mu$ is a (signed) measure on $\mathscr{S}$, we say that $\mu$ is a (signed) measure on the A-space $S$. A (signed) measure on an A-space is called regular if for any set $E \in \mathscr{S}$ and any $\varepsilon > 0$ there exists a closed set $F$ such that $F \in E$ and

$$|\mu(E \backslash F)| < \varepsilon. \tag{1}$$

This condition can also be given in the following form: For $E \in \mathscr{S}$ and $\varepsilon > 0$, there exists an open set $G$ such that $E \in G$ and

$$|\mu(G \backslash E)| < \varepsilon. \tag{2}$$

Indeed, for $S \backslash E$ we can find a closed set $F$ such that $|\mu(S - E - F)| < \varepsilon$.

Putting $S \backslash F = G$, we get (2). For a measure $\mu$ we can combine (1) and (2):

$$\mu(F) \leq \mu(E) \leq \mu(G), \qquad F \subset E \in G, \qquad \mu(G \backslash F) < 2\varepsilon. \tag{3}$$

**Theorem 1.** *If $\mu$ is a regular signed measure on an A-space, then the measures $\bar{\mu}$, $\underline{\mu}$, and $|\mu|$, determined according to Theorem 12.1 are regular measures.*

*Proof:* It follows by the definition of $\bar{\mu}$, $\underline{\mu}$ and $|\mu|$ that, for $E_0 \in \mathscr{S}$,

$$\bar{\mu}(E_0) = \sup_{F \in E_0} \mu(F), \qquad \mu(E_0) = \sup_{F \in E_0} -\mu(F),$$

where $F$ is a closed set. Hence there exist closed sets $F_1 \in E_0$ and $F_2 \in E_0$ for a given $\varepsilon > 0$ such that

$$\bar{\mu}(E_0 \backslash F_1) < \varepsilon, \qquad \mu(E_0 \backslash F_2) < \varepsilon;$$

hence

$$\bar{\mu}[E_0 \backslash (F_1 \cup F_2)] < \varepsilon, \quad \underline{\mu}[E_0 \backslash (F_1 \cup F_2)] < \varepsilon, \quad |\mu|(E_0 \backslash F_1 \cup F_2) < 2\varepsilon. \quad \square$$

A regular measure is then uniquely determined by its values on closed sets. As a complement to Theorem 10.2, we therefore get

**Theorem 2.** *If $\mu$ is a regular measure on an A-space, then $\mu$ is $\sigma$-smooth if and only if $\mu(F_n) \downarrow 0$ for any sequence $\{F_n\}$ of closed sets such that $F_n \downarrow \varnothing$.*

*Proof:* By Theorem 12.2, $\mu$ is $\sigma$-smooth if and only if $\mu(E_n) \downarrow 0$ for any sequence $\{E_n\}$, $E_n \in \mathscr{S}$, $E_n \downarrow \varnothing$. Hence it is sufficient to show that the condition $\mu(F_n) \downarrow 0$ for $F_n \downarrow \varnothing$, $F_n$ a closed set, implies $\mu(E_n) \downarrow 0$ for any sequence $\{E_n\}$ such that $E_n \in \mathscr{S}$, $E_n \downarrow \varnothing$. Let $E_n \downarrow \varnothing$, $E_n \in \mathscr{S}$. Since $\mu$ is regular, we may choose $F_n \subset E_n$ such that

$$\mu(E_n \backslash F_n) < \varepsilon/2^n. \tag{4}$$

Put $F^{(n)} = \bigcap_{n=1}^{n} F_n$. Then $F^{(n)} \downarrow \varnothing$ since $E_n \downarrow \varnothing$. Noting that $E_n = \bigcap_{k=1}^{n} E_k$, we can write

$$\begin{aligned} E_n \Big\backslash \bigcap_{k=1}^{n} F_k &= \bigcap_{k=1}^{n} E_k \Big\backslash \bigcap_{k=1}^{n} F_k \\ &= (E_1 \backslash F_1) \cap \left[\bigcap_{k=2}^{n} E_k\right] \\ &\quad \cup \left\{(E_2 \backslash F_2) \cap F_1 \cap \left[\bigcap_{k=3}^{n} E_k\right]\right\} \cup \cdots \cup \left\{(E_n \backslash F_n) \cap \left[\bigcap_{j=1}^{n-1} F_j\right]\right\}. \end{aligned}$$

From this relation and (4) we obtain

$$\mu[E_n \backslash F^{(n)}] \leq \varepsilon \cdot \sum_{j=1}^{n} 2^{-j} < \varepsilon.$$

Hence

$$\mu(E_n) \leq \mu(F^{(n)}) + \varepsilon \qquad \text{and} \qquad \mu(F^{(n)}) \downarrow 0 \qquad (n \to \infty)$$

imply $\mu(E_n) \downarrow 0$ $(n \to \infty)$ since $\varepsilon$ is arbitrary. □

A measure $\mu$ on an A-space is said to be tight if for any $\varepsilon > 0$ there exists a compact measurable set $K_\varepsilon$ such that

$$\mu(S \backslash K_\varepsilon) < \varepsilon. \tag{5}$$

**Theorem 3.** *If a regular measure $\mu$ on an A-space is tight, then $\mu$ is $\sigma$-smooth.*

*Proof:* Let the regular measure $\mu$ be tight. By Theorem 2, $\mu$ is $\sigma$-smooth if and only if $\mu\{F_n\} \downarrow 0$ $(n \uparrow +\infty$ for any sequence $\{F_n\}$ of closed sets such that $F_n \downarrow \varnothing$, hence, if and only if $\mu(G_n) \uparrow \mu(S)$ for any sequence $\{G_n\}$ of open sets such that $G_n \uparrow S$. For such a sequence $K_\varepsilon \subset \bigcup_{n=1}^m G_n = G_m$ for sufficiently large $m$, and

$$\mu(S) - \varepsilon < \mu(K_\varepsilon) < \mu(G_m) \leq \mu(S).$$

This is true for any $\varepsilon > 0$. Hence, $\mu(G_m) \uparrow \mu(S)$. □

Example: A $\sigma$-smooth measure on $R$ is tight. Indeed, as we have shown in Section 11, the closed set $\mathscr{T}_\alpha = \{x: |x| \leq \alpha\}$ is compact on $R$, and moreover we proved there that this is also true on $R^{(k)}$ for the set $\mathscr{T}_\alpha = \{x: |x^{(i)}| \leq \alpha, i = 1, 2, \ldots, k\}$ for any real number $\alpha$. Further,

$$\mu(S \backslash \mathscr{T}_\alpha] \downarrow 0, \qquad \alpha \uparrow \infty$$

both on $R$ and $R^{(k)}$ if $\mu$ is $\sigma$-smooth. □

**Theorem 4.** *Any $\sigma$-smooth measure on a completely normal A-space is regular.*

*Proof:* Let $\mathscr{S}$ be the algebra generated by the closed sets of the A-space $S$, and let $\mathscr{S}^*$ be the class of all sets $E \in \mathscr{S}$ such that for any $\varepsilon > 0$ there exists a closed set $F$ and an open set $G$ such that $F \subset E \subset G$ and

$$\mu(F) \leq \mu(E) \leq \mu(G), \qquad \mu(G \backslash F) < \varepsilon.$$

If $E$ is closed, we put $E = F$. It was shown in Lemma 6.5 that for a totally closed set $F$ there exists a sequence $G^{(n)}$ of open sets such that $G^{(n)} \backslash F \downarrow \varnothing$ $(n \to +\infty)$. Since the measure $\mu$ is $\sigma$-smooth, we have

$$\mu(G_n^{(n)} \backslash F) \downarrow 0.$$

Thus to $\varepsilon > 0$, we may choose $n_0 = n_0(\varepsilon)$ such that $\mu(G^{(n_0} \backslash F) \leq \varepsilon$. Hence $\mathscr{S}^*$ contains all totally closed sets and thus all closed sets, since the A-space is completely normal. In order to prove the regularity of $\mu$, it is then sufficient to prove that $\mathscr{S}^*$ is an algebra, since $\mathscr{S}$ is the algebra generated by the closed sets. Let $E_1$ and $E_2$ belong to $\mathscr{S}^*$. Then there exist $F_i$ and $G_i$ such that $F_i \subset E_i \subset G_i$, $\mu(G_i \backslash F_i) < (\varepsilon/2)$ for $i = 1, 2$. Then $F_1 \cup F_2 \subset E_1 \cup E_2 \subset G_1 \cup G_2$ and

$$\mu(G_1 \cup G_2 \backslash F_1 \cup F_2) \le \mu(G_1 \backslash F_1) + \mu(G_2 \backslash F_2) < \varepsilon.$$

Hence, $\mathscr{S}^*$ is additive, therefore an algebra, and identical with $\mathscr{S}$. □

## 14. EXTENSIONS OF MEASURES

Consider a set function $\mu$ on a class $\mathscr{E}$ of subsets of a set $S$ and assume that $\mathscr{E}$ is closed for finite unions and finite intersections, and that $\mathscr{E}$ contains $S$ and the empty set. We say that $\mu$ is nondecreasing on $\mathscr{E}$ if

$$\mu(E_1) \le \mu(E_2) \qquad \text{for} \quad E_1 \subset E_2, \tag{1}$$

subadditive on $\mathscr{E}$ if

$$\mu(E_1 \cup E_2) \le \mu(E_1) + \mu(E_2), \tag{2}$$

and superadditive on $\mathscr{E}$ if

$$\mu(E_1 \cup E_2) \ge \mu(E_1) + \mu(E_2). \tag{3}$$

Let $\mathscr{S}^*$ be an algebra or a $\sigma$-algebra of subsets of $S$.

A set function $\mu^{(i)}$ on $\mathscr{S}^*$ is called an inner measure on $\mathscr{S}^*$ if $\mu^{(i)}$ is nondecreasing and superadditive and takes the value 0 at the empty set $\varnothing$.

A set function $\mu^{(0)}$ on $\mathscr{S}^*$ is called an outer measure on $\mathscr{S}^*$ if it is nondecreasing and subadditive and takes the value 0 at $\varnothing$. For a class $\mathscr{E} \subset \mathscr{S}^*$ of subsets given as above and a nonnegative and subadditive set function $\mu$ on $\mathscr{E}$ such that $\mu(\varnothing) = 0$, we define the outer extension $\mu^{(0)}$ of $\mu$ on $\mathscr{S}^*$ by

$$\mu^{(0)}(E) = \inf_{E' \supset E,\, E' \in \mathscr{E}} (E'), \tag{4}$$

and for a nonnegative and superadditive set function $\mu$ on $\mathscr{E}$ such that $\mu(\varnothing) = 0$, we define the inner extension $\mu^{(i)}$ of $\mu$ on $\mathscr{S}^*$ by

$$\mu^{(i)}(E) = \sup_{E' \subset E,\, E' \in \mathscr{E}} \mu(E'). \tag{5}$$

If $\bar{\mu}$ is a set function on $\mathscr{S}^*$, we call a set $E \in \mathscr{S}^*$ $\bar{\mu}$-measurable if

$$\bar{\mu}(D) = \bar{\mu}(E \cap D) + \bar{\mu}(E^{c} \cap D). \tag{6}$$

**Lemma 1.** *Let $\mathscr{E}$ be a class of subsets of a set $S$ such that $\mathscr{E}$ is closed for finite unions and finite intersections and contains $\varnothing$ and $S$. Further, let $\mu$ be a nondecreasing set function on $\mathscr{E}$ such that $\mu(\varnothing) = 0$ and $\mu(S) < \infty$.*

*If $\mu$ is subadditive and $\mu^{(0)}$ is the outer extension of $\mu$, then the $\mu^{(0)}$-measurable sets form an algebra $\mathscr{S}^{(0)}$, and $\mu^{(0)}$ is a measure on $\mathscr{S}^0$. If $\mu$ is superadditive and $\mu^{(i)}$ is the inner extension of $\mu$, then the $\mu^{(i)}$-measurable sets form an algebra $\mathscr{S}^{(i)}$, and $\mu^{(i)}$ is a measure on $\mathscr{S}^{(i)}$.*

*Proof:* We first note that $\mu^{(0)}$ and $\mu^{(i)}$ both are nondecreasing, since $\mu$ is nondecreasing on $\mathscr{E}$. If, furthermore, $\mu$ is subadditive, we get for $E', E_1', E_2' \in \mathscr{E}$

$$\begin{aligned}\mu^{(0)}(E_1 \cup E_2) &= \inf_{E' \supset E_1 \cup E_2} \mu(E') \leq \inf_{E_1' \supset E_1,\, E_2' \supset E_2} \mu(E_1' \cup E_2') \\ &\leq \inf_{E_1' \supset E_1,\, E_2' \supset E_2} [\mu(E_1') + \mu(E_2')] \\ &= \inf_{E_1' \supset E_1} \mu(E_1') + \inf_{E_2' \supset E_2} \mu(E_2') = \mu^{(0)}(E_1) + \mu^{(0)}(E_2). \end{aligned} \tag{7}$$

Thus, $\mu^{(0)}$ is nondecreasing and subadditive on $\mathscr{S}^*$ when $\mu$ has these properties on $\mathscr{E}$. In the same way we find that $\mu^{(i)}$ is superadditive on $\mathscr{S}^*$ if $\mu$ has these properties on $\mathscr{E}$. To prove this, we change inf to sup, $\supset$ to $\subset$, and $\leq$ to $\geq$ in (7) when we change $\mu^{(0)}$ to $\mu^{(i)}$. Clearly $\mu(\varnothing) = 0$, $\mu^{(0)}(S) = \mu^{(i)}(S) = \mu(S) < \infty$.

In order to prove that $\mathscr{S}^{(0)}$ is an algebra, we have to show that $E_1 \cup E_2$ and $E_1 \cap E_2$ are $\mu^{(0)}$-measurable if $E_1$ and $E_2$ are $\mu^{(0)}$-measurable. But according to the definition of $\bar{\mu}$-measurable sets in (6), $E^c$ is $\mu^{(0)}$-measurable if $E$ is $\mu^{(0)}$-measurable. Hence if $E_1$, $E_2$, and $E_1 \cap E_2$ are $\mu^{(0)}$-measurable, $(E_1 \cap E_2)^c = E_1^c \cup E_2^c$ and $E_1 \cup E_2$ are also $\mu^{(0)}$-measurable. Thus it is sufficient to show that $E_1 \cap E_2$ is $\mu^{(0)}$-measurable. According to the subadditivity of $\mu^{(0)}$, for any sets $E$ and $D$ in $\mathscr{S}^*$ we have

$$\mu^{(0)}(D) = \mu^{(0)}(E \cap D \cup E^c \cap D) \leq \mu^{(0)}(E \cap D) + \mu^{(0)}(E^c \cap D). \tag{8}$$

Thus in order to show that $E_1 \cap E_2$ is a $\mu^{(0)}$-measurable set, it is sufficient to show that

$$\mu^{(0)}(D) \geq \mu^{(0)}(E_1 \cap E_2 \cap D) + \mu^{(0)}[(E_1 \cap E_2)^c \cap D]. \tag{9}$$

Now if $E_1$ and $E_2$ are $\mu^{(0)}$-measurable, we get

$$\mu^{(0)}(D) = \mu^{(0)}(E_1 \cap D) + \mu^{(0)}(E_1^c \cap D), \tag{10}$$

$$\mu^{(0)}(E_1 \cap D) = \mu^{(0)}(E_2 \cap E_1 \cap D) + \mu^{(0)}(E_2^c \cap E_1 \cap D). \tag{11}$$

Observing that

$$E_1^c \cup (E_2^c \cap E_1) = (E_1 \cap E_2)^c \tag{12}$$

and using the subadditivity of $\mu^{(0)}$, we get (9). Hence $E_1 \cap E_2$ is $\mu^{(0)}$-measurable. In the same way, we find that the $\mu^{(i)}$-measurable sets form an algebra. Indeed (10)–(12) hold for $\mu^{(i)}$, but according to the superadditivity of $\mu^{(i)}$, we have to change $\geq$ to $\leq$ in (9) if we change $\mu^{(0)}$ to $\mu^{(i)}$. On the other hand, we have to change $\leq$ to $\geq$ in (8) if we change $\mu^{(0)}$ to $\mu^{(i)}$, since $\mu^{(i)}$ is superadditive.

Using the fact that (6) is satisfied for any $D \in \mathscr{S}^*$, we obtain for $\bar{\mu} = \mu^{(0)}$ and disjoint sets $E_1$ and $E_2$ in $\mathscr{S}^{(0)}$

$$\mu^{(0)}(E_1 \cup E_2) = \mu^{(0)}(E_1) + \mu^{(0)}(E_2), \tag{13}$$

i.e., $\mu^{(0)}$ is a measure on $\mathscr{S}^{(0)}$. In the same way, we conclude that $\mu^{(i)}$ is a measure on $\mathscr{S}^{(i)}$. More generally, we get by (6) for any $D \subset \mathscr{S}^*$ and disjoint sets $E_1$ and $E_2$ in $\mathscr{S}^{(0)}$ and the outer extension $\mu^{(0)}$ of $\mu$

$$\mu^{(0)}(D \cap E_1 \cup E_2) = \mu^{(0)}(D \cap E_1) + \mu^{(0)}(D \cap E_2) \tag{14}$$

and the corresponding relation for disjoint sets $E_1$ and $E_2$ in $\mathscr{S}^{(i)}$ and the inner extension. By induction, we obtain from (14) for an outer extension $\mu^{(0)}$ and any $D \in \mathscr{S}^*$ and a finite disjoint sequence $\{E_i\}_{i=1}^m$, $E_i \in \mathscr{S}^{(0)}$:

$$\mu^{(0)}\left(D \cap \bigcup_{i=1}^{m} E_i\right) = \sum_{i=1}^{m} \mu^{(0)}(D \cup E_i), \tag{15}$$

which for $D = \bigcup_{i=1}^m E_i$ reduces; correspondingly for $\mu^{(i)}$. □

Let now $\mathscr{S}^*$ be the $\sigma$-algebra of all subsets of $\mathscr{S}$.

**Theorem 1** (Extension theorem). *If $\mathscr{S}$ is an algebra of subsets of a set $S$ and $\mu$ is a $\sigma$-smooth measure on $\mathscr{S}$, then there exists an extension $\bar{\mu}$ of $\mu$ on the $\sigma$-algebra $\mathscr{S}_\sigma$ generated by $\mathscr{S}$. [Extension of $\mu$ to $\bar{\mu}$ means that $\bar{\mu}(E) = \mu(E)$ for $E \in \mathscr{S}$.] Furthermore, $\mu$ is uniquely determined by $\mu$.*

*Proof:* Let $\mu^{(0)}$ be the outer measure determined by $\mu$ on $\mathscr{S}^*$ and $\mathscr{S}^{(0)}$, the algebra of $\mu^{(0)}$-measurable sets. By the lemma, $\mathscr{S}^{(0)}$ is an algebra and $\mu^{(0)}$ is a measure on $\mathscr{S}^{(0)}$. We shall show that $\mathscr{S}^{(0)}$ is even a $\sigma$-algebra and that $\mu^{(0)}$ is a $\sigma$-smooth measure on $\mathscr{S}^{(0)}$.

To begin with we prove that $\mu^{(0)}$ is sub–$\sigma$-additive on $\mathscr{S}^{(0)}$, i.e., that for any sequence $\{E_j\}_{j=1}^\infty$, $E_j \in \mathscr{S}^{(0)}$ we have

$$\mu^{(0)}\left(\bigcup_{i=1}^{\infty} E_j\right) \leq \sum_{i=1}^{\infty} \mu^{(0)}(E_j). \tag{16}$$

Then we observe that, by the definition of the outer measure, there exists for any $\varepsilon > 0$ a set $E_j'$ in $\mathscr{S}$ such that $E_j' \supset E_j$ and

$$\mu(E_j') \leq \mu^{(0)}(E_j) + \varepsilon/2^j.$$

Furthermore $\bigcup_{j=1}^{\infty} E_j \subset \bigcup_{j=1}^{\infty} E'_j$ and, hence, since $\mu$ is $\sigma$-additive on $\mathscr{S}$ and thus sub–$\sigma$-additive on $\mathscr{S}$,

$$\mu^{(0)}\left(\bigcup_{j=1}^{\infty} E_j\right) \leq \mu\left(\bigcup_{j=1}^{\infty} E'_j\right) \leq \sum_{j=1}^{\infty} \mu(E'_j) \leq \sum_{j=1}^{\infty} \mu^0(E_j) + \varepsilon.$$

This is true for any $\varepsilon > 0$. Thus (16) holds.

Consider now a disjoint sequence $\{E_j\}_{j=1}^{\infty}$ in $\mathscr{S}^0$ and put

$$\bigcup_{j=1}^{n} E_j = B_n, \qquad \bigcup_{j=1}^{\infty} E_j = B.$$

Since $B_n \in \mathscr{S}^0$, we get for any $D \in \mathscr{S}^0$

$$\mu^{(0)}(D) = \mu^{(0)}(B_n \cap D) + \mu^{(0)}(B_n^c \cap D) \geq \sum_{j=1}^{n} \mu^{(0)}(E_j \cap D) + \mu^{(0)}(B^c \cap D). \tag{17}$$

Letting $n \to \infty$ and taking note of (16), we obtain

$$\mu^{(0)}(D) \geq \mu^{(0)}(B \cap D) + \mu^{(0)}(B^c \cap D).$$

Hence, $B \subset \mathscr{S}^{(0)}$ and thus $\mathscr{S}^{(0)}$ is a $\sigma$-algebra. Choosing $D = B$ in (17) and letting $n \to \infty$, we get

$$\mu^{(0)}(B) = \sum_{i=1}^{\infty} \mu^{(0)}(E_j),$$

which proves that $\mu^{(0)}$ is $\sigma$-additive (is $\sigma$-smooth).

Clearly $\mathscr{S} \subset \mathscr{S}^{(0)}$ and thus $\mathscr{S}^{(0)}$ contains the $\sigma$-algebra $\mathscr{S}_\sigma$ generated by $\mathscr{S}$, and $\mu^{(0)}$ restricted to $\mathscr{S}_\sigma$ is a $\sigma$-smooth measure on $\mathscr{S}_\sigma$. It follows by Theorem 12.3 that this measure is uniquely determined by $\mu$, since $\mu^{(0)} = \mu$ on $\mathscr{S}$.

## 15. MEASURES ON INFINITE-DIMENSIONAL PRODUCT SPACES

In Section 10 we defined the infinite-dimensional product space $S_T$ of measurable spaces $S_t$, $t \in T$, where $T$ is a countable set on the line. For any projection $\pi_{\mathbf{t}}$ of $S_T$ onto the finite-dimensional product space $S_{\mathbf{t}} = \pi_{\mathbf{t}} S$, the set $\pi_{\mathbf{t}}^{-1}(E_{\mathbf{t}})$ is a measurable set on $S_T$ if $E_{\mathbf{t}}$ is a measurable set on $S_{\mathbf{t}}$.

Let measures $\mu^{\mathbf{t}}$ be given on $S_{\mathbf{t}}$ for any finite index vector $\mathbf{t}$ in $T$. We say that these measures satisfy the consistence conditions if

$$\mu^{\mathbf{s}} = \mu^{\mathbf{t}}(\pi_{\mathbf{ts}}^{-1}\cdot), \qquad \mu^{\mathbf{s}}(S_{\mathbf{s}}) = 1 \tag{1}$$

for all finite index vectors $\mathbf{s}$ and $\mathbf{t}$ where $\mathbf{s} \prec \mathbf{t}$ and $\pi_{\mathbf{ts}}$ denotes the projection of $S_{\mathbf{t}}$ onto $S_{\mathbf{s}}$.

**Theorem 1.** *If the measures $\mu^{\mathbf{t}}$ satisfy the consistence conditions, they determine a measure $\mu$ on the algebra $\mathscr{S}$ formed of the sets $\pi_{\mathbf{t}}^{-1}E^{\mathbf{t}}$, and $\mu$ satisfies the relations*

$$\mu^{\mathbf{t}} = \mu(\pi_{\mathbf{t}}^{-1}\cdot), \qquad \mu(S_T) = 1.$$

*Remark 1.* If measures $\mu^{\mathbf{t}'}$ are given only for $\mathbf{t}' \subset T'$, where $T'$ is a subset of $T$, such that for any $\omega \subset T$ there exists $\mathbf{t}$ in $T'$, $\omega \prec \mathbf{t}$, and if (1) holds for $s$, $t \in T'$, then defining $\mu^{\mathbf{u}} = \mu^{\mathbf{t}}(\pi_{\mathbf{tu}}^{-1}\cdot)$, we get consistent measures on all $S_{\mathbf{t}}$, $\mathbf{t} \in S_T$.

*Proof:* If $E \subset \mathscr{S}$, then $E = \pi_{\mathbf{t}}^{-1}E^{\mathbf{t}}$ for some $\mathbf{t} \in T$, where $E^{\mathbf{t}}$ is a measurable set in $S_{\mathbf{t}}$. We define

$$\mu(E) = \mu^{\mathbf{t}}(E^{\mathbf{t}}). \tag{2}$$

This definition is consistent, according to the consistence conditions (1). Indeed, if in addition $E = \pi_{\mathbf{s}}^{-1}E^{\mathbf{s}}$ for $\mathbf{s} \prec \mathbf{t}$, we have $S_{\mathbf{s}} = \pi_{\mathbf{ts}}S_{\mathbf{t}}$ with the projection $\pi_{\mathbf{ts}}$ and $\pi_{\mathbf{s}} = \pi_{\mathbf{ts}}\pi_{\mathbf{t}}$, and thus

$$\pi_{\mathbf{s}}^{-1}E^{\mathbf{s}} = \pi_{\mathbf{t}}^{-1}\pi_{\mathbf{ts}}^{-1}E^{\mathbf{s}}.$$

Hence, if $\mu(E)$ is given by (2) for $E^{\mathbf{t}} \subset S_{\mathbf{t}}$, by (1) we get

$$\mu(E) = \mu^{\mathbf{t}}(\pi_{\mathbf{ts}}^{-1}E^{\mathbf{s}}) = \mu^{\mathbf{s}}(E_{\mathbf{s}}).$$

Thus the definition of $\mu$ is unique. If $E_1$ and $E_2$ are two disjoint sets in $\mathscr{S}$, we have $E = \pi_{\mathbf{t}}^{-1}E_1^{\mathbf{t}}$, $E' = \pi_{\mathbf{t}}^{-1}(E_2^{\mathbf{t}})$ with $E_1^{\mathbf{t}}$ and $E_2^{\mathbf{t}}$ belonging to $S_{\mathbf{t}}$ for suitable $\mathbf{t}$, and hence by (2)

$$\mu(E \cup E') = \mu^{\mathbf{t}}(E_1^{\mathbf{t}} \cup E_2^{\mathbf{t}}) = \mu^{\mathbf{t}}(E_1^{\mathbf{t}}) + \mu^{\mathbf{t}}(E_2^{\mathbf{t}}) = \mu(E) + \mu(E').$$

Thus $\mu$ is a measure. Clearly $\mu(S_T) = 1$, and by the definition of $\mu$ we have $\mu^{\mathbf{t}}(E^{\mathbf{t}}) = \mu(\pi_{\mathbf{t}}^{-1}E^{\mathbf{t}})$ for any measurable set $E^{\mathbf{t}}$ in $S_{\mathbf{t}}$. □

If $\mu$, given by Theorem 1, is $\sigma$-smooth, then we can extend it to a probability measure on $S_T$. For metric spaces we have the following theorem, which is a generalization of a famous theorem of Kolmogorov.

**Theorem 2** (Kolmogorov's theorem). *Let $S_i$, $i = 1, 2, \ldots,$ be metric spaces, and $S^{\mathbf{i}}$ for the vector $\mathbf{i} = (1, 2, \ldots, i)$ the corresponding product space $S_1 \otimes S_2 \otimes \cdots \otimes S_i$ for any (i), and $S_T$ the infinite product space $S_1 \otimes S_2 \otimes \cdots$. Furthermore let $\mu^{\mathbf{i}}$ for any $\mathbf{i}$ be a tight probability measure on $S^{\mathbf{i}}$ and suppose that the sequence $\{\mu^{\mathbf{i}}\}$, $i = 1, 2, \ldots$ is a consistent class. Then these $\mu^{\mathbf{i}}$ determine a probability measure $\mu$ on $S_T$ such that $\mu^{\mathbf{i}} = \mu(\pi_{\mathbf{i}}^{-1}\cdot)$.*

*Proof:* By Theorem 1, the $\mu^{\mathbf{i}}$ determine a measure $\mu$ with $\mu(S_T) = 1$ on the algebra $\mathscr{S}$ consisting of all sets $\pi_{\mathbf{i}}^{-1}E^{\mathbf{i}}$, where $E^{\mathbf{i}}$ is any set in the $\sigma$-algebra generated by the closed sets in $S^{\mathbf{i}}$ and $i$ is any positive integer. In order to prove that $\mu$ is $\sigma$-smooth, we have to show that $\mu(E_n) \downarrow 0$ for any sequence $\{E_n\}$, $E_n \in \mathscr{S}$ and $E_n \downarrow \varnothing$. Hence it is sufficient to show that $\bigcap_{n=1}^{\infty} E_n$ is not empty if $\mu(E_n)$ does not tend to 0. By Theorem 13.2, we may choose the $E_n$ as closed sets.

Suppose that there exists $\varepsilon > 0$ such that

$$\mu(E_n) \geq \varepsilon, \qquad n = 1, 2, \ldots, \tag{3}$$

where $E_n \in \mathscr{S}$ and $E_n \subset E_{n-1}$. Since $E_n \in \mathscr{S}$, we have

$$E_n = \pi_{\mathbf{i}_n}^{-1} E_n^{\mathbf{i}_n}$$

for some $\mathbf{i}_n$, and we may choose $\mathbf{i}_n$ such that $\mathbf{i}_n > \mathbf{i}_{n-1}$ for $n = 2, 3, \ldots$. Then

$$\mu(E_n) = \mu^{\mathbf{i}_n}(E_n^{\mathbf{i}_n}) \geq \varepsilon.$$

Since $\mu^{\mathbf{i}_n}$ is tight, there exists a compact set $K_n^{\mathbf{i}_n}$ of $E_n^{\mathbf{i}_n}$ such that

$$\mu^{\mathbf{i}_n}(E_n^{\mathbf{i}_n} - K_n^{\mathbf{i}_n}) < \varepsilon 2^{-(n+1)}. \tag{4}$$

Putting

$$K_n = \pi_{\mathbf{i}_n}^{-1} K_n^{\mathbf{i}_n}, \qquad \mathscr{C}_n = \bigcap_{j=1}^{n} K_j, \tag{5}$$

we get

$$\mathscr{C}_1 \supset \mathscr{C}_2 \supset \cdots \supset \mathscr{C}_n \cdots$$

$$\mu(E_n - K_n) < \varepsilon 2^{-(n+1)}, \qquad \mu(E_n \backslash \mathscr{C}_n) < \varepsilon/2. \tag{6}$$

By (3) and (5) we obtain

$$\mu(\mathscr{C}_n) > \varepsilon/2 \qquad \text{for} \quad n = 1, 2, \ldots. \tag{7}$$

According to this inequality, $\mathscr{C}_n$ is nonempty. We choose a point $x_j$ in each $\mathscr{C}_j$ and so obtain a sequence $\{x_j\}$ of points in $S_T$. Note that

$$x_j = (x_j^{(1)}, x_j^{(2)} \ldots)$$

is a sequence and that $x_j \subset \mathscr{C}_n$ for $j \geq n$. Hence

$$\pi_n x_j = (x_j^{(1)}, \ldots, x_j^{(i_n)}) \in K_n^{i_n} \qquad \text{for} \quad j \geq n.$$

Since $K_1^{i_1}$ is compact, the sequence $\{x_j\}$ contains a subsequence such that $x_j^{(i)}$ converges for $i = 1, 2, \ldots, i_j$ for this subsequence. This means that there exists an infinite sequence $N_1$ of positive integers such that $\{x_j^{(i)}\}_{j \in N_1}$ converges for $i = 1, 2, \ldots, i_1$. In the same way, $N_1$ contains an infinite subsequence $N_2$ such that $\{x_j^{(i)}\}_{j\in N_2}$ converges for $i = 1, 2, \ldots, i_2$, and so on. Proceeding in this way, we obtain $N_1 \supset N_2 \supset \cdots$ such that $\{x_j^{(i)}\}_{j\in N_n}$ converges for $i = 1, 2, \ldots, i_n$. Using the selection principle, we let $\bar{N}$ be the sequence whose first number is the first number in $N_1$, whose second number is the second number in $N_2$, and so on. Then $\{x_j^{(i)}\}_{j\in \bar{N}}$ converges for all $i$ to an element $y^{(i)}$, which means that $\{x_j\}_{j\in N}$ converges to an element $y$ in $S_T$ according to the metric on the product space (compare Section 11). For any $j$ and $j_1$ in $\bar{N}$, we have $x_j \in \mathscr{C}_j$, for $j \geq j_1$, and hence $y \in \mathscr{C}_{j_1}$ since $\mathscr{C}_n$ is closed and thus contains the limit of $x_j$ ($K_n^{i_n}$, being a compact set, is closed, and then $\pi_{i_n}^{-1} K_n^{i_n}$ is closed since $\pi_{i_n}$ is continuous and $\mathscr{C}_n$ as intersection of closed sets is closed). Thus $y \subset \bigcap_{j=1}^{\infty} \mathscr{C}_j = \mathscr{C}$ and $\mathscr{C}$ is not empty; in other words $\bigcap_{n=1}^{\infty} E_n$ is not empty. □

Kolomogorov's theorem was first given on $R^{(\infty)}$. Note that the measures $\mu^{\mathbf{i}}$ on the projections $S^{\mathbf{i}}$ of $R^{(\infty)}$ are tight if the $\mu^{\mathbf{i}}$ are $\sigma$-smooth (see the example in Section 13).

## 16. COMPLETION OF MEASURES, CONTINUITY ALMOST SURELY AND ALMOST EVERYWHERE

If $(\Omega, \mathscr{B}, \mu)$ is measure space, where $\mathscr{B}$ denotes a $\sigma$-algebra and $\mu$ a $\sigma$-smooth measure on $\mathscr{B}$, and if $D$ is a measurable set with $\mu(D) = 0$ and $D \in \mathscr{B}$, we form a new $\sigma$-algebra $\bar{\mathscr{B}}$ and a measure $\bar{\mu}$ on $\bar{\mathscr{B}}$, defining $\bar{\mathscr{B}}$ and $\bar{\mu}$ as follows: $\bar{\mathscr{B}}$ contains any set $B \cup E$, $B \in \mathscr{B}$ $E \subset D$ and $\bar{\mu}(B \cup E) = \mu(B)$. It is easily seen that $\bar{\mathscr{B}}$ is a $\sigma$-algebra and $\bar{\mu}$ is a $\sigma$-smooth measure on $\bar{\mathscr{B}}$. We call $\bar{\mathscr{B}}$ the completion of $\mathscr{B}$ and $\bar{\mu}$ the completion of $\mu$ with respect to the null set $D$.

By definition (Section 3), a mapping $T$ from an A-space $S$ into an A-space $S'$ is continuous if $T^{-1}F'$ is a closed set in $S$ for any closed set $F'$ in $S'$, or equivalently $T^{-1}G'$ is an open set in $S$ for an open set $G'$ in $S'$. If $\mu$ is a measure on $S$, we say that a measurable mapping $T$ of $S$ into $S'$ is almost surely (abbreviated a.s.) continuous ($\mu$) if

$$\mu(\overline{T^{-1}F'}) = \mu(T^{-1}F')$$

for any closed set $F'$ in $S'$, or equivalently

$$\mu[(T^{-1}G')^0] = \mu(T^{-1}G')$$

for any open set $G'$ in $S'$.

For metric spaces we have introduced the concept "continuous at a point" and "continuous everywhere for a mapping $T$" (Section 8). We say that a measurable mapping $T$ from a metric space $S$ into a metric space $S'$

is continuous almost everywhere ($\mu$) [continuous a.e. ($\mu$)] if for the set $D_T$ of discontinuity points of $T$ we have $\mu(D_T) = 0$, where $D_T$ is measurable, i.e., that it belongs to the algebra generated by the closed sets. However, by Lemma 8.2, $D_T$ is always measurable with respect to the $\sigma$-algebra generated by the closed sets so that $\mu(D_T)$ is defined when we deal with $\sigma$-smooth measures.

**Theorem 1.** *A measurable mapping $T$ from a metric space $S$ into a metric space $S'$ is continuous a.s. ($\mu$) with respect to a $\sigma$-smooth measure if $T$ is continuous a.e. ($\mu$).*

*Proof:* By Lemma 8.2

$$\overline{T^{-1}F'} \subset T^{-1}F' \cup D_T$$

for any closed set $F'$ in $S$. Hence $\mu(D_T) = 0$ implies $\mu(\overline{T^{-1}F'}) = \mu(T^{-1}F')$. □

**Theorem 2.** *Let $\mu$ be a measure on an $A$-space $S$ that is mapped by a measurable mapping $T$ into an $A$-space $S'$ that is mapped by a measurable mapping $T'$ into an $A$-space $S''$. If $T$ is continuous a.s. ($\mu$) and $T'$ is continuous, then the mapping $T' \circ T$ of $S$ into $S''$ is continuous a.s. ($\mu$). If $T$ is continuous and $T'$ is continuous a.s. ($\mu(T^{-1}\cdot)$), then $T' \circ T$ is continuous a.s. $\mu$.*

*Proof:* Let $F''$ be a closed set in $S''$. If $T'$ is continuous and $T$ continuous a.s. ($\mu$), then $(T')^{-1}F''$ is closed and

$$\mu[(T)^{-1}(T')^{-1}F''] = \mu(\overline{T^{-1}(T')^{-1}F''}).$$

If $T$ is continuous and $T'$ continuous a.s. ($\mu(T^{-1}\cdot)$), *then*

$$\mu[T^{-1}(T')^{-1}F''] = \mu[T^{-1}\overline{(T')^{-1}F''}] = \mu(\overline{T^{-1}(T')^{-1}F''}). \quad \square$$

**Theorem 3.** *Let $\mu$ be a $\sigma$-smooth measure on an $A$-space $S$. The class $\Psi$ of all bounded, real-valued functions from $S$ that are continuous a.s. ($\mu$) form a Stone vector lattice.*

We first prove

**Lemma 1.** *$f \in \overline{\psi}$ if and only if*

$$\overline{\{x : f(x) \le a\}} = \{x : f(x) \le a\} \cup \Delta'_f(a), \tag{1}$$

$$\overline{\{x : f(x) \ge a\}} = \{x : f(x) \ge a\} \cup \Delta''_f(a) \tag{2}$$

*for any real number $a$, where*

$$\mu(\Delta'_f(a)) = \mu(\Delta''_f(a)) = 0.$$

*Proof:* Indeed, let $\tilde{\mathscr{F}}$ be the $\sigma$-topological class of closed sets on the real line. If $f$ is continuous a.s. ($\mu$), then clearly (1) and (2) are satisfied. Conversely, let (1) and (2) be satisfied for a real-valued bounded function $f$ from $S$. Let $\tilde{\mathscr{F}}_0$ be the class of all sets $\tilde{F}$ in $\tilde{\mathscr{F}}$ for which $\overline{f^{-1}(\tilde{F})} = F + \Delta_F$ with $\mu(\Delta_F) = 0$. It is easy to show that $\tilde{\mathscr{F}}_0$ is a $\sigma$-topological class of closed sets and it contains the intervals $(-\infty, \mathrm{a}]$ and $[\mathrm{a}, \infty)$. But these intervals generate $\tilde{\mathscr{F}}$. Hence $\tilde{\mathscr{F}}$ is identical with $\tilde{\mathscr{F}}_0$. □

Applying the lemma, we immediately find that $f \in \bar{\psi} \Rightarrow \alpha f \in \bar{\psi}$ for any real number $\alpha$, $f, g \in \bar{\psi} \Rightarrow \max(f, g) \in \bar{\psi}$, and $f + g \in \bar{\psi}$ we get

$$\{x : \max[f(x), g(x)] \leq a\} = \{x : f(x) \leq a\} \cap \{x : g(x) \leq a\},$$
$$\{x : \max[f(x), g(x)]\} \geq a = \{x : f(x) \geq a\} \cup \{x : g(x) \geq a\},$$

$$\{x : f(x) + g(x) \leq a\} = \bigcap_{r=1}^{\infty} \{x : f(x) \leq r\} \cap \{x : g(x) \leq a - r\},$$

$$\{x : f(x) + g(x) \geq a\} = \bigcap_{r=1}^{\infty} \{x : f(x) \geq r\} \cap \{x : g(x) \geq a - r\},$$

where on the right-hand side we may change all sets to the corresponding closed sets when we disregard null sets. □

CHAPTER II

# INTEGRALS, BOUNDED, LINEAR FUNCTIONALS, AND MEASURES

## 1. INTEGRALS AS NONNEGATIVE, BOUNDED, LINEAR FUNCTIONALS

Though the reader may be well acquainted with different integral concepts, we shall nevertheless give a rather detailed introduction to the abstract integral in order to give it a suitable place in our presentation of weak convergence.

A set $E$ in the algebra, generated by the closed sets in an A-space $S$, is called a continuity set for the measure $\mu$ on $S$ if $\mu(\bar{E}) - \mu(E^\circ) = 0$, where $\bar{E}$ is the closure and $E^\circ$ the interior of $E$. Note that $E^\circ$ is open and $\bar{E}$ closed, and $E^\circ \subset E \subset \bar{E}$. Now let $f$ be a bounded, continuous function from $S$ into $R$, and put $F_t = \{x : f(x) \geq t\}$. Then $F_t$ is a closed set. The function $g$, defined by

$$g(t) = \mu(F_t),$$

is nonincreasing, and thus has, at most, countably many discontinuity points. We claim

**Theorem 1.** *If $t_0$ is a continuity point of $g$, then $F_{t_0}$ is a continuity set for $\mu$.*

*Proof:* For any $t > t_0$, the set $G_t = \{x : f(x) > t\}$ is open, and clearly $G_t \subset F_{t_0}^\circ \subset F_{t_0} = \bar{F}_{t_0}$, and

$$\mu(G_t) \leq \mu(F_{t_0}^\circ) \leq \mu(F_{t_0}),$$

where $\mu(G_t) \downarrow \mu(F_{t_0})$ as $t \downarrow t_0$, since $t_0$ is a continuity point of $g$. □

In Section I.4, we introduced various Stone vector lattices, such as the class of bounded, continuous functions and the class of bounded, measurable functions. By Theorem I.4.2, these are complete and even rings. We call a real-valued function from a Stone vector lattice $\Psi$ a functional, and we call it *linear* if

$$\text{(i)} \qquad L(af) = aL(f) \qquad \text{for} \quad f \in \Psi, \qquad a \text{ real number,}$$

$$\text{(ii)} \qquad L(f_1 + f_2) = L(f_1) + L(f_2),$$

*nonnegative* if

$$\text{(iii)} \qquad L(f) \geq 0 \qquad \text{for} \quad f \geq 0,$$

*bounded* if

$$\text{(iv)} \qquad |L(f)| < \infty \qquad \text{for} \quad \|f\| < \infty.$$

Let $[S, \mathscr{S}, \mu]$ be a measure space; $\mathscr{S}$, algebra of subsets of $S$; and $\mu$, a measure. Consider the Stone vector lattice $\Psi_0$ of all simple functions

$$f = \sum_{i=1}^{n} a_i 1_{E_i},$$

where $\{E_i\}$, $i = 1, \ldots, n$ is a partition of $S$ such that $E_i \in \mathscr{S}$, $E_i \cap E_j = \varnothing$, $\bigcup_{i=1}^{n} E_i = S$, and the $a_i$ are real numbers. We define the functional $L_0$ on $\Psi_0$ by

$$L_0(f) = \sum_{i=1}^{n} a_i \mu(E_i).$$

Consider the two representations

$$f = \sum_{i=1}^{m} \sum_{j=1}^{m'} a_i 1_{E_i \cap E'_j}, \qquad g = \sum_{i=1}^{m} \sum_{j=1}^{m'} b_j 1_{E_i \cap E'_j} \tag{1}$$

[compare I.4(4)]. By these we find that $L_0(f) = L_0(g)$ for $f = g$. It follows immediately that $L_0$ satisfies the conditions (i), (iii), and (iv) on $\Psi_0$, and the relation (ii) follows by the help of the representations of $f$ and $g$ given above. Hence $L_0$ is a nonnegative, bounded, linear functional on $\Psi_0$. We use the notation

$$L_0(f) = \int f(x)\, \mu(dx), \qquad f \in \Psi_0, \tag{2}$$

and call $L_0(f)$ the abstract integral of $f$ with respect to $\mu$. (We say integral for short when there is no need to point out its character.) We extend the functional $L_0$ on $\Psi_0$ to a functional on the closure $\bar{\Psi}_0$ of $\Psi_0$ with respect to the uniform metric by putting

$$L(f) = \lim_{n \to \infty} L_0(f_n) \qquad \text{for} \quad \|f_n - f\| \to 0 \quad \text{as} \quad n \to \infty, \tag{3}$$

where $\{f_n\}$ is a sequence in $\Psi_0$ and $f \in \bar{\Psi}_0$. The limit (2) exists. Indeed, using representations (1) for $f_n$ and $g_n$, we get

$$|L_0(f_n) - L_0(f_m)| \leq \|f_n - f_m\| \mu(S), \tag{4}$$

where the right-hand side tends to 0 as $n, m \to +\infty$ since $\{f_n\}$ is convergent in the uniform metric. Hence $\{L_0(f_n)\}$ is Cauchy convergent and thus convergent. Clearly, $L(f)$ only depends on $f$. By the properties of limits, $L$ is a nonnegative, bounded, linear functional on $\bar{\Psi}_0$. Using the integral notation, we write

$$L(f) = \int f(x)\,\mu(dx), \qquad f \in \bar{\Psi}_0. \tag{5}$$

The integral in (4) is an integral on the given set $S$. Instead of the notation (4) of the integral, and hence of the linear functional, we use the notation

$$\mu(f) = \int f(x)\,\mu(dx). \tag{6}$$

Hence, $\mu(f)$ means the integral (6) for a function $f$, and $\mu(E)$ the measure of the set $E$ for a set $E$. Note that then

$$\mu(1_E) = \mu(E).$$

The integral of a function $f$ on a set $E \in \mathscr{S}$ is defined by $\mu(f \cdot 1_E)$, also written

$$\int_E f(x)\,\mu(dx).$$

This integral has a significance, since $f \cdot 1_E$ belongs to $\bar{\Psi}_0$ if $f \in \bar{\Psi}_0$. It follows from the definition of the integral that

$$\mu(f \cdot 1_{E_1 \cup E_2}) = \mu[(f \cdot {}^1E_1)] + \mu(f \cdot 1_{E_2}) \tag{7}$$

for disjoint sets $E_1, E_2$ in $\mathscr{S}$, the functional $\mu$ being linear.

**Theorem 2.** *The set function $\lambda$, defined on $\mathscr{S}$ for a given $f \in \bar{\Psi}_0$ by*

$$\lambda(E) = \mu(1_E f),$$

*is a signed measure on $\mathscr{S}$, and a measure if $f \geq 0$. Further $\lambda$ is $\sigma$-smooth if $\mu$ is $\sigma$-smooth.*

*Proof:* The additivity of $\lambda$ follows by (6), and clearly $\lambda(\varnothing) = 0$, $\lambda(E) < +\infty$.

Put

$$f^+(x) = \begin{cases} f(x) & \text{for } f(x) \geq 0, \\ 0 & \text{for } f(x) < 0, \end{cases}$$

$$f^-(x) = \begin{cases} 0 & \text{for } f(x) > 0, \\ -f(x) & \text{for } f(x) \leq 0. \end{cases}$$

Then $f(x) = f^+(x) - f^-(x)$, and, by the linearity of the functional $\mu$,

$$\lambda(E) = \mu(1_E \cdot f) = \mu(1_E \cdot f^+) - \mu(1_E \cdot f^-) = \lambda^+(E) - \lambda^-(E),$$

where $\lambda^+(E) = \mu(1_E \cdot f^+)$ and $\mu(1_E \cdot f^-)$ are nonnegative and thus measures. Hence $\lambda$ is a signed measure, and a measure if $f^-(x) = 0$ for all $x$. If $\mu$ is $\sigma$-smooth, then $E_n \in \mathscr{S}$, $E_n \downarrow \varnothing$ implies $\mu(E_n) \downarrow 0$, and hence

$$|\lambda(E_n)| = |\mu(1_{E_n} \cdot f)| \leq \|f\| \mu(E_n) \downarrow 0. \quad \square$$

By Theorem I.4.2, the class $\bar{\Psi}_0$ contains all real-valued, bounded, continuous functions from an A-space $S$ ($\mathscr{S}$ algebra generated by the closed sets). By the definition of the abstract integral, we get

**Theorem 3.** *To any real-valued, bounded, continuous function $f$ from an A-space $S$ and given $\varepsilon > 0$, let the real numbers $a$, $b$, and $a_i$, $i = 0, \ldots, n$, be chosen such that $a < \inf_{x \in S} f(x)$, $b > \sup_{x \in S} f(x)$, $a = a_0 < a_1 < \cdots < a_n = b$, $a_i - a_{i-1} < \varepsilon$, and put*

$$F_i = [x : f(x) \geq a_i],$$

$$\underline{g} = \sum_{i=1}^{n} a_{i-1} 1_{F_{i-1} - F_i}, \qquad \bar{g} = \sum_{i=1}^{n} a_i 1_{F_{i-1} - F_i}.$$

*Then the $F_i$ are closed sets, $\underline{g} \leq f \leq \bar{g}$:*

$$\mu(\underline{g}) \leq \mu(f) \leq \mu(\bar{g}), \qquad \mu(\bar{g}) - \mu(\underline{g}) \leq \varepsilon \mu(S).$$

**Corollary.** *For any real-valued, bounded, continuous function $f$ from $S$ we have*

$$|\mu(f)| \leq \mu(|f|) \leq \|f\| \mu(S).$$

*Remark:* The theorem and its corollary hold for bounded, measurable functions, but then the $F_i$ are measurable. (The integrals of measurable functions are defined below.)

Now let $(S, \mathscr{S}_\sigma)$ be a measurable space, $\mathscr{S}_\sigma$ a $\sigma$-algebra, $\Psi_1$ the class of simple measurable functions, $\bar{\Psi}_1$ its closure in the uniform metric, and $\mu$ a measure on $\mathscr{S}_\sigma$. The definitions and theorems for integrals given above remain true if we change $\Psi_0$ to $\Psi_1$ and $\bar{\Psi}_0$ to $\bar{\Psi}_1$. Thus we define the abstract integral of any bounded, measurable function with respect to a measure.

## 2. GENERALIZATIONS OF THE ABSTRACT INTEGRAL

In Section 1 we introduced abstract integrals in a rather limited sense. This integral concept can be generalized in different directions. The following two generalizations follow readily. At first we reject the restriction that the function $f$ is bounded, but we require that it still be measurable. Then for a nonnegative function $f$ we define $f_m$ by $f_m(x) = \min(f(x), m)$ for a positive integer $m$, and

$$\mu(f) = \lim_m \mu(f_m). \tag{1}$$

This limit always exists, but it may be infinite. For an arbitrary measurable function $f$, we then define

$$\mu(f) = \mu(f^+) - \mu(f^-),$$

provided that one of these integrals in finite.

In the next generalization, we pass over to $\sigma$-finite measures, but then require that they be $\sigma$-smooth. For a representation

$$\mu = \sum_{i=1}^{\infty} \mu_i, \qquad \mu_i(E) = \mu E \cap S_i, \qquad \bigcup (S_i) = S \tag{2}$$

of pairwise disjoint subsets $S_i$ in the $\sigma$-algebra $\mathscr{S}_\sigma$, we define

$$\mu(f) = \sum_{i=1}^{\infty} \mu_i(f), \tag{3}$$

when this sum has a sense, the terms being defined by (1). The sum has a meaning if it is convergent to a finite value, to $\infty$, or to $-\infty$. It is easy to show that (2) is independent of the particular representation (3). If $\mu(f)$ is finite, we say that $f$ is *integrable.*

The integral, defined by (3), has the following property.

**Theorem 1** (Monotone convergence theorem). *Let $\{f_n\}$ be a nondecreasing sequence of nonnegative, measurable functions and define $f = \lim_n f_n$. Then $\mu(f) = \lim_n \mu(f_n)$ for any $\sigma$-smooth measure $\mu$.*

*Proof:* As limit of a sequence of measurable functions, $f$ is measurable. It follows from (2) that it is sufficient to prove the theorem for (finite) $\sigma$-smooth measures. First consider the case when $f$ is bounded. As we have remarked in Theorem 1.2, there then exists a sequence $\{\underline{g}_n\}$ of elementary functions such that $\underline{g}_n \uparrow f$, $\|\underline{g}_n - f\| \to 0$, $\mu(f) = \lim_n \mu(\underline{g}_n)$. Put $E_n = \{x : \underline{g}_n(x) > f_n(x)\}$. Then

$$\mu(\underline{g}_n) - \mu(\underline{g}_n \cdot 1_{E_n}) \leq \mu(f_n) \leq \mu(f)$$

since $\underline{g}_n(x) \leq f_n(x)$ for $x \in S \backslash E_n$. Now

$$\mu(\underline{g}_n \cdot 1_{E_n}) \leq \|f\| \mu(E_n) \downarrow 0 \qquad \text{as} \quad n \uparrow +\infty,$$

since $E_n \downarrow \varnothing$ and $\mu$ is $\sigma$-smooth. Hence $\mu(f) = \lim_n \mu(f_n)$. □

We say that a sequence $\{f_n\}$ of measurable, real-valued functions from a measure space $(S, \mathscr{S}_\sigma, \mu)$ converges almost everywhere $(\mu)$ to a function $f$, if

$\lim_n f_n(x) = f(x)$ at all points $x \in S \setminus E_0$, where $\mu(E_0) = 0$. If then $f$ is not defined on $E_0$, we may define it as 0 at $x \in E_0$. Then $f$ is a measurable function. If $f$ is a measurable function on $S$ and we define $\bar{f}$ by $\bar{f}(x) = f(x)$ on $S \setminus E_0$ and $\bar{f}(x) = 0$ on $E_0$, we have $\mu(\bar{f}) = \mu(f)$. This follows by the definition of the integrals.

**Theorem 2** (Dominated convergence theorem). *Let $\mu$ be a $\sigma$-smooth, $\sigma$-finite measure on a $\sigma$-algebra $\mathcal{S}_\sigma$ on S. If a sequence $\{f_n\}$ of real-valued, measurable functions from S converges almost everywhere $\mu$ to a measurable function $f$, and if $|f_n(x)| \leq g(x)$ for $x \in S$, where $g$ is integrable with respect to $\mu$, then $f$ is integrable with respect to $\mu$ and*

$$\lim_n \mu(f_n) = \mu(f).$$

*Proof:* Putting $f = f^+ - f^{-1}$, we find by the linearity of the integral as a functional that it is sufficient to prove the theorem for a nonnegative, measurable function and a $\sigma$-smooth finite measure $\mu$. Hence consider this case and define $g_n$ and $h_n$ by $g_n(x) = \inf_{j \geq n} f_j(x)$, $h_n(x) = \sup_{j \geq n} f_j(x)$. Then $g_n$ and $h_n$ are measurable, and $0 \leq g_n \leq f_n \leq h_n \leq g$, $g_n \uparrow f$, $h_n \downarrow f$,

$$\mu(g_n) \leq \mu(f_n) \leq \mu(h_n) \leq \mu(g). \tag{4}$$

By this relation and the monotone convergence theorem, we obtain $\mu(g_n) \uparrow \mu(f)$, $\mu(g - h_n) \uparrow \mu(g - f)$, and thus $\mu(h_n) \downarrow \mu(f)$. Hence, by (4), $\lim_n \mu(f_n) = \mu(f)$. □

## 3. THE REPRESENTATIONS OF BOUNDED, LINEAR FUNCTIONALS BY INTEGRALS

We saw in Section 1 that integrals are bounded, linear functionals on a Stone vector lattice ring $\Psi$ of real-valued, bounded, continuous functions from an A-space. Now consider this lattice. Let the A-space be normal, and let $L$ be a bounded, nonnegative, linear functional on $\Psi$, and $\mathcal{F}$ the $\sigma$-topological class of closed sets on $S$. We then define a set function $\mu$ on the algebra $\mathcal{S}$, generated by the closed sets in $S$, as follows. Let $\mathcal{D}(F)$ for given $F \in \mathcal{F}$ denote the subclass of $\Psi$ that consists of all nonnegative functions $f \in \Psi$ for which $f(x) \geq 1$ for $x \in F$. Put

$$\mu(F) = \inf_{f \in \mathcal{D}(F)} L(f), \qquad \mu(E) = \sup_{F \in E,\, F \in \mathcal{F}} \mu(F) \qquad \text{for } E \in \mathcal{S}. \tag{1}$$

In the next section we shall show that for certain Stone vector lattices the set function $\mu$ defined in this way is a measure determined uniquely by $L$ and $\Psi$. However, it is possible that $L$ on a smaller lattice than $\Psi$, say $\Psi_0$, determines $\mu$; then, of course, $\mathcal{D}(F)$ denotes the corresponding subclass of

$\Psi_0$. We may thus ask: Which lattices determine a given measure $\mu$ on $\mathscr{S}$ by a bounded, linear functional $L$ and by (1)? Since an integral on a Stone vector lattice $\Psi_0$ of bounded, continuous functions is a bounded, linear functional, we may also ask if and under which conditions the converse holds. We answer these questions by

**Theorem 1.** *Let $\Psi$ be a Stone vector lattice of real-valued, bounded, continuous functions on a normal space $S$ such that any two disjoint closed sets are connected by a function in $\Psi$, and $\Psi$ contains the function $x \to 1$ for $x \in S$. If a regular measure $\mu$ on $S$ (i.e., on the algebra generated by the closed sets) is determined by a nonnegative, bounded, linear functional $L$ on $\Psi$ according to* (1)*, then*

$$L(f) = \mu(f) \qquad \text{for} \quad f \in \Psi.$$

*Furthermore, $\mu$ is uniquely determined by $L$.*

*Remark:* The question of whether $L$ determines a measure will be answered in the next section.

For the proof we need

**Lemma 1.** *If $L_1$ and $L_2$ are bounded, linear functionals on a Stone vector lattice $\Psi$ of bounded, real-valued functions containing the function $x \to 1$, and $L_1(f) \leq L_2(f)$ for all nonnegative functions $f \in \Psi$, and $L_1(1) = L_2(1)$ for the function $x \to 1$, then $L_1(f) = L_2(f)$ for $f \in \Psi$.*

*Proof:* Put $a = \inf_{x \in S} f(x)$ for given $f \in \Psi$. Then $f - a \in \Psi$ and $f(x) - a \geq 0$ for $x \in S$. Hence, by this assumption,

$$L_1(f) - aL_1(1) = L_1(f - a) \leq L_2(f - a) = L_2(f) - aL_2(1)$$

and

$$L_1(f) \leq L_2(f).$$

We also get this inequality for $-f$, and thus $L_1(-f) \leq L_2(-f)$, i.e., $L_1(f) \geq L_2(f)$, and then $L_1(f) = L_2(f)$. □

*Proof of Theorem 1:* It follows by the definition of $\mu$ that it is a regular measure, if it is a measure, as we have assumed. Put

$$L_1(f) = \mu(f) \qquad \text{for} \quad f \in \Psi.$$

Then $L_1$ is a bounded, linear functional on $\Psi$ (See Section 1). We have $L_1(1) = \mu(S)$, and by (1), $L(1) = \mu(S)$. Applying the lemma, we conclude that it is sufficient to show that $L_1(f) \geq L(f)$ for any nonnegative function $f$ in $\Psi$. We shall show this. It is then sufficient to deal with $f$ satisfying $0 \leq f(x) \leq 1$ for $x \in S$. Applying Theorem 1.3, we approximate $f$ by a sequence

$\{\bar{f}_n\}$ of functions, which we define as

$$\bar{f}_n = \sum_{i=1}^{n+1} \frac{i-1}{n} 1_{F_{i-1}-F_i}, \qquad F_i = \left\{x : f(x) \geq \frac{i}{n}\right\},$$

which we may so define, since $0 \leq f(x) \leq 1$ for $x \in S$. Then $\bar{f}_n \leq f$ and

$$\mu(f) \geq \mu(\bar{f}_n) = \sum_{i=1}^{n+1} \frac{i-1}{n} [\mu(F_{i-1}) - \mu(F_i)] = -\frac{1}{n}\mu(S) + \frac{1}{n}\sum_{i=1}^{n} \mu(F_i).$$

Furthermore, $\bar{f}_n$ tends uniformly to $f$. By the definition of $\mu$, there exists a function $f_i \in \mathscr{D}(F_i)$ such that $f_i \geq 1$ on $F_i$, $f_i \geq 0$ on $S$, and $\mu(F_i) \geq L(f_i) - 1/n$. Hence we get by the relation above

$$\mu(f) \geq -\frac{1}{n}\mu(S) + \frac{1}{n}\sum_{i=1}^{n} L(f_i) - \frac{1}{n}. \tag{2}$$

On $F_i$ we have $(i/n) \leq f(x) < [(i+1)/n]$ for $x \in F_i \backslash F_{i+1}$. On the other hand, $f_j \geq 1$ on $F_j$, and thus, since $F_{i+1} \subset F_i$, $f_j(x) \geq 1$ for $x \in F_i$ for all $j \leq i$. Hence

$$\frac{1}{n}\sum_{i=1}^{n} f_i(x) \geq \frac{i}{n} \geq f(x) - \frac{1}{n}$$

on $F_i \backslash F_{i+1}$ for $i = 0, \ldots, n$, and so the left-hand side is at least equal to $f - 1/n$ on $S$, and

$$L\left(\frac{1}{n}\sum_{i=1}^{n} f_i\right) \geq L(f) - \frac{1}{n}L(1).$$

By (2) we then get

$$\mu(f) \geq L(f) - \frac{1}{n}\mu(S) - \frac{1}{n} - \frac{1}{n}L(1).$$

Letting $n \to +\infty$, we obtain $\mu(f) \geq L(f)$.

Since $\mu$, by the definition, is regular, and obviously any regular measure is uniquely determined by its values on the closed sets, the measure $\mu$ is uniquely determined by $L$ and $\Psi$. □

## 4. MEASURES BELONGING TO A NONNEGATIVE, BOUNDED, LINEAR FUNCTIONAL ON A NORMAL A-SPACE

The main result of this section is given by

**Theorem 1** (Alexandrov's first theorem). *Let $\Psi$ be a Stone vector lattice ring of real-valued, bounded, continuous functions from a normal space $S$,* *and suppose that any two disjoint sets are connected by a function in* $\Psi$ *and that* $\Psi$ *contains the function* $x \to 1$ *for* $x \in S$. *Then a given nonnegative, bounded, linear functional* $L$ *on* $\Psi$ *determines uniquely a regular measure* $\mu$ *on* $S$ *such that*

$$L(f) = \mu(f) \qquad \text{for} \quad f \in \Psi.$$

**Corollary 1.** *For a metric space* $S$, *we may choose* $\Psi$ *as the class of real-valued, bounded, uniformly continuous functions, provided that* $S$ *is pseudo-compact, and particularly if* $S$ *is compact.*

*Remark:* Note that a regular measure on a compact space is $\sigma$-smooth (according to Theorem I.12.3).

**Corollary 2.** *Let* $\Psi$ *be the class of real-valued, bounded, measurable functions from a measurable space* $S$. *Then a given nonnegative, bounded, linear functional* $L$ *on* $\Psi$ *determines uniquely a measure on* $S$ *such that* $L(f) = \mu(f)$ *for* $f \in \Psi$.

*Remark:* Note that $\mu(f)$ for a measure $\mu$ is defined for all real-valued, bounded, continuous functions and, moreover, for all real-valued, bounded functions, measurable with respect to the algebra generated by the closed sets.

In order to prove the theorem, we define a set function on the $\sigma$-topological class $\mathscr{F}$ of closed sets according to (3.1), and let $\mathscr{S}$ be the algebra generated by $\mathscr{F}$.

Define $\mu(F)$ for $F \in \mathscr{F}$ and $\mu(E)$ for $E \in \mathscr{S}$ by (3.1). Note then that

$$\mu(E) = \sup_{F \subset E,\ F \in \mathscr{F}} \mu(F)$$

holds for all sets $E$, including the closed sets $F_0$, since $f \in \mathscr{D}(F_0)$ implies $f \in \mathscr{D}(F)$. For the proof of the theorem, we need some lemmas.

**Lemma 1.** *If* $E_1$ *and* $E_2$ *are disjoint sets in* $\mathscr{S}$, *then* $\mu(E_1 \cup E_2) \geq \mu(E_1) + \mu(E_2)$.

*Proof:* By the definition of $\mu$, for any given closed, disjoint sets $F_1$ and $F_2$ there exists $f \in \mathscr{D}(F_1 \cup F_2)$ for given $\varepsilon > 0$ such that

$$L(f) < \mu(F_1 \cup F_2) + \varepsilon.$$

Let $g \in \Psi$ be a function that connects $F_1$ with $F_2$: $g(x) = 0$ on $F_1$ and 1 on $F_2$; $0 \leq g(x) \leq 1$ on $S$. Put $f_1 = f(1 - g)$, $f_2 = fg$. The functions $f_1$ and $f_2$ belong to $\Psi$, since $\Psi$ is some vector lattice ring. We have

$$f = f_1 + f_2, \qquad f_1 \in \mathscr{D}(F_1), \qquad f_2 \in \mathscr{D}(F_2).$$

Hence we get

$$\mu(F_1 \cup F_2) > L(f) - \varepsilon = L(f_1) + L(f_2) - \varepsilon \geq \mu(F_1) + \mu(F_2) - \varepsilon,$$

or, since $\varepsilon$ is arbitrary,

$$\mu(F_1 \cup F_2) \geq \mu(F_1) + \mu(F_2).$$

Thus we have proved the lemma for disjoint closed sets. It then follows for any disjoint sets, since the sets $\mu(E)$ for any set can be approximated arbitrarily closely by $\mu(F)$ for closed sets according to the definition of $\mu$. □

**Lemma 2.** *For $E_1, E_2 \in \mathscr{S}$, we have*

(i) $$\mu(E_1 \cup E_2) = \mu(E_1) + \mu(E_2),$$

*provided that there exists a closed set $F \supset E_1$, where $F \cap E_2 = \varnothing$.*

*Proof:* We may suppose that $E_1 \cup E_2$ is not empty. Since $\mathscr{S}$ is generated by the closed sets, there exists at least one closed set $F_1 \subset E_1 \cup E_2$. To a given $f \in \mathscr{D}(F_1 \cap F)$ and $0 < \varepsilon < 1$, there belongs the closed set $\{x : f(x) \leq 1 - \varepsilon\} = F_2$. Choose $g \in \mathscr{D}(F_1 \cap F_2)$. If $x \in F_1$, then either $x \in F_1 \cap F_2$ and $g(x) \geq 1$, or $x \in F_1 \backslash F_2$ and $f(x) > 1 - \varepsilon$ by the definition of $F_2$. Hence, for any $x \in F_1$, we have $f(x) + g(x) > 1 - \varepsilon$, and thus $(f + g)/(1 - \varepsilon) \in \mathscr{D}(F_1)$ and

$$\frac{1}{1-\varepsilon}[L(f) + L(g)] = L\left(\frac{f+g}{1-\varepsilon}\right) \geq \mu(F_1). \tag{1}$$

Now $F \supset E_1$ and $F_1 \subset E_1 \cup E_2$, where $E_2 \cap F = \varnothing$. Hence $F \cap F_1 \subset F \cap E_1 = E_1$, and thus $F_1 \backslash (F_1 \cap F) \subset E_2$. Remembering that $f \in \mathscr{D}(F_1 \cap F)$, and thus $f \geq 1$ on $F_1 \cap F$, we conclude that $F_2$ has no points in common with $F_1 \cap F$, since $f(x) \leq 1 - \varepsilon$ for $x \in F_2$. Consequently,

$$F_1 \cap F_2 \subset F_1 \backslash (F \cap F_1) \subset E_2$$

and

$$\mu(F_1 \cap F_2) \leq \mu(E_2). \tag{2}$$

Since $g$ can be chosen arbitrarily in $\mathscr{D}(F_1 \cap F_2)$, by (1) and (2) we get

$$[1/(1 - \varepsilon)]\{L(f) + \mu(E_2)\} \geq \mu(F_1).$$

This relation holds for any $f \in \mathscr{D}(F_1 \cap F)$ and for any $\varepsilon$, where $0 < \varepsilon < 1$. Hence

$$\mu(F_1 \cap F) + \mu(E_2) \geq \mu(F_1). \tag{3}$$

Since $F_1 \subset E_1 \cap E_2$, we have $F_1 \cap F \subset E_1$, and thus

$$\mu(E_1) + \mu(E_2) \geq \mu(F_1).$$

This is true for any $F_1 \subset E_1 \cap E_2$. Hence, by the definition of $\mu$,

$$\mu(E_1) + \mu(E_2) \geq \mu(E_1 \cup E_2)$$

On the other hand, since $E_1$ and $E_2$ are disjoint, we obtain by Lemma 1

$$\mu(E_1 \cup E_2) = \mu[E_1 \cup (E_2 \backslash E_1)] \geq \mu(E_1) + \mu(E_2 \backslash E_1) = \mu(E_1) + \mu(E_2). \quad \square$$

*Proof of the Theorem:* We apply Lemma I.14.1 to $\mathscr{E} = \mathscr{F}$ and $\mu$ on $\mathscr{F}$ defined by (3.1). Then let $\mu^{(i)}$ be the inner extension of $\mu$ on the algebra $\mathscr{S}$ generated by $\mathscr{F}$. By the definition (3.1) of $\mu$ on $E \in \mathscr{S}$, we get $\mu^{(i)}(E) = \mu(E)$ for $E \in \mathscr{S}$. We have shown in Section I.12 that $\mu^{(i)}$ is superadditive, i.e.,

$$\mu^{(i)}(E_1 \cup E_2) \geq \mu^{(i)}(E_1) + \mu^{(i)}(E_2)$$

for any $E_1, E_2 \in \mathscr{S}$. By Lemma I.12.1, the $\mu^{(i)}$-measurable sets form an algebra $\mathscr{S}^{(i)}$, and $\mu^{(i)}$ is a measure on $\mathscr{S}^{(i)}$. This algebra contains $\mathscr{F}$. Indeed, by Lemma 2 we get for $F \in \mathscr{F}$ and any $D \in \mathscr{S}$,

$$\mu^{(i)}(D) = \mu^{(i)}(F \cap D) + \mu^{(i)}(F^c \cap D),$$

since $F \supset F \cap D$, $F \cap F^c \cap D = \varnothing$. Hence $\mathscr{F} \subset \mathscr{S}^{(i)}$ and $\mathscr{S} = \mathscr{S}^{(i)}$, since $\mathscr{S}$ is generated by $\mathscr{F}$. Thus $\mu^{(i)}$ is a measure on $S$ and $\mu^{(i)}(F) = \mu(F)$ for $F \in \mathscr{F}$. The uniqueness of $\mu^{(i)}$ follows by Theorem 3.1. $\square$

*Proof of Corollary 1:* The class $\Psi_0$ of real-valued, bounded, uniformly continuous functions is obviously a Stone vector lattice ring. However, any two disjoint closed sets need not be connected by a function in $\Psi_0$. If the distance between any two disjoint closed sets is positive, say $\rho(F_1, F_2) \geq 2\alpha$ for the disjoint closed sets $F_1$ and $F_2$ in the metric $\rho$, then they are connected by a real-valued, bounded, uniformly continuous function according to the remark on Theorem I.7.2, and by the same remark any two disjoint compact sets have a positive distance. $\square$

*Proof of Corollary 2:* The corollary follows from the theorem if we consider the $\sigma$-algebra $\mathscr{S}_0$ on $S$ as a $\sigma$-topological class of closed sets. Note that two disjoint sets are then connected by the indicator function of one of the sets. $\square$

## 5. TRANSFORMATIONS OF MEASURES AND INTEGRALS

Let $S$ and $S'$ be normal A-spaces, and $\mathscr{S}$ and $\mathscr{S}'$ the algebras generated by the closed sets on $S$ and $S'$, respectively. Denote the class of all bounded, real-valued, continuous functions from $S$ and $S'$ by $\Psi$ and $\Psi'$, respectively. Consider a continuous mapping $T$ of $S$ into $S'$ and a measure $\mu$ on $\mathscr{S}$. If

$f' \in \Psi'$, then $x \to f'(Tx)$ is a continuous function from $S$, and the integral

$$L'(f') = \int_S f'(Tx)\,\mu(dx)$$

is defined for all $f' \in \Psi'$. By Theorem 4.1, the bounded, linear functional $L'$ on $\Psi'$ determines a regular measure $\mu'$ on $\mathscr{S}'$ such that

$$\int_S f'(Tx)\,\mu(dx) = \int_{S'} f'(x')\,\mu'(dx') = \mu'(f'). \tag{1}$$

Note that $\{f'(T\cdot)\}$ for $f' \in \Psi'$ is a subclass $\tilde{\Psi}$ of $\Psi$. Since $\Psi'$ is a Stone vector lattice ring, it follows that $\tilde{\Psi}$ also is such a ring and the integral on the left-hand side of (1) determines a bounded, linear functional $\tilde{L}$ on $\tilde{\Psi}$. For any closed set $F'$ on $S'$, the set $T^{-1}F'$ is closed since $T$ is continuous. By (1), we have

$$\tilde{L}[f'(T\cdot)] = L'(f'), \qquad \text{where} \quad f' \in \Psi'. \tag{2}$$

By the definition of regular measures in Section 3, it then follows that (2), and hence (1), implies

$$\mu(T^{-1}F') = \mu'(F'). \tag{3}$$

We observe that the topology in $S'$ and $T$ determine a topology in $S$, since any closed set on $S'$ determines the closed set $T^{-1}F'$ on $S$. If $\mathscr{S}$ and $\mathscr{S}'$ are $\sigma$-algebras of subsets on sets $S$ and $S'$, respectively, and $T$ is a measurable mapping of $S$ into $S'$, we consider $\mathscr{S}$ and $\mathscr{S}'$ as $\sigma$-topological classes of subsets, and so make $S$ and $S'$ to normal spaces. Then (1) follows as before. Any indicator function $1_{E'}$ from $S'$ is continuous (i.e., measurable), and by (1), we get for any measurable set $E'$ on $S$,

$$\int_S 1_{E'}(Tx)\,\mu(dx) = \int_{S'} 1_{E'}(x')\,\mu(T^{-1}\,dx) = \mu(T^{-1}E),$$

and hence

$$\mu'(E') = \mu(T^{-1}E'),$$

since $1_{E'}(Tx) = 1$ if and only if $Tx \in E'$, i.e., $E' = T^{-1}E$. Clearly $\mu'$ is $\sigma$-smooth if $\mu$ is $\sigma$-smooth. Thus we have proved

**Theorem 1.** *If $S$ and $S'$ are A-spaces and $T$ is a continuous mapping of $S$ into $S'$, then a measure $\mu$ on $S$ determines a measure $\mu'$ on $S'$ such that*

(i) $$\mu'(F') = \mu(T^{-1}F')$$

*for any closed set $F'$ on $S'$ and*

(ii) $$\int_{S'} f'(x')\,\mu'(dx') = \int_S f'(Tx)\,\mu(dx)$$

*for any bounded, continuous function $f'$ from $S'$ into $R$.*

*Remark:* If $\mu$ is regular, then (i) also implies this relation for any set $E'$ in the algebra generated by the closed sets on $S'$. This follows by the definition of regular measures.

**Corollary.** *If $S$ and $S'$ are measurable spaces with $\sigma$-algebras $\mathscr{S}_\sigma$ and $\mathscr{S}'_\sigma$, respectively, and $\mu$ is a measure on $\mathscr{S}_\sigma$, then $T$ and $\mu$ determine a measure $\mu'$ on $\mathscr{S}'_\sigma$ such that* (i) *holds for any $F' \in \mathscr{S}'_\sigma$ and* (ii) *for any bounded, measurable function from $S'$ into $R$. If $\mu$ is $\sigma$-smooth, then $\mu'$ is $\sigma$-smooth.*

## 6. CONSTRUCTIONS OF MEASURES ON METRIC SPACES BY RIEMANN–STIELTJES INTEGRALS

Let $E^{(m)}$ for $m = 1, 2, \ldots,$ where $E^{(m)} \subset E^{(m+1)}$, be subsets of a metric space $S$, and let $N^{(m)}$ be a partition of $E^{(m)}$ into disjoint subsets $E_j^{(m)}$ $j = 1, 2, \ldots, r_m$ for each $E^{(m)}$, i.e., $N^{(m)}$ is the sequence of subsets $E_j^{(m)}$. We call $N^{(m)}$ a net on $E^{(m)}$, and the $E_j^{(m)}$, meshes of the net, and say that $\{N^{(m)}\}_{m=1}^{\infty}$ is a dense, disjoint, directed sequence of nets on $S$ if

(1) the distance between any two points in $E_j^{(m)}$ is smaller than $\varepsilon_m$ for $j = 1, \ldots, r_m$, where $\varepsilon_m \to 0$ as $m \to \infty$;
(2) $E_j^{(m)} \cap E_j^{(m)} = \varnothing$ for $i \neq j$; and
(3) any $E_i^{(m+1)}$ is a subset of some $E_j^{(m)}$.

Furthermore, we call the sequence of nets measurable if the $E_j^{(m)}$ belong to the algebra generated by the closed sets in $S$.

A dense, disjoint, and directed sequence of nets on a space $S$ determines a class $\mathscr{E}$ of subsets such that $\mathscr{E}$ is closed for finite unions and finite intersections and is generated by the class consisting of all meshes of the nets of $S$, and of the empty set. The sets in $\mathscr{E}$ are finite unions of disjoint meshes.

**Theorem 1.** *Let $\{N^{(m)}\}$ be a dense, disjoint, and directed sequence of nets on a pseudocompact metric space $S$ and let $\mathscr{E}$ be the class of subsets belonging to $\{N^{(m)}\}$, as above. A nonnegative, additive set function $\mu$ on $\mathscr{E}$ such that $\mu(S) < \infty$,*

$$\mu(S) - \mu(E^{(m)} \to 0) \qquad (m \to \infty),$$

*determines a regular measure $\lambda$ on $S$. If $S$ is compact, then $\lambda$ is $\sigma$-smooth.*

*Proof:* Let $\Psi$ be the class of real-valued, bounded, uniformly continuous functions. To the net $N^{(m)}$, $f \in \Psi$, and $\mu$, there belongs the sum

$$L_m(f) = \sum_{1 \leq j \leq r_m} f[x_j^{(m)}]\mu[E_j^{(m)}], \tag{1}$$

where $x_j^{(m)}$ is some point in $E_j^{(m)}$. For a dense, disjoint, directed sequence of nets, any $E_i^{(m)}$ is the union of meshes $E_j^{(m+k)}$ for any positive integer $k$. Hence we get, for the metric $\rho$,

$$|L_{m+k}(f) - L_m(f)| \leq \sup_{\rho(x,y) < \varepsilon_m} |f(x) - f(y)|\mu(S) + \|f\| \sum_{E_j^{(m+k)} \not\subset E^{(m)}} \mu(E_j^{m+k}) \quad (2)$$

The first term on the right-hand side tends to 0 as $m \to \infty$, and so does the second term since it is at most equal to

$$\mu(S) - \mu[E^{(m)}].$$

Thus (5) implies Cauchy convergence and, hence, the convergence of $L_m(f)$ to a finite value $L(f)$. It easily follows by the properties of limits that $L(f)$ is a nonnegative, bounded, linear functional on $\Psi$, and, by the corollary of Theorem 4.1, this functional determines a regular measure $\lambda$ on $S$. If $S$ is compact, then $\lambda$ is $\sigma$-smooth according to Theorem I.13.3. □

If $\{K^{(m)}\}$ is an increasing sequence of compact sets, we may consider a dense, disjoint, directed sequence $\{N^{(m)}\}$ of nets such that

$$E^{(m)} = \bigcup_{j=1}^{r_m} E_j^{(m)} = K^{(m)}. \quad (3)$$

Let $\mathscr{E}$ be the class of subsets of $S$ that belongs to $\{N^{(m)}\}$ as above, and consider a nonnegative, additive set function $\mu$ on $E$. We say that $\mu$ is tight on the sequence $\{N^{(m)}\}$ if

$$\mu(S) - \mu(K^{(m)}) \to 0 \qquad (m \to \infty), \quad (\mu(S) < \infty).$$

Since for any positive integer $r$ the sequence of nets with meshes $E_j^{(m)} \cap K^{(r)}$, where $m > r$, form a dense, disjoint, directed sequence of nets, it follows by the theorem above that $\mu$ and the sequence $\{N^{(m)}\}$ of nets determine a $\sigma$-smooth measure $\lambda_r$ on $K^{(r)}$ for $r = 1, 2, \ldots$ . Define the $\sigma$-smooth measure $\overline{\lambda}_r$ on $S$ by

$$\overline{\lambda}_r(E) = \lambda_r(E \cap K_r)$$

for any measurable set $E$ on $S$. Then

$$\lambda = \overline{\lambda}_1 + \sum_{r=1}^{\infty} (\overline{\lambda}_{r+1} - \overline{\lambda}_r) \quad (4)$$

is a $\sigma$-smooth measure on $S$. Indeed, it is easily seen that $\overline{\lambda}_{r+1}(E) - \overline{\lambda}_r(E)$ is nonnegative and equal, at most, to $\overline{\lambda}_{r+1}(S) - \overline{\lambda}_r(S)$ and $\lambda(S) \leq \mu(S)$. Thus the series is uniformly convergent with respect to $E \subset S$.

If $\{K^{(m)}\}$ is an increasing sequence of compact sets, then we can construct a dense, disjoint, directed sequence of nets satisfying (3). We proceed as follows.

The open balls $\{x:\rho(x, y)\}$, where $y$ is any point in $K^{(m)}$, cover $K^{(m)}$, and since $K^{(m)}$ is compact, it is covered by finitely many such balls, say $B_1^{(m)}$, $B_2^{(m)}, \ldots, B_{r_m}^{(m)}$. Putting

$$\tilde{E}_1^{(m)} = B_1^{(m)} \cap K^{(m)}, \qquad \tilde{E}_2^{(m)} = B_2^{(m)} \cap K^{(m)} \backslash E_2^{(m)}, \ldots,$$

$$\tilde{E}_j^{(m)} = B_j^{(m)} \cap K^{(m)} \Big\backslash \bigcup_{i=1}^{j-1} E_i^{(m)}$$

for $j = 2, \ldots, r_m$, we get a disjoint net $\tilde{N}^{(m)}$ on any $K^{(m)}$ such that $\rho(x, y) < 1/m$ for $x$ and $y$ belonging to a mesh in this net. From these nets we get a directed, dense, disjoint sequence of nets simply by choosing $E_j^{(1)} = \tilde{E}_j$ for all $j$, the meshes of $N^{(2)}$ as all sets $E_j^{(1)} \cap \tilde{E}_i^{(2)}$ for $j = 1, \ldots, r_1$, $i = 1, \ldots, r_2$, and in the same manner all the meshes in $N^{(m)}$ as the union of all intersections of $\tilde{E}_i^{(m)}$ with the meshes in $E_j^{(m-1)}$ for all $m$.

In Chapter III we shall deal with tight set functions on dense, disjoint, directed sequences of nets with the property (3).

## 7. MEASURES ON PRODUCT SPACES

Product spaces were introduced in Section I.9. We considered there the class $\Psi_0$ of simple functions

$$f = \sum_{i=1}^{m} \alpha_i 1_{E_i' \otimes E_i''}, \qquad \alpha_i \text{ real number},$$

$$E_i' \in \mathscr{S}', \qquad E_i'' \in \mathscr{S}'', \qquad \bigcup E_i' \otimes E_i'' = S' \otimes S'', \tag{1}$$

$$E_i' \otimes E_i'' \cap E_j' \otimes E_j'' = \varnothing \qquad \text{for} \quad i \neq j,$$

where $\mathscr{S}'$ and $\mathscr{S}''$ are the algebras generated by the closed sets in $S'$ and $S''$, respectively. We proved that the closure $\bar{\Psi}_0$ with respect to the uniform metric is a complete Stone vector lattice ring in this metric. The closure $\bar{\bar{\Psi}}_0$ of $\Psi_0$ with respect to pointwise bounded limits is also a Stone vector lattice ring, and it is closed with respect to pointwise bounded limits. The classes of continuous functions in $\bar{\Psi}_0$ and $\bar{\bar{\Psi}}_0$ are denoted by $\bar{\Psi}$ and $\bar{\bar{\Psi}}$, respectively. We observed in Section I.9 that any closed set $F$ in $S' \otimes S''$, when $S'$ and $S''$ are completely normal, has a representation $F = \{x: f(x) = 0\}$ with $f \in \bar{\Psi}$, $\|f\| = 1$, and we observe that $\bar{\bar{\Psi}}$ is the class of all bounded, continuous functions.

Let $\mu'$ and $\mu''$ be $\sigma$-smooth measures on completely normal spaces $S'$ and $S''$. The algebras $\mathscr{S}'$ and $\mathscr{S}''$ and the $\sigma$-smooth measures can then be extended to $\sigma$-algebras $\mathscr{S}_\sigma'$ and $\mathscr{S}_\sigma''$ and $\sigma$-smooth measures on these $\sigma$-algebras. Then in (1), we let $E'$ and $E''$ be sets in the $\sigma$-algebras and $\Psi_0$ be the Stone vector lattice ring of the elementary functions formed in this way. Furthermore, we still let $\bar{\Psi}_0$ be the closure of $\Psi_0$ with respect to the uniform metric and $\bar{\bar{\Psi}}_0$ be the closure of $\Psi_0$ with respect to pointwise bounded limits. As we have remarked in Section I.9, $\bar{\bar{\Psi}}_0$ then contains all bounded, measurable functions, and consequently the Stone vector lattice $\bar{\bar{\Psi}}$ of bounded, continuous functions is contained in $\bar{\bar{\Psi}}_0$. To $f \in \Psi_0$ belongs the functional

$$L(f) = \sum_{i=1}^{m} \alpha_i \mu'(E_i')\mu''(E_i''). \tag{2}$$

It is easy to show that $L$ so defined is a nonnegative, bounded, linear functional on $\Psi_0$. Indeed, any two functions in $\Psi_0$ have representations of the form (1) with the same sets $E_i' \times E_i''$. We can write (2) as repeated integrals

$$\begin{aligned} L(f) &= \int_{S'}\left[\int_{S''} f(x)\,\mu''(dx'')\right]\mu'(dx') \\ &= \int_{S''}\left[\int_{S'} f(x)\,\mu'(dx')\right]\mu''(dx''), \qquad x = (x', x''). \end{aligned} \tag{3}$$

Now any $f \in \bar{\bar{\Psi}}_0$ is a pointwise bounded limit of a sequence $\{f_n\}$, $f_n \in \Psi_0$ and (3) holds for $f_n$. Applying the dominated convergence theorem to the repeated integral and observing that the sections of $f$ are measurable (Theorem I.9.2), we find first that the inner integrals in (3) exist, are bounded and, as pointwise bounded limits, are measurable. Hence, the repeated integrals exist. Since $L$ is a nonnegative, bounded, linear functional on $\Psi_0$, it follows by the properties of limits that (3) holds for any $f \in \bar{\bar{\Psi}}_0$ and that $L$ is a nonnegative, bounded, linear functional on $\bar{\bar{\Psi}}_0$ and, hence, also on $\bar{\bar{\Psi}}$. By Theorem 1.4, this functional on $\bar{\bar{\Psi}}$ determines a regular measure on the algebra generated by the closed sets on $S' \otimes S''$. By the definition of integrals,

$$\mu(F) = \mu(1_F)$$

and

$$\mu(1_F) = L(1_F) = \int_{S'}\left[\int_{S''} 1_F(x)\,\mu''(dx'')\right]\mu'(dx'), \tag{4}$$

since this relation holds for functions in $\Psi_0$. Consider now any nonincreasing sequence $\{F_n\}$ such that $F_n \downarrow \varnothing$ and hence $1_{F_n} \downarrow 0$ as $n \uparrow \infty$. By the integral representation (4) of $\mu(1_{F_n})$, we conclude that $\mu(F_n) \downarrow 0$ as $n \uparrow \infty$; but then $\mu$ is $\sigma$-smooth according to Theorem I.13.2. We may now consider the nonnegative, bounded, linear functional $\mu(f)$ on $\Psi_0$ and $\bar{\bar{\Psi}}_0$, and get

$$\mu(f) = \int_{S'}\left[\int_{S''} f(x)\,\mu''(dx'')\right]\mu'(dx') = \int_{S''}\left[\int_{S'} f(x)\,\mu'(dx')\right]\mu''(dx'') \tag{5}$$

for any $f \in \bar{\bar{\Psi}}_0$. Indeed, this relation holds for $f \in \Psi_0$, and if $\{f_n\}$ tends pointwise bounded to $f$, then $\mu(f_n)$ and the corresponding repeated integrals in (5) tend to $\mu(f)$ and the repeated integrals in (5).

We call $\mu$ the product measure of $\mu'$ and $\mu''$ and use the notation $\mu = \mu' \otimes \mu''$. Clearly $\mu' \otimes \mu'' = \mu'' \otimes \mu'$. By the definition of $\mu$, we could consider the $\sigma$-algebras $\mathscr{S}'_\sigma$ and $\mathscr{S}''_\sigma$ as $\sigma$-topological classes of closed sets; then, as we have remarked in Section I.9, any "closed" set in $S' \otimes S''$ means a measurable set, and by (3) the functional $L(f)$ defined by (3) determines a measure $\mu''$ on $S' \otimes S''$ such that $\mu(f) = L(f)$ for $f \in \bar{\bar{\Psi}}_0$. It follows as above that $\mu$ is $\sigma$-smooth. For this definition we may consider only measurable spaces $S'$ and $S''$ (not necessarily measurable spaces generated by the closed sets).

We may form product measures as above of several $\sigma$-smooth measures on corresponding A-spaces (measurable spaces). Note that then

$$\mu' \otimes (\mu'' \otimes \mu''') = (\mu' \otimes \mu'') \otimes \mu'''$$

on $S' \otimes S'' \otimes S'''$. Indeed, the integral of any bounded, measurable function with respect to $\mu' \otimes (\mu'' \otimes \mu''')$ and $(\mu' \otimes \mu'') \otimes \mu'''$ are pointwise bounded limits of sums

$$\sum_{i=1}^{m} \alpha_i \mu'(E_i')\mu''(E_i'')\mu'''(E_i'''),$$

where $E_i'$, $E_i''$, and $E_i'''$ are measurable sets on $S'$, $S''$, and $S'''$, respectively, and

$$E_j' \otimes E_i'' \otimes E_i''' \cap E_j' \otimes E_j'' \otimes E_j''' = \varnothing \qquad \text{for} \quad i \neq j.$$

Let us return to the case where $E_i'$ and $E_i''$ in (1) are sets in the algebra generated by the closed sets in the completely normal spaces $S'$ and $S''$, respectively, and $S' \otimes S''$ is, therefore, a completely normal space. Then $\Psi_0$ is the corresponding class of simple functions (1) and $\bar{\Psi}_0$ the closure with respect to the uniform metric. Let $\mu'$ and $\mu''$ be measures (not necessarily $\sigma$-smooth) on $S'$ and $S''$, respectively. We still get, using the definition of integrals,

$$L(f) = \int_{S'} \left[ \int_{S''} f(x)\, \mu''(dx'') \right] \mu'(dx') = \int_{S''} \left[ \int_{S'} f(x)\, \mu'(dx') \right] \mu''(dx'')$$

for $f \in \Psi_0$, and then for $f \in \bar{\Psi}_0$, by passing to limits in the uniform metric of sequences $\{f_n\}$ to $f$. It also follows that $L(f)$ is a nonnegative, bounded, linear functional on $\bar{\Psi}_0$, and hence on the class $\bar{\Psi}$ of continuous functions. The class $\bar{\Psi}$ contains the function $x \to 1$ from $S' \otimes S''$ into $R$. However, $\bar{\Psi}$ does not necessarily contain a function $f$ to any disjoint closed sets $F_1$ and $F_2$ such that $f$ connects these sets. The application of Theorem 4.1 requires that $\bar{\Psi}$ contains such a function. If any disjoint closed sets are connected by a function in $\bar{\Psi}$, then the functional $L(f)$ on $\bar{\Psi}$ determines a measure $\mu$ on $S' \otimes S''$ such that

$$L(f) = \mu(f) \qquad \text{where} \quad f \in \bar{\Psi} \quad \text{and} \quad \mu = \mu' \otimes \mu'' = \mu'' \otimes \mu'.$$

If the product space is a metric pseudocompact space, then, by the corollary of Theorem 1.4, the functional $L$ on $\bar{\Psi}$ determines a measure $\mu$ on $S' \otimes S''$. We call this measure the product measure of $\mu'$ and $\mu''$ and denote it by $\mu' \otimes \mu''$.

## 8. CONVOLUTIONS OF MEASURES

In the preceding sections we have not required any algebraic structure in the A-spaces under consideration. Now we shall require such an algebraic structure. We say that an A-space $S$ is a topological group if it is a group and if the mapping $(x, y) \to xy$ of $S \otimes S$ into $S$ is continuous, composition in the group being written multiplicatively, and moreover the mapping $x \to x^{-1}$ of $S$ onto $S$ is continuous. If the A-space is a commutative topological group, we write the composition additively.

**Theorem 1.** *If the A-space $S$ is a topological group (composition multiplicative), then the mapping $(x, y) \to yx$ from $S \otimes S$ onto $S$ and the mappings $x \to xy$ and $x \to yx$ for given $y \in S$ into $S$ are continuous.*

*Proof:* By $(x, y) \to (x^{-1}, y^{-1})$, a mapping $\upsilon$ of $S \otimes S$ onto $S \otimes S$ is defined. For any closed sets $F'$ and $F''$ in $S$, we have

$$\upsilon^{-1}[F' \otimes F''] = \{x : x^{-1} \in F'\} \otimes \{y : y^{-1} \in F''\}, \tag{1}$$

and the right-hand side is a closed rectangle in $S \otimes S$, since $x \to x^{-1}$ is continuous. Since the rectangles $F' \otimes F''$ generate the $\sigma$-topological class of closed sets, it follows by (1) that $\upsilon$ is continuous (compare the proof of Theorem I.3.2). Moreover, using the fact that the mapping $(x^{-1}, y^{-1}) \to x^{-1}y^{-1} = (yx)^{-1}$ and the mapping $(yx)^{-1} \to yx$ are continuous, we conclude that $(x, y) \to yx$ is continuous.

Since $(x, y) \to xy$ is continuous and the section of this function at $y$ maps $x$ on $xy$, the section being continuous, we conclude that $x \to yx$ is continuous.

Consider now a normal A-space $S$, that is a topological group, and let $\Psi$ be the class of bounded, real-valued, measurable functions from $S$. We form the product A-space $S \otimes S$. If $f \in \Psi$, and $x = (x', x'')$ denotes a point in $S \otimes S$, then $x \to f(x'x'')$ is a real-valued, bounded, measurable function from $S \otimes S$ into $R$. For $\sigma$-smooth measures $\mu_1$ and $\mu_2$ on $S$, there exists, according to Theorem 7.1, a product measure $\mu_1 \otimes \mu_2$ such that

$$L(f) = \int_S \left[ \int_S f(x'x'')\mu_1(dx') \right] \mu_2(dx'')$$
$$= \int_S \left[ \int_S f(x'x'')\mu_2(dx') \right] \mu_1(dx''). \tag{2}$$

According to Theorem 4.1, the bounded, linear functional $L$ on $\Psi$ defined by (2) determines a regular measure $\mu$ on $S$. For any measurable set $E$ in $S$ we get, by (2),

$$\mu(E) = \mu(1_E) = \int_S \left[ \int_S 1_E(x_1x_2)\mu_2(dx_2) \right] \mu_1(dx_1)$$
$$= \int_S \left[ \int_S 1_E(x_1x_2)\mu_1(dx_1) \right] \mu_2(dx_2). \tag{3}$$

If $\{E_n\}$ is any sequence of measurable sets such that $E_n \downarrow \varnothing$, we first find by this relation, using the monotone convergence theorem and the fact that $\mu_1$ and $\mu_2$ are $\sigma$-smooth, that the inner integral tends nonincreasingly to 0, and then also that the outer integral tends to 0. Hence, $\mu$ is $\sigma$-smooth. Note that the inner integrals in (3) are measurable since they are limits of elementary functions. We call $\mu$ the convolution of $\mu_1$ and $\mu_2$ and use the notation $\mu = \mu_1 * \mu_2$.

For an A-space $S$ that is a topological group and for $\sigma$-smooth measures $\mu_1$, $\mu_2$, and $\mu_3$, we form successively the convolutions $(\mu_1 * \mu_2) * \mu_3$ and $\mu_1 * (\mu_2 * \mu_3)$. These are equal and determine the same functional

$$L(f) = \int_S \left\{ \int_S \left[ \int_S f(x_1x_2x_3)\mu_1(dx_1) \right] \mu_2(dx_2) \right\} \mu_3(dx_3). \tag{4}$$

Thus the convolution is associative. If $S$ is a commutative topological group, we write the composition additively, i.e., the continuous mapping from $S \otimes S$ into $S$ is denoted by $(x_1, x_2) \to x_1 + x_2$. Then it follows by (3) that the convolution is commutative and by (4) that it is associative. Note that the inner integrals in (3) can be written $\mu_2(E - x_1)$ and $\mu_1(E - x_2)$, which are measurable functions. We collect our results. □

**Theorem 2.** *The convolution of $\sigma$-smooth measures, determined by* (2) *on an A-space, which is also a topological group, is an associative composition. If the topological group is commutative, then the convolution is also commutative.*

**Theorem 3.** *Let $S$ and $S'$ be A-spaces that are also topological groups, and let $\pi$ be a measurable mapping of $S$ into $S'$. If $\mu_1$ and $\mu_2$ are $\sigma$-smooth measures on $S$ and $\mu = \mu_1 * \mu_2$, then $\mu(\pi^{-1}\cdot) = \mu_1(\pi^{-1}\cdot) * \mu_2(\pi^{-1}\cdot)$ on $S'$, provided $\pi x y = \pi x \pi y$ for $x, y \in S$.*

*Proof:* Let $\Psi'$ be the class of all real-valued, bounded, continuous functions from $S'$ into $R$. By the definition of convolution and Fubini's relation, we get, putting

$$\mu' = \mu_1(\pi^{-1}\cdot) * \mu_2(\pi^{-1}\cdot)$$

for any $f \in \Psi'$, $x', y' \in S'$ that

$$L(f) = \int_{S'} f(z)\,\mu'(dz') = \int_{S'}\left\{\int_{S'} f(x'y')\,\mu_2(\pi^{-1}\,dy')\right\}\mu_1(\pi^{-1}\,dx').$$

By $y' = \pi y$, $x' = \pi x$, the second integral is transformed into

$$L(f) = \int_S\left\{\int_S f(\pi x\cdot\pi y)\,\mu_2(dy)\right\}\mu_1(dx) = \int_S f(\pi z)\,\mu(dz) = \int_{S'} f(z')\,\mu(\pi^{-1}\,dz').$$

This is true for any $f \in \Psi'$, and $L$ on $\Psi'$ determines the measure on $S'$ uniquely. □

## 9. PROBABILITY SPACES AND RANDOM VARIABLES

A set $\Omega$ together with a $\sigma$-algebra $\mathscr{B}$ and a probability measure $P$ on $\mathscr{B}$ form a probability space. For this we use the notation $[\Omega, \mathscr{B}, P]$. By definition, a probability measure is a $\sigma$-smooth measure with $\mu(\Omega) = 1$. Members of $\mathscr{B}$ are called events.

A family $\{E_i, i \in I\}$ of events is said to be independent if for any finite sets $(i_1, i_2, \ldots, i_k)$ of indices $i_\nu \in I$, $\nu = 1, \ldots, k$, we have

$$P\left(\bigcap_{\nu=1}^{k} E_{i_\nu}\right) = \prod_{\nu=1}^{k} P(E_{i_\nu}).$$

**Theorem 1.** *If the family $\{E_i, i \in I\}$ is independent, then also the family $\{E_i : i \in I\} \cup \{\complement E_i, i \in I\}$ is independent.*

*Proof:* The statement follows by induction from the following fact: If $A$ and $E$ are independent, then $A$ and $\complement E$ are independent. This statement follows by

$$\begin{aligned} P(A \cap \complement E) &= P[A \cap (S - E)] = P(A) - P(A \cap E) \\ &= P(A)[1 - P(A)] = P(A)P(\complement A). \quad \square \end{aligned}$$

**Lemma 1** (Borel–Cantelli lemma). *For a sequence $\{E_n\}_{n=1}^{\infty}$, where $E_n \in \mathscr{B}$, we have $P(\limsup E_n) = 0$ if $\sum_{n=1}^{\infty} P(E_n) < \infty$; and $P(\limsup E_n) = 1$ if $\sum_{n=1}^{\infty} P(E_n) = \infty$, and moreover the sequence $\{E_n\}_{n=1}^{\infty}$ is independent.*

*Remark:* By this lemma, we get the Borel–Cantelli zero–one law: If the sequence $\{E_n\}_{n=1}^{\infty}$, $E_n \in \mathscr{B}$ is independent, then $P\{\limsup E_n\} = 0$ or 1, depending on whether $\sum_{n=1}^{\infty} P(E_n)$ is convergent or divergent.

*Proof:* By definition, we have

$$\limsup_{n\to+\infty} \mathrm{E}_n = \bigcap_{n=1}^{\infty} \bigcup_{k=n}^{\infty} E_k,$$

and thus

$$P(\limsup E_n) \leq \bigcup_{k=n}^{\infty} P(E_k)$$

for any $n$. The right-hand side tends to 0 as $n \to \infty$ if $\sum_{n=1}^{\infty} P(E_n) < \infty$. Now let $\{E_n\}_{n=1}^{\infty}$ be an independent sequence and $\sum_{n=1}^{\infty} P(E_n) = \infty$. By Theorem 1, then, the sequence $\{\complement E_n\}_{n=1}^{\infty}$ is also independent. By the use of de Morgan's rules, we get

$$\complement \limsup E_n = \bigcup_{n=1}^{\infty} \bigcap_{k=n}^{\infty} \complement E_k.$$

Using the elementary inequality $1 - x \leq \exp(-x)$, where $0 \leq x \leq 1$, we get $P(\complement E_k) \leq \exp[-P(E_k)]$ and, since the sequence $\complement E_k$ is independent,

$$P\left(\bigcap_{k=n}^{\infty} \complement E_n\right) \leq \exp\left[-\sum_{k=n}^{\infty} P(E_n)\right],$$

where the right-hand side is 0, since $\sum_{k=1}^{\infty} E_n = \infty$. Hence $P(\complement \limsup E_n) = 0$ and $P(\limsup E_n) = 1$. □

Let $(S, \mathscr{S})$ be a measurable space, $\mathscr{S}$ denoting a $\sigma$-algebra on $S$. A measurable function $X$ from a probability space $(\Omega, \mathscr{B}, P)$ into $S$ determines the element $X(\omega)$ in $S$. We call $X$ a random variable from $\Omega$ into $S$ (often a random function if $S$ is a function space). The probability $P$ and $X$ determine the probability measure $P_X = P(X^{-1}\cdot)$ on $\mathscr{S}$ and thus determine a probability space $(S, \mathscr{S}, P_X)$. We call $P_X$ the distribution function of $X$. If $f$ is any measurable, real-valued function from $S$, we get by transformation of the integral

$$\int_S f(x)\, P_X(dx) = \int_\Omega f(X(\omega))\, P(d\omega), \tag{1}$$

whenever one of these integrals is defined.

Observe that $X$ is a mapping with the map $X(\omega)$ of $\omega$. However, often the same notation is used for the mapping and the map; so we may write $\int X\, P(d\omega)$ instead of $\int X(\omega)\, P(d\omega)$.

Consider a sequence $\{(\Omega_i, \mathscr{B}_i, P_i)\}$ of probability spaces and random variables $X_i$ from $(\Omega_i, \mathscr{B}_i, P_i)$ into $(S, \mathscr{S})$, $i = 1, 2, \ldots$. We form the product measurable space $(\Omega, \mathscr{B})$ of these spaces, and the product probability measure $P$ on $\mathscr{B}$ of these $P_i$ (compare Sections I.9 and I.15). Then by Theorem I.9.3, the mapping $\gamma_i$ of $(\Omega, \mathscr{B})$ onto $(\Omega_i, \mathscr{B}_i)$ is measurable. Since $X_i$ is a measurable mapping of $(\Omega_i, \mathscr{B}_i)$ into $(S, \mathscr{S})$, the mapping $\bar{X}_i = X_i \circ \gamma_i$ is a measurable mapping of $(\Omega, \mathscr{B}, P)$ into $(S, \mathscr{S})$ and hence a random variable. Note that $\bar{X}_i$ determines the measure

$$P(\bar{X}_i^{-1} \cdot) = P(\gamma_i^{-1} X_i^{-1} \cdot) = P_i(X_i^{-1} \cdot)$$

on $(S, \mathscr{S})$. It follows by this relation that we may consider countably many random variables from different probability spaces as random variables from the same probability space when we deal with the distributions of the random variables.

Observing that the random variables $X_i$ determine the probability spaces $[S, \mathscr{S}, P_i(X_i^{-1})]$, we may in the same way as for the $(\Omega_i, \mathscr{B}_i, P_i)$ form the product probability space of the probability spaces $(S, \mathscr{S}, P_{X_i})$. The probability measure on this product probability space is called the distribution of the vector $X_1, \ldots, X_r$ when the product space has the dimension $r$.

Random variables $X_i$, $i = 1, \ldots, r$ of random variables $X_i$ from a probability space $(\Omega, \mathscr{B}, P)$ into $(S, \mathscr{S})$ are said to be independent if the family $X_i^{-1}E$ is independent for any $E \in \mathscr{S}$.

## 10. EXPECTATIONS, CONDITIONAL EXPECTATIONS, AND CONDITIONAL PROBABILITIES

For a random variable $\xi$ from a probability space $(\Omega, \mathscr{B}, P)$ into $R$, the integral

$$\boldsymbol{E}(\xi) = \int_\Omega \xi(\omega)\, P(d\omega) \tag{1}$$

is called the expectation when it is defined. The notation $\boldsymbol{E}(\xi)$ is common, though, with our general integral notation, we could also write $P(\xi)$. The existence of (1) implies the existence of

$$\mu(E) = \int_E \xi(\omega)\, P(d\omega) \tag{2}$$

for any $E \in \mathscr{B}$. By (2), a $\sigma$-smooth, $\sigma$-finite signed measure $\mu$ is defined on $\mathscr{B}$, provided that (1) exists. This follows by Theorem 1.1 and the generalizations in Section 2. Indeed, $\mu$ is a $\sigma$-smooth, $\sigma$-finite measure if $\xi$ is nonnegative; and moreover $\mu$ is a $\sigma$-smooth measure if $\xi$ is also bounded.

A random variable $\eta$ from a probability space $(\Omega, \mathscr{B}_0, P)$, where $\mathscr{B}_0 \subset \mathscr{B}$, is called the conditional expectation of $\xi$, given $\mathscr{B}_0$, if

$$\int_E \eta(\omega)\, P(d\omega) = \int_E \xi(\omega)\, P(d\omega) \tag{3}$$

for any $E \in \mathscr{B}_0$. We use the notation

$$\eta = \boldsymbol{E}[\xi \,|\, \mathscr{B}_0] \qquad \text{or} \qquad \boldsymbol{E}^{\mathscr{B}_0}(\xi).$$

Now $\eta$ is not uniquely determined by $\xi$ and $\mathscr{B}_0$ according to (3), but if (3) is satisfied by $\eta_1$ and $\eta_2$ for all $E \in \mathscr{B}_0$, we have

$$\int_E [\eta_1(\omega) - \eta_2(\omega)]\, P(d\omega) = 0, \qquad E \in \mathscr{B}_0, \tag{4}$$

i.e., $\eta_1 = \eta_2$ almost surely $(P_{\mathscr{B}_0})$, if $P_{\mathscr{B}_0}$ denotes the restriction of $P$ to $\mathscr{B}_0$. We call any $\eta$ satisfying (3) a version of the conditional expectation of $\xi$, given $\mathscr{B}_0$, and write $\eta = \boldsymbol{E}[\xi \,|\, \mathscr{B}_0]$ a.s., if $\eta$ is such a version.

**Theorem 1.** *If a real-valued random variable $\xi$ from a probability space $(\Omega, \mathscr{B}, P)$ has a finite expectation, and if $\mathscr{B}_0$ is a sub–$\sigma$-algebra of $\mathscr{B}$, then the conditional expectation $\boldsymbol{E}^{\mathscr{B}0}(\xi)$ exists and is almost surely uniquely determined.*

**Theorem 2** (Radon–Nikodym theorem). *If $(\Omega, \mathscr{B}, P)$ is a probability space and $\nu$ a $\sigma$-smooth measure on a sub–$\sigma$-algebra $\mathscr{B}_0$ of $\mathscr{B}$, then there exists a real-valued function $\eta$ from $\Omega$, measurable $\mathscr{B}_0$, such that*

(i) $$\int_E \eta(\omega)\, \nu(d\omega) = P(E) \qquad \textit{for} \quad E \in \mathscr{B}_0,$$

(ii) $$\int_\Omega \zeta(\omega)\eta(\omega)\, \nu(d\omega) = \int_\Omega \zeta(\omega)\, P(d\omega)$$

*for any real-valued function $\zeta$ from $\Omega$ with finite expectation and measurable $(\mathscr{B}_0)$. Furthermore, $\eta$ is uniquely determined a.s. by $\nu$.*

*Remark:* We call $\eta$ the Radon–Nikodym density.

*Proof:* We shall prove the two theorems simultaneously. Consider first the case when $\xi(\omega) \geq 0$ for $\omega \in \Omega$. Let $C$ be the class of all real-valued functions $\zeta$, measurable $\mathscr{B}_0$, and satisfying the inequality

$$\int_E \zeta(\omega)\, P(d\omega) \leq \int_E \xi(\omega)\, P(d\omega). \tag{5}$$

The class $C$ is not empty since it contains the function $\zeta \to 0$. If $\zeta_1$ and $\zeta_2$ belong to $C$, then $\max(\zeta_1, \zeta_2) \in C$. Indeed, put

$$E_1 = \{\omega : \zeta_1(\omega) \geq \zeta_2(\omega)\}.$$

Then $E_1 \in \mathscr{B}_0$, $E_1^c \in \mathscr{B}_0$, and $\max(\zeta_1, \zeta_2)$ is measurable $\mathscr{B}_0$. Furthermore, for $E \in \mathscr{B}_0$, we get by (5)

$$\int_E \max[\zeta_1(\omega)\zeta_2(\omega)]\, P(d\omega) = \int_{E \cap E_1} \zeta_1(\omega)\, P(d\omega) + \int_{E \cap E_1^c} \zeta_2(\omega)\, P(d\omega)$$

$$\leq \int_{E \cap E_1} \xi(\omega)\, P(d\omega) + \int_{E \cap E_1^c} \xi(\omega)\, P(d\omega)$$

$$= \int_E \xi(\omega)\, P(d\omega).$$

Put

$$\sup_{\zeta \in C} \int_\Omega \zeta(\omega)\, P(d\omega) = \alpha. \tag{6}$$

Then $\alpha < \infty$ and thus there exists a sequence $\{\zeta_n\}$, $\zeta_n \in C$ such that $E[\zeta_n] \to \alpha$. Define $\zeta_n^*$ by $\zeta_n^* = \max_{i \leq n} \zeta_i$. Then $\zeta_n^* \uparrow \eta$ with $\eta$ measurable $\mathscr{B}_0$ and $\eta \in C$, and $E(\eta) = \alpha$. We claim that $\eta$ satisfies (3). To prove this we first note that $\mu$, given by

$$\mu(E) = \int_E \xi(\omega)\, P(d\omega) - \int_E \eta(\omega)\, P(d\omega), \tag{7}$$

is a $\sigma$-smooth measure, absolute continuous with respect to $P$. Suppose that $\mu(E_0) \geq 2\varepsilon > 0$ for some $E_0 \in \mathscr{B}_0$ and consider the signed measure

$$\mu_n(E) = \mu(E) - (1/n)P(E \cap E_0).$$

By the Hahn decomposition (see [19]) there exists a decomposition of the set $E_0$ into two disjoint sets, $E_0 = A_n \cup B_n$, such that $\mu_n(E) \geq 0$ for any $E \subset A_n$ and $\mu_n(E) \leq 0$ for any $E \in B_n$. We choose $n$ so large $P(E_0) \leq n\varepsilon$. Then $\mu(A_n) \geq \varepsilon$ and

$$\nu_n(E) = \mu(E) - (1/n)P(E \cap A_n)$$

is a $\sigma$-smooth measure on $\mathscr{B}_0$. Indeed $\nu_n(E) = \mu(E)$ for $E \cap A_n = \varnothing$, $\nu_n(E) \geq 0$ for $E \in A_n$. Thus (7) remains true if we change $\eta$ to $\eta + 1_{A_n}$. This contradicts the maximal property, $E(\eta) = \alpha$, since $\mu(A_n) \geq \varepsilon > 0$ implies $P(A_n) > 0$, and $E[\eta + (1/n)1_{A_n}] = E[\eta] + (1/n)P(A_n) > 0$. Hence (3) holds.

If $\eta_1$ and $\eta_2$ are $\mathscr{B}_0$-measurable and satisfy (3) then we find by (4) that $\eta_1 = \eta_2$ a.s.

If $\xi$ is integrable but not necessarily nonnegative, we put $\xi = \xi^+ - \xi^-$ and conclude that $\xi^+$ and $\xi^-$ have conditional expectations $E^{\mathscr{B}_0}(\xi^+)$ and $E^{\mathscr{B}_0}(\xi^-)$ and then clearly

$$E^{\mathscr{B}_0}(\xi) = E^{\mathscr{B}_0}(\xi^+) - E^{\mathscr{B}_0}(\xi^-).$$

To prove Theorem 2, we have only to change $P$ on the left-hand side of (5) to $\nu$ and $\xi(\omega)$ on the right-hand side of (5) to 1 and define $C$ as the class of functions satisfying this new inequality (5). Then it follows, as in the proof of Theorem 1, that there exists a real-valued function $\eta$ from $\Omega$, measurable

$(\mathscr{B}_0)$, such that (i) holds. To prove (ii), we observe that (ii) holds for any indicator function $\zeta = 1_E$, where $E \in \mathscr{B}_0$, by (i). Then it follows for any simple function $\zeta$. Considering nondecreasing sequences of simple functions, we then find by the monotone convergence theorem that it holds for any nonnegative function $\zeta$ measurable $\mathscr{B}_0$. Since any $\zeta$, measurable $\mathscr{B}_0$, with finite expectation, has a representation $\zeta = \zeta^+ - \zeta^-$, we conclude at last that (ii) holds for any $\zeta$, measurable $\mathscr{B}_0$ with finite expectation. □

We shall now give some fundamental properties of conditional expectations. Note that the existence of a conditional expectation $\boldsymbol{E}^{\mathscr{B}_0}(\xi)$ requires the existence of $\boldsymbol{E}(\xi)$. This follows by the definition (3), which we now write in the form

$$\int_E \boldsymbol{E}^{\mathscr{B}_0}(\xi)\, P(d\omega) = \int_E \xi\, P(d\omega), \qquad E \in \mathscr{B}_0. \tag{8}$$

(In the following we drop $\omega$ in $\xi(\omega)$ when we consider integrals, remembering that the random variables depend on $\omega \in \Omega$.) Observe that the existence of $\boldsymbol{E}(\xi)$ requires the existence of $\boldsymbol{E}(\xi^+)$ and $\boldsymbol{E}(\xi^-)$, and finiteness of one of these expectations.

**Theorem 3.** *Consider real-valued random variables $\xi, \zeta, \ldots$ from a probability space $(\Omega, \mathscr{B}, P)$, and let $\mathscr{B}_0$ be a sub–$\sigma$-algebra of $\mathscr{B}$. Then the following relations hold, provided that the conditional expectations under consideration exist:*

$$\begin{array}{lll}
\text{(i)} & \boldsymbol{E}^{\mathscr{B}_0}(\xi + \zeta) = \boldsymbol{E}^{\mathscr{B}_0}(\xi) + \boldsymbol{E}^{\mathscr{B}_0}\zeta & a.s.\ (P_{\mathscr{B}_0}), \\
\text{(ii)} & \boldsymbol{E}^{\mathscr{B}_0}c = c & a.s.\ (P_{\mathscr{B}_0}), \qquad \xi \geq \zeta \quad a.s. \\
 & \Rightarrow \boldsymbol{E}^{\mathscr{B}_0}(\xi) \geq \boldsymbol{E}^{\mathscr{B}_0}(\zeta) & a.s.\ (P_{\mathscr{B}_0}), \quad particularly \quad \xi \geq 0 \\
 & a.s.\ (P) \Rightarrow \boldsymbol{E}^{\mathscr{B}_0}(\xi) \geq 0 & a.s.\ \ P_{\mathscr{B}_0}, \\
\text{(iii)} & \boldsymbol{E}^{\mathscr{B}_0}(\xi\zeta) = \zeta \boldsymbol{E}^{\mathscr{B}_0}\xi & a.s.\ (P_{\mathscr{B}_0}) \quad if\ \zeta\ is\ measurable\ \mathscr{B}_0, \\
\text{(iv)} & \boldsymbol{E}^{\mathscr{B}_0}[\boldsymbol{E}^{\mathscr{B}_1}(\xi)] = \boldsymbol{E}^{\mathscr{B}_0}(\xi) & a.s.\ (P_{\mathscr{B}_0}),
\end{array}$$

*if $\mathscr{B}_1$ is a sub–$\sigma$-algebra of $\mathscr{B}$ and $\mathscr{B}_0 \subset \mathscr{B}_1$.*

*Proof:* For the proof we shall mainly use the relation in (8), which transforms integrals over conditional expectations $\boldsymbol{E}^{\mathscr{B}_0}\xi$ to integrals over $\xi$.

(i) It follows directly by (8).
(ii) The first relation follows directly by (5). Put

$$B = \{\omega : \boldsymbol{E}^{\mathscr{B}_0}(\zeta) - \boldsymbol{E}^{\mathscr{B}_0}\xi \geq \varepsilon\}.$$

By (8) we get

$$P(B) \leq \frac{1}{\varepsilon} \int_{\mathscr{B}} (\zeta - \xi)\, P(d\omega) = 0$$

for any $\varepsilon > 0$.

(iii) According to (i), it is sufficient to consider nonnegative random variables $\xi$ and $\zeta$ (but we must check that the conditional expectations under consideration exist). If $\xi$ has finite expectation, then $1_{B'}\xi$ has finite expectation and $E^{\mathscr{B}_0}(1_{B'}\xi)$ is defined. Let $B, B' \in \mathscr{B}^0$. By (5) we obtain

$$\int_B E^{\mathscr{B}_0}(1_{B'}\xi)\, P(d\omega) = \int_B 1_{B'}\xi\, P(d\omega) = \int_{BB'} \xi\, P(d\omega)$$
$$= \int_B E^{\mathscr{B}_0}(\xi)\, P(d\omega) = \int_B 1_{B'} E^{\mathscr{B}_0}(\xi)\, P(d\omega).$$

This is true for any $B \in \mathscr{B}_0$. Hence, (iii) holds for indicator functions $\zeta$, measurable $\mathscr{B}_0$. Then, however, it follows for simple functions by the monotone convergence theorem for nonnegative, measurable functions.

(iv) Since $B \in \mathscr{B}_0 \Rightarrow B \in \mathscr{B}_1$ we get, by (8),

$$\int_B E^{\mathscr{B}_0}[E^{\mathscr{B}_1}(\xi)]\, P(d\omega) = \int_B E^{\mathscr{B}_1}(\xi)\, P(d\omega) = \int_B \xi\, P(d\omega) = \int_B E^{\mathscr{B}}(\xi)\, P(d\omega). \qquad \square$$

**Theorem 4.** *Let $\{\xi_n\}$ be a nondecreasing sequence of nonnegative, random variables from a probability space $(\Omega, \mathscr{B}, P)$, and let $\mathscr{B}_0$ be a sub–$\sigma$-algebra of $\mathscr{B}$. Furthermore, suppose that $E^{\mathscr{B}_0}(\xi_n)$ exists for $n = 1, 2, \ldots$. Then the pointwise limit $\zeta$ of $\{\xi_n\}$ has a conditional expectation and*

$$E^{\mathscr{B}_0}(\xi_n) \uparrow E^{\mathscr{B}_0}(\zeta) \qquad \text{a.s.}$$

*Proof:* By (ii) in Theorem 1, $\{E^{\mathscr{B}_0}(\xi_n)\}$ is nondecreasing and thus has a limit $\bar{\zeta}$, measurable $\mathscr{B}_0$,

$$E^{\mathscr{B}_0}(\xi_n) \uparrow \bar{\zeta},$$

where $E^{\mathscr{B}_0}(\xi_n) \geq 0$ a.s. Furthermore, $\zeta$ is measurable $\mathscr{B}$. By (8) and the monotone convergence theorem, we obtain, for any $E \in \mathscr{B}_0$,

$$\int_E E^{\mathscr{B}_0}(\xi_n)\, P(d\omega) \uparrow \int_E \bar{\zeta}\, P(d\omega),$$
$$\int_E E^{\mathscr{B}_0}(\xi_n)\, P(d\omega) = \int_E \xi_n\, P(d\omega) \uparrow \int_E \zeta\, P(d\omega).$$

Hence,

$$\int_E \bar{\zeta}\, P(d\omega) = \int_E \zeta\, P(d\omega) \qquad \text{for} \quad E \in \mathscr{B}_0,$$

which implies

$$\bar{\zeta} = E^{\mathscr{B}_0}(\zeta)$$

according to the definition (8).

Using (8), we may obtain other relations for conditional expectations that hold true for unconditional expectations. The expectation $E^{\mathscr{B}_0}(1_E)$ for any $E \in \mathscr{B}$, $\mathscr{B}_0 \subset \mathscr{B}$, is called the conditional probability of $E$ given $\mathscr{B}_0$ and is denoted by $P^{\mathscr{B}_0}(E)$. Note that

$$\boldsymbol{E}[P^{\mathscr{B}_0}E] = P(E). \tag{9}$$

Hence $P^{\mathscr{B}_0}(E)$ has many properties of a measure. If $E_1$ and $E_2$ are disjoint and belong to $\mathscr{B}$, we get

$$P^{\mathscr{B}_0}(E_1 \cup E_2) = P^{\mathscr{B}_0}(E_1) + P^{\mathscr{B}_0}(E_2) \qquad \text{a.s.} \quad (P_{\mathscr{B}_0}).$$

If $E_n \uparrow E$, we get

$$P^{\mathscr{B}_0}(E_n) \uparrow P^{\mathscr{B}_0}(E) \qquad \text{a.s.} \quad P_{\mathscr{B}_0} \tag{10}$$

by the monotone convergence theorem for integrals.

If $E' \in \mathscr{B}$, $E \in \mathscr{B}_0$, we also use the notation $P(E'\,|\,E)$ instead of $P(E'\,|\,\mathscr{B}_0)$. This notation is often suitable, since

$$P(E)P(E'\,|\,E) = P(E \cap E'), \tag{11}$$

according to Theorem 3(iii).

## 11. THE JENSEN INEQUALITY

In Chapters IV and V we shall use the famous Jensen inequality, and we shall show it to be an application of the theorems in Section 10. This inequality concerns real-valued convex functions $q$ from $R$. Such a function is characterized by the property

$$q[\alpha x_1 + (1 - \alpha)x_2] \leq \alpha q(x_1) + (1 - \alpha)q(x_2) \tag{1}$$

for $0 \leq \alpha \leq 1$ and any $x_1, x_2 \in R$. This relation implies

$$q(x) \leq \frac{x_2 - x}{x_2 - x_1} q(x_1) + \frac{x - x_1}{x_2 - x_1} q(x_2) \tag{2}$$

for $x_1 \leq x$, $x \leq x_2$, $x_1 < x_2$, since

$$\frac{x_2 - x}{x_2 - x_1} x + \frac{x - x_1}{x_2 - x_1} x = x.$$

Often (2) is used as a definition of a convex function. By (2), we conclude that a convex function is continuous at any point. Indeed, if $x \downarrow x_1$, we obtain $q(x_1 + 0) \leq q(x_1)$, and if $x_2 \downarrow x$, $q(x) \leq q(x + 0)$. Hence, $q(x) = q(x + 0)$. In the same way, we find that $q(x) = q(x - 0)$. The inequality (2) may be written

$$\frac{q(x) - q(x_1)}{x - x_1} \leq \frac{q(x_2) - q(x)}{x_2 - x}, \tag{3}$$

provided $x > x_1, x < x_2$. By (3), we further obtain

$$\frac{q(x_2) - 1(x_1)}{x_2 - x_1} \le \frac{q(x_2) - q(x)}{x_2 - x}. \tag{4}$$

It follows by this inequality that the right-hand side tends nondecreasingly to a limit as $x \uparrow x_2$, and this limit is the left derivative $q_1'(x_2)$ at $x_2$. By (3), we then find that $q_1'(x_2)$ is finite and nondecreasing at any point. Letting $x \uparrow x_2$ in (4), we get

$$q(x_1) \ge q(x_2) + (x_1 - x_2)q_1'(x_2), \qquad x_1 \le x_2. \tag{5}$$

By (3) we also obtain

$$\frac{q(x_2) - q(x_1)}{x_2 - x_1} \ge \frac{q(x) - q(x_1)}{x - x_1}, \qquad x_1 < x < x_2, \tag{6}$$

which shows that the quotient tends nonincreasingly to a limit, namely the right derivative $q_r'(x_1)$, as $x \downarrow x_1$. It follows, as above, that $q_r'$ is nondecreasing and finite at any point $x_1$. Letting $x_1 \uparrow x$, $x_2 \downarrow x$ in (3), we conclude that $q_1'(x) \le q_r'(x)$. We then get by (6), letting $x \downarrow x_1$,

$$q(x_2) \ge q(x_1) + (x_2 - x_1)q_r'(x_1) \ge q(x_1) + (x_2 - x_1)q_1'(x_1).$$

According to this relation and (5), we thus have

$$q(x) \ge q(x_0) + (x - x_0)q_1'(x_0) \tag{7}$$

for any $x$ and $x_0$ on $R$. By induction, we extend (1) to

$$q\left(\sum_{i=1}^{k} \alpha_i x_i\right) \le \sum_{i=1}^{k} \alpha_i q(x_i) \qquad \text{for} \quad \alpha_i \ge 0, \quad \sum_{i=1}^{k} \alpha_i = 1, \quad x_i \in R.$$

We shall now state the Jensen inequality for convex functions. It does not depend on all the properties of a convex function that we have given above, but the proof can be carried through with more or fewer of these properties.

**Theorem 1** (Jensen inequality). *Let $\xi$ be a random variable from a probability space $(\Omega, \mathscr{B}, P)$ into a finite closed interval on $R$ and let $q$ be a convex function on $R$. Then for any sub–$\sigma$-algebra $\mathscr{B}_0$ of $\mathscr{B}$,*

(i) $$q[\mathbf{E}^{\mathscr{B}_0}(\xi)] \le \mathbf{E}^{\mathscr{B}_0} q(\xi).$$

*Particularly*

(ii) $$q[\mathbf{E}(\xi)] \le \mathbf{E}[q(\xi)].$$

**Corollary.** (i) *and* (ii) *hold for any random variable $\xi$ into any closed subset of $R$ provided that the conditional expectations $\mathbf{E}^{\mathscr{B}_0}(\xi)$ and $\mathbf{E}^{\mathscr{B}_0}[q(\xi)]$ exist. (Note that these existences require the existences of $\mathbf{E}(\xi)$ and $\mathbf{E}[q(\xi)]$.)*

*First proof of (i):* We first prove (i) for simple probability measures

$$\xi = \sum_{i=1}^{k} x_i 1_{E_i}, \qquad \sum_{i=1}^{k} 1_{E_i} = 1, \qquad x_i \in [a,b].$$

Then a.s. $P$, $\{E_i\}$ disjoint,

$$\boldsymbol{E}^{\mathscr{B}_0}(\xi) = \sum_{i=1}^{k} x_i \boldsymbol{E}^{\mathscr{B}_0}(1_{E_i}), \qquad \sum_{i=1}^{k} \boldsymbol{E}^{\mathscr{B}_0}(1_{E_i}) = 1,$$

$$q(\xi) = \sum_{i=1} q(x_i) 1_{E_i}, \qquad \boldsymbol{E}^{\mathscr{B}_0} q(\xi) = \sum_{i=1}^{k} q(x_i) \boldsymbol{E}^{\mathscr{B}_0}(1_{E_i}) \qquad \text{a.s.} \quad (P).$$

By (8), we then obtain

$$q[\boldsymbol{E}^{\mathscr{B}_0}\xi] \le \sum_{i=1}^{k} q(x_i) \boldsymbol{E}^{\mathscr{B}_0}(1_{E_i}) = \boldsymbol{E}^{\mathscr{B}_0}(q(\xi)) \qquad \text{a.s.} \quad (P).$$

Thus (i) follows for simple random variables.

By the remark on Theorem 1.3, any random variable $\xi$ from $(\Omega, \mathscr{B}, P)$ into $[a,b]$ can be approximated by a simple random variable $\xi_n$ from $(\Omega, \mathscr{B}, P)$ into $[a,b]$ such that

$$\|\xi - \xi_n\| \le 1/n,$$

and thus, by (ii) in Theorem 10.3,

$$\|\boldsymbol{E}^{\mathscr{B}_0}(\xi - \xi_n)\| \le 1/n.$$

Since $q$ is continuous on $R$ and, hence, uniformly continuous on $[a,b]$, we get

$$\|\boldsymbol{E}^{\mathscr{B}_0} q(\xi) - \boldsymbol{E}^{\mathscr{B}_0} q(\xi_n)\| \to 0, \qquad n \to \infty, \quad \text{a.s.} \quad (P),$$
$$\|q(\boldsymbol{E}^{\mathscr{B}_0}\xi) - q(\boldsymbol{E}^{\mathscr{B}_0}\xi_n)\| \to 0, \qquad n \to \infty, \quad \text{a.s.} \quad (P).$$

Thus (i) follows for any random variable into $[a,b]$. □

*Second proof of (i):* We apply the inequality (7) and get

$$q(\xi) \ge q[\boldsymbol{E}^{\mathscr{B}_0}(\xi)] + \{\xi - \xi[\boldsymbol{E}^{\mathscr{B}_0}(\xi)]\} q_1'[\boldsymbol{E}^{\mathscr{B}_0}(\xi)] \quad \text{a.s.} \quad (P).$$

Note that all random variables in this inequality are finite a.s. $(P)$. By conditioning and using Theorem 1.3, we obtain

$$\boldsymbol{E}^{\mathscr{B}_0} q(\xi) \ge q[\boldsymbol{E}^{\mathscr{B}_0}(\xi)]. \quad \square$$

*Proof of the corollary:* Let $D_a$ be any set in $\mathscr{B}_0$ such that $|\xi| \le a$ a.s. on $D_a$. If there is no such set, then $\mathscr{B}_0$ only consists of $S$ and $\varnothing$ and there is nothing to prove. We now apply (i) in the theorem to the random variable $1_{D_a}\xi$ and observe that $1_{D_a}$ is measurable $\mathscr{B}_0$. We then obtain

$$q[\boldsymbol{E}^{\mathscr{B}_0}(1_{D_a}\xi) \le \boldsymbol{E}^{\mathscr{B}_0} q(1_{D_a}\xi) \qquad \text{a.s.} \quad (P).$$

Since

$$1 = 1_{D_a} + 1_{D_a^c},$$

this relation may be written

$$1_{D_a} q(1_{D_a} \boldsymbol{E}^{\mathscr{B}_0}\xi) + 1_{D_a^c} q(1_{D_a} \boldsymbol{E}^{\mathscr{B}_0}\xi) \leq \boldsymbol{E}^{\mathscr{B}_0}[1_{D_a} q(1_{D_a}\xi)] + \boldsymbol{E}^{\mathscr{B}_0}[1_{D_a^c} q(1_{D_a}\xi)].$$

Here

$$\begin{aligned} 1_{D_a^c} q(1_{D_a} \boldsymbol{E}^{\mathscr{B}_0}\xi) &= 1_{D_a^c} q(0) \qquad \text{a.s.} \quad (P), \\ 1_{D_a^c} q(1_{D_a}\xi) &= 1_{D_a^c} q(0), \\ 1_{D_a} q(1_{D_a} \boldsymbol{E}^{\mathscr{B}_0}\xi) &= 1_{D_a} q(\boldsymbol{E}^{\mathscr{B}_0}\xi) \qquad \text{a.s.} \quad (P), \\ 1_{D_a} q(1_{D_a}\xi) &= 1_{D_a} q(\xi). \end{aligned}$$

Thus we obtain from (9)

$$1_{D_a} q(\boldsymbol{E}^{\mathscr{B}_0}(\xi)) \leq 1_{D_a} \boldsymbol{E}^{\mathscr{B}_0} q(\xi) \quad \text{a.s.} \quad (P).$$

CHAPTER III

# WEAK CONVERGENCE IN NORMAL SPACES

## 1. WEAK CONVERGENCE OF SEQUENCES OF MEASURES ON NORMAL SPACES

Let $\Psi$ be the Stone vector lattice ring of real-valued, bounded, continuous functions from a normal space $S$, and let $\{\mu_n\}$ be a sequence of measures on $S$, i.e., on the algebra, generated by the closed sets in $S$. Then

$$L_n(f) = \mu_n(f), \qquad f \in \Psi$$

defines a nonnegative, bounded, linear functional on $\Psi$. If

$$L(f) = \lim_{n \to +\infty} L_n(f) \qquad \text{for} \quad f \in \Psi, \quad |L(f)| < \infty \tag{1}$$

exists, it is easily seen that $L$ is also a nonnegative, bounded, linear functional on $\Psi$. According to Theorem II.4.1, the functional $L$ uniquely determines a measure $\mu$ on $S$ such that $L(f) = \mu(f)$ for $f \in \Psi$. We say that $\{\mu_n\}$ converges weakly to $\mu$ when (1) holds and use the notation $\mu_n \underset{w}{\to} \mu$ $(n \to \infty)$.

The weak convergence of $\mu_n$ to $\mu$ does not mean that $\mu_n(E)$ converges to $\mu(E)$ for all sets $E \in \mathscr{S}$. However, we shall show that this holds for the continuity sets for $\mu$, and we shall also give important limit relations for all open and closed sets.

**Theorem 1** (Alexandrov's second theorem). *Let $\Psi_0$ be a Stone vector lattice of real-valued, bounded, continuous functions from a normal A-space $S$, and suppose that $\Psi_0$ contains the function $x \to 1$ from $S$. Furthermore, let $\mu$ be a given measure on $S$ such that for any closed set $F$ and any $\varepsilon > 0$ exists an open set $G_\varepsilon \supset F$ with $\mu(G_\varepsilon \backslash F) < \varepsilon$ and $S \backslash G_\varepsilon$, and $F$ are connected by a function in $\Psi_0$. For a sequence $\{\mu_n\}$ of measures on $S$, the following conditions* *imply each other*:

(i) $\mu_n \underset{w}{\rightarrow} \mu \quad (n \to \infty)$,

(ii) $\mu_n(f) \to \mu(f) \quad (n \to \infty) \qquad$ *for* $\; f \in \Psi_0$,

(iii) $\limsup_{n\to\infty} \mu_n(F) \le \mu(F)$ *for any closed set* $F$ *and* $\mu_n(S) \to \mu(S)$ $(n \to \infty)$,

(iv) $\liminf_{n\to\infty} \mu_n(G) \ge \mu(G)$ *for any open set* $G$ *and* $\mu_n(S) \to \mu(S)$ $(n \to \infty)$,

(v) $\mu_n(E) \to \mu(E)$ $(n \to \infty)$ *for any continuity set* $E$ *for* $\mu$, *and* $\mu_n(S) \to \mu(S)$.

*Remark:* The condition $\mu(G_\varepsilon \backslash F) < \varepsilon$ with an open set $G_\varepsilon > F$ for given $F$ and suitable $G_\varepsilon$ is fulfilled if $\mu$ is regular. The requirement that $F$ and $S \backslash G_\varepsilon$ be connected by a function in $\Psi_0$ imposes a stronger regularity on $\mu$, if not any two disjoint closed sets are connected by a function in $\Psi_0$.

**Corollary.** *In a pseudocompact metric space, we may choose* $\Psi_0$ *as the class of all real-valued, bounded, uniformly continuous functions, and if* $\mu$ *is* $\sigma$*-smooth we may make this choice for any metric space.*

*Proof:* We shall show the implications in the order

$$\begin{array}{c}(i) \Rightarrow (ii) \Rightarrow (iii) \Rightarrow (v) \Rightarrow (i)\\ \Updownarrow \\ (iv)\end{array}$$

(a) (i) $\Rightarrow$ (ii) is obvious.

(b) (ii) $\Rightarrow$ (iii): Let (ii) hold, and choose $G_\varepsilon$ to $F$ and $\varepsilon$ such that $\mu(G_\varepsilon \backslash F) < \varepsilon$, and then choose $f \in \Psi_0$ such that it connects $S \backslash G_\varepsilon$ with $F$, $f(x) = 1$ for $x \in F$, $f(x) = 0$ for $x \in S \backslash G_\varepsilon$, $0 \le f(x) \le 1$ for $x \in S$. Then we get

$$\mu(f) \le \mu(1_F \cdot f) + \mu(1_{G \backslash F}) \le \mu(F) + \varepsilon, \qquad \mu_n(f) \ge \mu_n(1_F \cdot f) = \mu_n(F),$$

and thus

$$\limsup_{n\to\infty} \mu_n(F) \le \limsup_{n\to\infty} \mu_n(f) = \mu(f) \le \mu(F) + \varepsilon.$$

Since $\varepsilon > 0$ is arbitrary, (ii) $\Rightarrow$ (iii) for any $F$ (also for $F = S$ if we choose $f = 1$).

(c) The implications (iii) $\Leftrightarrow$ (iv) follow since $S \backslash F$ is open if $F$ is closed, and $S \backslash G$ is closed if $G$ is open.

(d) (iii) $\Rightarrow$ (v): Let $E$ be a continuity set for $\mu$. Then for the interior $E^\circ$ and the closure $\bar{E}$ of $E$ we have $\mu(\bar{E} \backslash E^\circ) = 0$. Since $E^\circ$ is open and $\bar{E}$ closed we get, by (iii),

$$\mu(E^\circ) \le \liminf \mu_n(E^\circ) \le \liminf \mu_n(E) \le \limsup \mu_n(E) \le \limsup \mu_n(\bar{E}) \le \mu(\bar{E}).$$

(e)  (v) $\Rightarrow$ (i): In order to prove the relation $\lim \mu_n(f) = \mu(f)$ for any real-valued, bounded, continuous function from $S$, it is sufficient to deal with such functions for which $0 \le f(x) \le 1$ for all $x \in S$. By Theorem II.1.3, we can choose real numbers $a_i$, where $a_0 < a_i < \cdots < a_m$ to $f$, these $a_i$ forming a net $N$, and give approximations uniformly with respect to different measures $\mu$ and $\mu_n$ under consideration:

$$\underline{\sigma}_N(f, \mu) \le \mu(f) \le \bar{\sigma}_N(f, \mu), \tag{2}$$

where

$$\underline{\sigma}_N(f, \mu) = a_0 \mu(S) + \sum_{i=1}^{m} (a_i - a_{i-1}) \mu(F_i),$$

$$\bar{\sigma}_N(f, \mu) = a_0 \mu(S) + \sum_{i=1}^{m} (a_i - a_{i-1}) \mu(F_{i-1}),$$

$$F_i = [x : f(x) \ge a_i].$$

We choose the points $a_i$ such that $F_i$ is a continuity set for $\mu$ or $F_i = \varnothing$ or $S$. Then, besides (2), we have a corresponding inequality with $\mu$ changed to $\mu_n$. Now (v) implies

$$\bar{\sigma}_N(f, \mu_n) \to \bar{\sigma}_N(f, \mu), \qquad \underline{\sigma}_N(f, \mu_n) \to \underline{\sigma}_N(f, \mu).$$

Since we may choose the points in the net $N$ such that $\underline{\sigma}_N(f, \mu)$ and $\bar{\sigma}_N(f, \mu)$ are arbitrarily close, uniformly with respect to different measures $\mu$, the relation (i) follows.  □

*Proof of the corollary:*  In a metric space $S$, the class $\Psi_0$ of real-valued, bounded, uniformly continuous functions is a Stone vector lattice containing the function $x \to 1$ from $S$. If $S$ is pseudocompact, any two closed sets in $S$ are connected by a function $f \in \Psi_0$ (Theorem I.7.2). If $\mu$ is $\sigma$-smooth, we let $f = h \circ (1/\varepsilon)\rho(x, F)$, where $h(t) = 1$ for $t < 0$, $h(t) = 1 - t$ for $0 \le t \le 1$, and $h(t) = 0$ for $t \ge 1$. Consider the open set

$$G_\varepsilon = \{x : \rho(x, F) < \varepsilon\}.$$

We have $f(x) = 1$ for $x \in F$, and $f(x) = 0$ for $x \in S \backslash G_\varepsilon$. Furthermore, since $G_\varepsilon \backslash F \downarrow \varnothing$ as $\varepsilon \downarrow 0$, and $\mu$ is $\sigma$-smooth,

$$\mu(G_\varepsilon \backslash F) \downarrow 0 \qquad \text{as} \quad \varepsilon \downarrow 0.$$

Hence the requirements in the theorem are fulfilled.  □

Let $\mu_n$ and $\nu$ be $\sigma$-smooth, $\sigma$-finite measures on an A-space $S$. Suppose that there exists a partition $S = \{S_i\}_{i=1}^{\infty}$ of $S$ into disjoint measurable subsets

$S_i$ such that $\mu_n(S_i) < \infty$, $\nu(S_i) < \infty$. Define the measures $\mu_{ni}$ and $\nu_i$ on $S$ by $\mu_{ni}(E) = \mu_n(E \cap S_i)$, $\nu_i(E) = \nu(E \cap S_i)$ for any measurable set $E$ in $S$. If $\{\mu_{ni}\}$ converges weakly to $\nu_i$ for all $i$, we say that $\{\mu_n\}$ converges weakly to $\nu$.

Now consider a measure space $(\Omega, \mathscr{B}, P)$, where $\mathscr{B}$ is a $\sigma$-algebra, and let $S$ be an A-space. A measurable function $X$ from $(\Omega, \mathscr{B}, P)$ into $S$ determines the measure $P(X^{-1}\cdot)$ on $S$. We say that a sequence $\{X_n\}_{n=1}^{\infty}$ of measurable functions $X_n$ from $\Omega$ into $S$ converges weakly if $\{P(X^{-1}\cdot)\}$ converges weakly. If there exists a measurable function $X$ from $\Omega$ into $S$ such that $\{P(X_n^{-1}\cdot)\}$ converges weakly to $P(X^{-1}\cdot)$, we say that $\{X_n\}$ converges weakly to $X$.

**Theorem 2.** *If a sequence $\{X_n\}$ of random variables from a probability space $(\Omega, \mathscr{B}, P)$ into a completely normal space $S$ converges a.s. $(P)$ to a random variable $X$, then $\{P(X_n^{-1}\cdot)\}$ converges weakly.*

*Remark:* If $\{X_n\}$ converges a.s. $P$ to $X$ and thus determines a probability measure $\mu$, we may complete $(\Omega, \mathscr{B}, P)$ to $(\Omega, \bar{\mathscr{B}}, \bar{P})$ with respect to $\mu$ and extend $X_n$ and $X$ to random variables $\bar{X}_n$ and $\bar{X}$ such that $\{\bar{P}(X_n^{-1}\cdot)\}$ converges weakly to $\bar{P}(X^{-1}\cdot)$.

*Proof:* By transforming the integrals and applying the dominated convergence theorem for integrals, we get

$$\int_S f(x)\,P(X_n^{-1}\,dx) = \int_\Omega f(X_n(\omega))\,P(d\omega) \to \int_\Omega f(X)\,P(d\omega) \tag{3}$$

on the class $\Psi$ of real-valued, bounded, continuous functions from $S$ into $R$. The last integral determines a bounded, linear functional and thus a measure $\mu$ on $S$. Since the measures $P(X_n^{-1}\cdot)$ are probability measures, $\mu$ is also a probability measure. This follows by Theorem 4.1. (See Section 4.) However, using the statement in the remark, we can write the last integral in (3)

$$\int_\Omega f(\bar{X})\,\bar{P}(d\omega) = \int_S f(x)\,\bar{P}(\bar{X}^{-1}\,d\bar{\omega}),$$

where $P(\bar{X}^{-1}\cdot)$ is a probability measure, since $\bar{X}$ is a measurable mapping into $S$ ($X$ may not be a mapping into $S$). Let us then prove the statement in this remark. Suppose that $X_n$ converges to $X$ on the measurable set $D$, where $\mu(D^c) = 0$. The complete $\sigma$-algebra $\bar{\mathscr{B}}$ on $\Omega$ is formed of all sets $E \cup \triangle$, where $\triangle$ is any subset of $D^c$ and $E$ any set in $S$. Furthermore, we put

$$\bar{P}(E \cup \triangle) = P(E).$$

For some $x_0 \in S$, we define $\bar{X}$ by

$$\bar{X}(\omega) = X(\omega) \quad \text{for} \quad \omega \in D, \qquad \bar{X}(\omega) = x_0 \quad \text{for} \quad \omega \in D^c.$$

Then $X$ is a random function from $(\Omega, \bar{\mathscr{B}}, \bar{P})$ into $S$, and

$$\lim_{n \to +\infty} \int f(X_n(\omega))\, P(d\omega) = \int f(X(\omega))\, P(d\omega)$$
$$= \int_\Omega f(\bar{X}(\omega))\, \bar{P}(d\omega) = \int_S f(x)\, \bar{P}(\bar{X}^{-1}\, d\omega),$$

according to the dominated convergence theorem for integrals.

## 2. WEAK CONVERGENCE OF SEQUENCES OF INDUCED MEASURES AND TRANSFORMED MEASURES

Let $F_0$ be a closed set on a normal A-space $S$. Then any closed set $F'$ on $S$ determines the closed set $F = F_0 \cap F'$ and hence the $\sigma$-topological class of closed sets on $S$ *induces* a $\sigma$-topological class of closed sets on $F_0$ and hence an A-space on $F_0$. A continuous function from $S$ into $R$ thus induces a continuous function from $F_0$ into $R$, and a measure on $S$, i.e., on the algebra generated by the closed sets, induces a measure $v$ on the A-space $F_0$, where $v$ is determined by $v(E) = \mu(E \cap F_0)$ for any $E$ belonging to the algebra generated by the closed sets on $S$. We shall use the same notation for the induced functions and measures on $S$ and on $F_0$.

**Theorem 1.** *If the sequence $\{\mu_n\}$ of measures on the normal A-space $S$ converges weakly to the regular measure $\mu$ on $S$, and if $F_0$ is a closed continuity set for $\mu$ on $S$, then the sequence $\{\mu_n\}$ of induced measures on $F_0$ converges weakly to the induced measure $\mu$ on $F_0$.*

**Corollary** (Helly's theorem). *If the sequence $\{\mu_n\}$ of measures on the normal A-space $S$ converges weakly to a regular measure $\mu$, and if $F_0$ is a closed continuity set for $\mu$ in $S$, then*

$$\lim_{n \to +\infty} \int_{F_0} f(x)\, \mu_n(dx) = \int_{F_0} f(x)\, \mu(dx)$$

*for any function $f$ from $F_0$ into $R$, that is bounded and continuous with respect to the induced $\sigma$-topology, hence for any bounded, continuous function $f$ from $S$ into $R$.*

*Proof:* The closed sets in the induced $\sigma$-topological space $F_0$ are the closed sets $F \subset F_0$, where $F$ is a closed set in $S$. Clearly a continuity set $F$ for $\mu$ on $F_0$ is a continuity set for $\mu$ on $S$. The weak convergence of $\{\mu_n\}$ to $\mu$ on $S$ implies the convergence $\mu_n(F) \to \mu(F)$ $(n \to +\infty)$ for any continuity set $F$. Thus the theorem follows by Theorem 1.1(iv), and the corollary, by the definition of weak convergence.

If $\Upsilon$ is a measurable mapping of an A-space $S$ into an A-space $S'$ and $\mu$ is is a measure on $S$, then $\mu(\Upsilon^{-1}\cdot)$ is a measure on $S'$. We call $\mu(\Upsilon^{-1}\cdot)$ a transformed measure.

**Theorem 2.** *If $\Upsilon$ is a mapping, continuous a.s. $(\mu)$ for a measure $\mu$, of a normal A-space $S$ into a normal A-space $S'$, and if a sequence $\{\mu_n\}$ of measures on $S$ converges weakly to $\mu$, then $\{\mu_n(\Upsilon^{-1}\cdot)\}$ converges weakly to $\mu(\Upsilon^{-1}\cdot)$.*

*Proof:* Applying Theorem 1.1(ii), and the relations

$$\limsup_n \mu_n(\Upsilon^{-1}F') \leq \limsup_n \mu_n(\overline{\Upsilon^{-1}F'}) \leq \mu(\overline{\Upsilon^{-1}F'}) = \mu(\Upsilon^{-1}F'),$$

which hold for any closed set $F'$ when $\Upsilon$ is continuous a.s. $(\mu)$, we get the statement. □

The assertion of Theorem 2 that the weak convergence of $\mu_n$ to $\mu$ implies the weak convergence of $\mu_n(\Upsilon^{-1}\cdot)$ to $\mu(\Upsilon^{-1}\cdot)$ for any mapping $\Upsilon$ of $S$ into a normal A-space $S'$ (which may be $S$) by a mapping continuous a.s. $(\mu)$, is called the invariance principle. This principle can be used in order to determine distributions $\mu(\Upsilon^{-1}\cdot)$.

## 3. UNIFORMLY $\sigma$-SMOOTH SEQUENCES OF MEASURES

By Theorem I.13.2, a measure $\mu$ on an A-space is $\sigma$-smooth if and only if $\mu(F_m) \downarrow 0$ as $m \uparrow +\infty$ for any nonincreasing sequence $\{F_m\}$ tending to $\varnothing$ as $m \to +\infty$. We call a sequence $\{\mu_n\}_{n=1}^{\infty}$ of measures on an A-space uniformly $\sigma$-smooth if $\{\mu_n(F_m)\}$ tends to 0 uniformly with respect to $n$ as $m \to +\infty$ for any nonincreasing sequence $\{F_m\}$ of closed sets tending to $\varnothing$ as $m \to +\infty$.

**Theorem 1.** *If a sequence $\{\mu_n\}$ of $\sigma$-smooth measures on a completely normal A-space converges weakly to a measure $\mu$, then $\{\mu_n\}$ is uniformly $\sigma$-smooth.*

*Proof:* We shall use an indirect proof, assuming that $\{\mu_n\}$ converges weakly but $\{\mu_n(F_m)\}$ does not converge to 0 uniformly with respect to $n$ for some nonincreasing sequence $\{F_m\}$ tending to $\varnothing$ as $m$ tends to $\infty$. Then $\mu_{n_k}(F_{n_k}) \geq \varepsilon$ for $k = 1, 2, \ldots,$ and some $\varepsilon > 0$. Changing the notation, we may assume that

$$\mu_k(F_k) \geq \varepsilon, \qquad k = 1, 2, \ldots, \qquad F_k \downarrow \varnothing. \tag{1}$$

Observing that the A-space is completely normal, we find by Lemma I.6.5 that there exists to a sequence $\{F_k\}$ of closed sets such that $F_k \downarrow \varnothing$, a sequence

$\{G_k\}$ of open sets $G_k \downarrow$, $F_k \subset G_n$, $\bigcap_{k=1}^{\infty} F_k = \bigcap_{k=1}^{\infty} G_k$ and hence $G_k \downarrow \varnothing$. Since $\mu_1$ is a $\sigma$-smooth measure and $G_n \downarrow \varnothing$, it follows from (1) that $\mu_1(F_1 \backslash G_{n_1}) > \varepsilon/2$ for sufficiently large $n_1$. In the same way, we can determine $n_2 > n_1$ such that $\mu_{n_1}(F_{n_1} \backslash G_{n_2}) > \varepsilon/2$. Using induction, we conclude that there exists a sequence $\{n_k\}_{n=1}^{\infty}$, $n_1 < n_2 < \cdots$ such that $\mu_{n_k}[F_{n_k} \backslash G_{n_{k+1}}] > \varepsilon/2$. By Lemma I.6.1, the sequence $\{F_{n_k} \backslash G_{n_{k+1}}\}$ is divergent. Thus we have reduced the proof to the proof of the following statement.

**Lemma 1.** *If a sequence of $\sigma$-smooth measures on a completely normal A-space converges weakly to a measure $\mu$, then $\lim_{n \to +\infty} \mu_n(F_n) = 0$ for any divergent sequence $\{F_n\}$.*

We call attention to the definition of a divergent sequence $\{F_n\}$ as a disjoint sequence of closed sets such that every union $\bigcap_{j=1}^{\infty} F_{n_j}$ is a closed set.

*Proof:* Our proof is indirect. Suppose that the statement in the lemma is not true. Then for some $\varepsilon > 0$ and some divergent sequence $\{F_n\}$, there exists an increasing sequence $\{n_k\}$ of positive integers such that

$$\mu_{n_k}(F_{n_k}) > 2\varepsilon.$$

Since $\{F_n\}$ is divergent, the sequence $\{F_{n_k}\}$ is also divergent. Changing the notation, we may assume that

$$\mu_k(F_k) > 2\varepsilon, \qquad n = 1, 2, \ldots . \tag{2}$$

Clearly $\mu(F_k) \to 0$ $(k \to \infty)$ for the weak limit $\mu$ of $\{\mu_k\}$, since $\{F_k\}$ is divergent, implying that $\mu(S) \geq \sum_{k=1}^{\infty} \mu(F_k)$. Thus $\mu(F_n) < \varepsilon$ for $n > n_0(\varepsilon)$, and we get, by (2),

$$\mu_n(F_n) - \mu(F_n) > \varepsilon \qquad \text{for} \quad n > n_0(\varepsilon). \tag{3}$$

Now $\hat{\mu}_n = \mu_n - \mu$ is a signed measure. Let $|\hat{\mu}_n|$ be its total variation. It is a regular measure, according to Theorem I.13.1, since $\mu$ is regular by definition and $\mu_n$ is regular by Theorem I.13.4. Hence there exists an open set $G_n'$ to $F_n$ and given $\varepsilon_n > 0$ such that $F_n \subset G_n'$ and $|\hat{\mu}_n|(G_n' \backslash F_n) < \varepsilon_n$. Furthermore, by Lemma I.6.3 there exists a disjoint sequence $\{G_n''\}$ to $\{F_n\}$ such that $F_n \subset G_n''$. Putting $G_n = G_n' \cap G_n''$, we then have $|\hat{\mu}_n|(G_n \backslash F_n) < \varepsilon_n$. We choose $\varepsilon_n = \hat{\mu}_n(F_n) - \varepsilon$ and observe that then, according to (3),

$$\begin{gathered} \varepsilon_n = \mu_n(F_n) - \mu(F_n) - \varepsilon > 0, \\ |\hat{\mu}_n|(G_n \backslash F_n) < \hat{\mu}_n(F_n) - \varepsilon. \end{gathered} \tag{4}$$

Since $\{F_n\}$ is divergent and $\{G_n\}$ a disjoint sequence of closed sets and $F_n \subset G_n$ for all $n$, it follows by Lemma I.6.4 that there exist functions $f_n$ connecting $S \backslash G_n$ and $F_n$ such that any sum (finite or infinite) of distinct functions $f_n$ is continuous. Note that the connecting functions satisfy the relations $0 \leq f_n(x) \leq 1$ on $S$,

$$f_n(x) = 0 \quad \text{on} \quad S \setminus G_n, \qquad f_n(x) = 1 \quad \text{on} \quad F_n.$$

Consider now the bounded linear functional $L_n$,

$$L_n(f) = \int_S f(x)\, \hat{\mu}_n(dx)$$

on the Stone vector lattice $\Psi$ of real-valued, bounded, continuous functions from $S$. By the definition of the $f_n$, we have

$$\begin{aligned} L_n(f_n) &= \int_{F_n} f_n(x)\, \hat{\mu}_n(dx) + \int_{G_n \setminus F_n} f_n(x)\, \hat{\mu}_n(dx) \\ &= \hat{\mu}_n(F_n) + \int_{G_n \setminus F_n} f_n(x)\, \hat{\mu}_n(dx). \end{aligned}$$

Regarding (4), we then get

$$L_n(f_n) \geq \varepsilon. \tag{5}$$

We shall use this inequality in order to construct a sequence $\{n_k\}$ of positive integers such that

$$\left| \sum_{i=1}^{k-1} L_{n_k}(f_{n_i}) \right| < \frac{\varepsilon}{4}, \qquad \left| \sum_{i=k+1}^{\infty} L_{n_k}(f_{n_i}) \right| < \frac{\varepsilon}{4}. \tag{6}$$

Let us suppose that we have constructed this sequence $\{n_k\}$ and put

$$g = \sum_{i=1}^{\infty} f_{n_i},$$

where $g$ is continuous. Remembering that the sequence $\{G_n\}$ is disjoint and that $f_n(x) = 0$ if $x \notin G_n$ and $0 \leq f_n(x) \leq 1$ for $x \in G_n$, we conclude that $0 \leq g(x) \leq 1$ for all $x$. Furthermore, by (5) and (6),

$$L_n(g) > \varepsilon/2. \tag{7}$$

But by the weak convergence of $\mu_n$ to $\mu$, we get

$$\lim_{n\to\infty} L_n(f) = \lim_{n\to\infty} \int g(x)\, \mu_n(dx) - \int g(x)\, \mu(dx) = 0,$$

in contradiction to (7). Thus Lemma 1 and hence also Theorem 1 follow. However, it remains to construct the sequence $\{n_k\}$. Choose $n_1 = 1$. Since $L_n(f_{n_1}) \to 0$ as $n \to \infty$ according to the weak convergence of $\{\mu_n\}$ to $\mu$, we may choose $n_0(\varepsilon)$ so large that $|L_n(f_{n_1})| < \varepsilon/4$ for $n > n_0(\varepsilon)$. Furthermore, $|\hat{\mu}_n|$ is a measure and, the sequence $\{G_n\}_{n=1}^{\infty}$ being disjoint, we have

$\sum_{i=1}^{\infty} |u_n|(G_i) < \infty$. Thus we may choose $n_2$ such that

$$\left|\sum_{i \geq n_2}^{\infty} L_{n_1}(f_i)\right| \leq \sum_{i \geq n_2}^{\infty} |\hat{\mu}_n|(G_i) < \frac{\varepsilon}{4}. \tag{8}$$

Suppose that we have chosen $n_1 < n_2 < \cdots < n_r$ such that (6) is satisfied for $k \leq r$. Then we choose $n_{r+1}$ so large that

$$|L_{n+1}(f_{n_i})| \leq \frac{\varepsilon}{4r} \qquad \text{for} \quad i = 1, 2, \ldots, n_r, \quad n = n_{r+1},$$

which is possible since $L_n(f) \to 0\ (n \to +\infty)$ for any bounded, continuous function $f$. Furthermore, by an estimation analogous to (8), we find that we may choose $n_{r+2}$ so large that

$$|L_{n_{r+1}}(f_{n_i})| < \varepsilon/4 \qquad \text{for} \quad i \geq n_{r+2}.$$

Then (8) also holds for $k = r + 1$. Thus the construction is verified by induction. □

## 4. WEAK LIMITS OF SEQUENCES OF $\sigma$-SMOOTH MEASURES ON COMPLETELY NORMAL A-SPACES

The following theorem is an important result in Alexandrov's research on weak convergence.

**Theorem 1** (Alexandrov's third theorem). *If the sequence $\{\mu_n\}$ of $\sigma$-smooth measures on a completely normal A-space converges weakly to $\mu$, then $\mu$ is $\sigma$-smooth.*

The proof will follow from Theorem 3.1 and the following lemma.

**Lemma 1.**† *If $\mu$ is a regular measure on a completely normal A-space and $\{F_m\}$ is a nonincreasing sequence of closed sets such that $F_m \downarrow \varnothing\ (m \uparrow \infty)$ and $\lim_{m \to +\infty} \mu(F_m) > 0$, then there exists a corresponding nonincreasing sequence $\{\hat{F}_m\}$ of closed continuity sets, $\{\hat{F}_m\} \downarrow \varnothing\ (m \uparrow \infty)$ such that $\lim_{m \to \infty} \mu(\hat{F}_m) > 0$.*

Before proving this lemma, we use it as an indirect proof of Theorem 1. Suppose that $\mu_n \underset{w}{\to} \mu\ (n \to \infty)$ and $\mu$ is not $\sigma$-smooth. By Theorem I.13.2 there exists at least one sequence $\{F_m\}$, $F_m \downarrow \varnothing\ (m \uparrow \infty)$ of closed sets such that $\lim_{m \to \infty} \mu(F_m) > 0$, and then, by Lemma 1, a sequence $\{\hat{F}_m\}$ of closed continuity sets, $\hat{F}_m \downarrow \varnothing\ (m \uparrow \infty)$ such that $\mu(\hat{F}_m) \geq \varepsilon_0$ for some $\varepsilon_0 > 0$ and all $m$. But since $\mu_n(\hat{F}_m) \to \mu(\hat{F}_m)\ (n \to \infty)$, for a continuity set, this contradicts Theorem 3.1. □

† The lemma holds also for signed regular measures.

For the proof of Lemma 1 we need

**Lemma 2.** *If $\mu$ is a regular measure on a completely normal A-space S, then to any pair $F_0$ and $F_1$ of disjoint closed subsets of S there exists a continuous function f with the following properties:*

(i) $0 \leq f(x) \leq 1$ *for* $x \in S$, $F_0 = \{x : f(x) = 0\}$, $F_1 = \{x : f(x) = 1\}$.
(ii) *For* $F_t = [x : f(x) \leq t]$ *and* $0 \leq t \leq 2^{-(m+1)}$, *the inequality* $\mu(F_t \backslash F_0) < 1/m$ *holds* $(m = 1, 2, \ldots)$.

*Proof:* Since $\mu$ is regular, for $n = 1, 2, \ldots$, there exist open sets $G_n \supset F_0$ such that $\mu(G_n \backslash F_0) < 1/n$. Put $G'_n = G_n \cap (S \backslash F_1)$, $\hat{G}_n = \bigcap_{k=1}^{n} G'_k$. Then $\hat{G}_n$ is nonincreasing and $F_0 \subset \hat{G}_n$. Since $\bigcap_{n=1}^{\infty} F_{1/n} = F_0$, Lemma I.6.5 asserts the existence of a nonincreasing sequence $\{\tilde{G}_n\}$ of open sets such that $F_{1/n} \subset \tilde{G}_n$, $\bigcap_{n=1}^{\infty} \tilde{G}_n = \bigcap_{n=1}^{\infty} F_{1/n} = F_0$. Put $\hat{G}_n \cap \tilde{G}_n = G^{(n)}$ for $n = 1, 2, \ldots$, $G^0 = S \backslash F_1$. Then $F_0 \subset G^{(n)}$ and $\mu(G^{(n)} \backslash F_0) < 1/n$. Since $S$ is completely normal, there exists a continuous function $g_n$ such that $0 \leq g_n(x) \leq 1$ for $x \in S$, $F_0 = \{x : g_n(x) = 1\}$, $S \backslash G^{(n)} = \{x : g_n(x) = 0\}$. The function

$$g = \sum_{n=0}^{\infty} 2^{-(n+1)} g_n$$

is continuous, and $0 \leq g(x) \leq 1$ for $x \in S$. We claim that $f = 1 - g$ is the function required in the lemma. Indeed,

$$\{x : f(x) = 0\} = \{x : g(x) = 1\} = \bigcap_{n=0}^{\infty} \{x : g_n(x) = 1\} = F_0.$$

If $x \in F_1$, then $g_n(x) = 0$ for all $n$ according to the definition of $G'_n$, $\hat{G}_n$, $\tilde{G}_n$, and $G^{(n)}$, and thus $g(x) = 0$ and $f(x) = 1$ for $x \in F_1$. Furthermore, $f(x) = 1$ implies $g(x) = 0$ and $x \in F_1$, since $G^{(0)} = S \backslash F_1$. Thus $f$ satisfies the condition (i). If $x \notin G^{(m)}$, then $x \in S \backslash G^{(n)}$ for $n \geq m$, and hence $g_n(x) = 0$ for $n \geq m$, and $g(x) < 1 - 2^{-(m+1)}$, i.e., $f(x) > 2^{-(m+1)}$. Thus, if $0 \leq t < 2^{-(m+1)}$, we have

$$F_0 \subset F_t \subset \{x : f(x) \leq 2^{-(m+1)}\} \subset G^{(m)}.$$

It follows from the definition of $G^{(m)}$ that $\mu(G^{(m)} \backslash F_0) < 1/m$, and hence $(F_t \backslash F_0) < 1/m$. □

We shall use Lemma 2 for the construction of the sequence $\{\hat{F}_n\}$ to the given sequence $\{F_n\}$ in accordance with the requirements in Lemma 1.

Assume that

$$\mu(F_n) \geq 2\varepsilon \qquad \text{for} \quad n = 1, 2, \ldots . \tag{1}$$

By Lemma I.6.5, there exists a nonincreasing sequence $\{G_n\}$ such that $F_n \subset G_n$, $\varnothing = \bigcap_{n=1}^{\infty} F_n = \bigcap_{n=1}^{\infty} G_n$.

By Lemma 2, there exists a continuous function $h_n$ to $F_n$ and $G_n$ such that $0 \leq h_n(x) \leq 1$, $F_n = \{x : h_n(x) = 0\}$, $\{S \backslash G_n\} = \{x : h_n(x) = 1\}$, $\mu(F_{n,t} \backslash F_n) < 1/m$ for $F_{n,t} = \{x : h_n(x) \leq t\}$, and $0 \leq t \leq 2^{-(m+1)}$.

Next we determine continuous functions $f_n$, $0 \leq f_n(x) \leq 1$ for $x \in S$ such that the sets $\tilde{F}_{n,t} = \{x : f_n(x) \leq t\}$ have the same properties as the $F_{n,t}$, but are nonincreasing in the sense $\tilde{F}_{n,t} \subset \tilde{F}_{n-1,t}$ for every $t$. Indeed, put

$$f_1 = h_1, \qquad f_n = \max(f_{n-1}, h_n) \qquad \text{for} \quad n = 2, 3, \ldots .$$

Then the functions $f_n$ are continuous and $0 \leq f_n(x) \leq 1$ for $x \in S$. If $f_n(x) = 0$, then $h_k(x) = 0$ for $k \leq n$ and thus $x \in F_n$. If $x \in F_n$, then $x \in F_k$ for $k \leq n$ and hence $h_k(x) = 0$ for $k \leq n$, which implies $f_n(x) = 0$. Thus $F_n = [x : f_n(x) = 0]$. If $f_n(x) = 1$, then $h_m(x) = 1$ for some $m \leq n$ and $x \in S \backslash G_m \subset S \backslash G_n$. If $x \in S \backslash G_n$, then $h_n(x) = 1$, and thus $f_n(x) = 1$. Hence $S \backslash G_n = [x : f_n(x) = 1]$. Noting that $f_n \geq h_n$, $f_n \geq f_{n-1}$, we get $\tilde{F}_{n,t} \subset F_{n,t}$, $\tilde{F}_{n,t} \subset \tilde{F}_{n-1,t}$ and

$$\mu(\tilde{F}_{n,t} \backslash F_n) \leq \mu(F_{n,t} \backslash F_n) < 1/m \tag{2}$$

for $0 \leq t \leq 2^{-(m+1)}$. This inequality will now be used for our construction of the continuity sets $\hat{F}_m$ in Lemma 1.

Let $\varepsilon > 0$ be given. Then choose $t_1$ such that $F_{1,t_1}$ is a continuity set for $\mu$ and $\mu(\tilde{F}_{1,t_1} \backslash F_1) < \varepsilon$.

We may choose $0 < t_i < t_{i-1}$ such that $F_{2,t_2}$ is a continuity set for $\mu$ and $\mu(\tilde{F}_{2,t_2} \backslash F_2) < \varepsilon$. Generally, we may choose $\tilde{F}_{i,t_i}$ as a continuity set for $\mu$ such that $\mu(\tilde{F}_{i,t_i} \backslash F_i) < \varepsilon$. Note that we then have $\tilde{F}_{i,t_i} \subset \tilde{F}_{i-1,t_i} \subset \tilde{F}_{i-1,t_{i-1}}$. Regarding (1), we get[†]

$$\mu(\tilde{F}_{i,t_i}) \geq \varepsilon.$$

Since $\tilde{F}_{i,t_i} \subset F_{i,t_i} \subset G_i$ and $\bigcap_{i=1}^{\infty} G_i = \varnothing$, we have $\bigcap_{i=1}^{+\infty} \tilde{F}_{i,t_i} = \varnothing$. Thus $\tilde{F}_i = \tilde{F}_{i,t_i}$ satisfies the requirements in Lemma 1. □

In Chapter I we introduced the concept "tight measure." We say that a sequence $\{\mu_n\}$ of measures on an A-space $S$ is tight if for any $\varepsilon > 0$ there exists a $\sigma$-compact set $K_\varepsilon$ such that

$$\mu_n(S \backslash K_\varepsilon) < \varepsilon \qquad n = 1, 2, \ldots . \tag{3}$$

**Theorem 2.** *If a sequence $\{\mu_n\}$ of measures on an A-space is tight, then it is uniformly $\sigma$-smooth.*

*Proof:* Let $\{F_m\}$ be a sequence of closed sets and $F_m \downarrow \varnothing$ $(m \uparrow \infty)$. Put $G_m = S \backslash F_m$. Since $F_m \downarrow \varnothing$, $G_m \uparrow S$, i.e., $K_\varepsilon \subset \bigcup_{m=1}^{\infty} G_m$, and then $\sigma$-compactness implies $K_\varepsilon \subset G_{m_\varepsilon}$ for some $\varepsilon$ and $\mu_n(F_{m_\varepsilon}) = \mu_n(S \backslash G_{m_\varepsilon}) < \varepsilon$ for all $n$. □

[†] Also, for a signed regular measure $\mu$ since then (2) holds for $|\mu|$.

In Section II.6 we considered directed dense sequences $\{N^{(m)}\}$ of nets on a metric space $S$. We say that a sequence $\{\mu_n\}$ of measures is tight on the sequences of nets if

$$\lim_{m\to+\infty} \limsup_{n\to+\infty} \mu_n(S\backslash E^{(m)}) = 0, \tag{4}$$

where $\{E^{(m)}\}$ is the nondecreasing sequence of primary meshes in the nets. Furthermore, $\{\mu_n\}$ is called bounded if

$$\sup_n \mu_n(S) < +\infty. \tag{5}$$

**Theorem 3.** *If a sequence $\{\mu_n\}$ of measures on a metric space is tight, then there exists a directed dense sequence of nets on which $\{\mu_n\}$ is tight.*

*Proof:* Let $\{\mu_n\}$ be tight. For a given $m$ we choose $K^{(m)}$ such that $\mu_n(S\backslash K^{(m)}) < 1/m$, and let $E^{(m)} = K^{(m)}$ be a primary sequence of meshes in a net, which we construct as in Section II.6. □

## 5. REDUCTION OF WEAK LIMIT PROBLEMS BY TRANSFORMATIONS

In many situations, the convergence in weak sense of a sequence $\{\mu_n\}$ of measures on an A-space $S$ is partly determined by corresponding convergences in other spaces into which $S$ is mapped. We shall consider such situations, which occur in many applications. Observe that $\mu$ is called a measure on an A-space if $\mu$ is a measure on the algebra $\mathscr{S}$ generated by the closed sets. If $\mu$ is $\sigma$-smooth, we can consider as well the $\sigma$-algebra generated by $\mathscr{S}$, since $\mu$ can be extended to this $\sigma$-algebra. If $S$ and $S'$ are A-spaces with algebras $\mathscr{S}$ and $\mathscr{S}'$, we say that a function $\pi$ from $S$ into $S'$ is measurable if $\pi^{-1}E' \in S$ for any $E' \in S'$.

The method indicated above is given by

**Lemma 1.** *Let the following conditions hold for a completely normal A-space $S$:*

(i) *$S$ is mapped onto a normal A-space $\tilde{S}^{(r)}$ by a measurable mapping $\pi_r$, and $\tilde{S}^{(r)}$ is mapped onto $S^{(r)} \subset S$ by a continuous mapping $V_r$ for $r = 1, 2, \ldots$ .*

(ii) *$\psi_0$ is a Stone vector lattice ring of real-valued, bounded, continuous functions from $S$, such that $\psi_0$ contains the function $x \to 1$, and for any disjoint closed sets $F_1$ and $F_2$, $\psi_0$ contains a function which connects $F_1$ and $F_2$.*

(iii) *$\{\mu_n\}$ is a sequence of measures on S such that $\sup_n \mu_n(S) < \infty$ and $\{\mu_n(\pi_r^{-1}\cdot)\}$ converges weakly to a measure $\tilde{\mu}^{(r)}$* on $\tilde{S}^{(r)}$ for $r = 1, 2, \ldots$ .

(iv)
$$\lim_{r\to\infty} \limsup_{n\to\infty} \mu_n\{x : |f(x) - f(V_r\pi_r x)| \geqq \varepsilon\} = 0$$

*for any $\varepsilon > 0$ and any $f \in \psi_0$. Then we have the relations*

(1°) *$\{\mu_n\}$ converges weakly to a measure $\mu$, which is $\sigma$-smooth if the $\mu_n$ are $\sigma$-smooth.*

(2°) *If $\pi_r$ is continuous a.s. ($\mu$), then $\tilde{\mu}^{(r)} = \mu(\pi_r^{-1}\cdot)$.*

*Conversely, if $\{\mu_n\}$ converges weakly to a measure $\mu$ and the $\mu_n$ are $\sigma$-smooth and $\pi_r$ is continuous a.s. $\mu$ for all r, and*

(v)
$$|f(x) - f(V_r\pi_r x)| \to 0 \qquad \textit{as} \quad r \to \infty,$$

*then* (i) *and* (ii) *imply* (iii) *and* (iv) *and $\mu$ is $\sigma$-smooth.*

*Proof:* For $f \in \psi_0$, put

$$\begin{aligned} L_n(f) &= \int_S f(x)\,\mu_n(dx), \\ L_n^{(r)}(f) &= \int_{\tilde{S}^{(r)}} f(V_r\tilde{x}^{(r)})\,\mu_n(\pi_r^{-1}\,d\tilde{x}^{(r)}). \end{aligned} \tag{1}$$

By transforming the integral, we can write

$$L_n^{(r)}(f) = \int_S f(V_r\pi_r x)\,\mu_n(dx), \tag{2}$$

$$L_n(f) - L_n^{(r)}(f) = \int_S [(f(x) - f(V_r\pi_r x)]\,\mu_n(dx). \tag{3}$$

Then (iv) implies

$$\lim_{r\to\infty} \limsup_{n\to\infty} |L_n^{(r)}(f) - L_n(f)| = 0. \tag{4}$$

Furthermore, (iii) implies

$$\lim_{n\to\infty} L_n^{(r)}(f) = L^{(r)}(f) = \int_{\tilde{S}^{(r)}} f(V_r\tilde{x}^{(r)})\,\tilde{\mu}^{(r)}(d\tilde{x}^{(r)}) \tag{5}$$

since $f(V_r\cdot)$ is a continuous function from $\tilde{S}^{(r)}$ into $R$. By (5), we obtain for positive integers $r_1$ and $r_2$ and $f \in \psi_0$

$$\begin{aligned} |L^{(r_1)}(f) - L^{(r_2)}(f)| &= \lim_{n\to\infty} |L_n^{(r_1)}(f) - L_n^{(r_2)}(f)| \\ &\leqq \limsup_{n\to\infty} |L_n^{(r_1)}(f) - L_n(f)| + \limsup_{n\to\infty} |L_n(f) - L_n^{(r_2)}(f)|. \end{aligned}$$

As $r_1 \to \infty$, $r_2 \to \infty$, the right-hand side of the inequality tends to 0 according to (4). Hence $\{L^{(r)}(f)\}$ is Cauchy convergent and thus convergent:

$$\lim_{r\to\infty} L^{(r)}(f) = L(f). \tag{6}$$

It follows from the properties of limits that $L$ is a nonnegative, bounded, linear functional on $\psi_0$. Furthermore, (4)–(6) imply that

$$\lim_{n\to\infty} L_n(f) = L(f). \tag{7}$$

By Theorem II.4.1, the functional $L$ on $\psi_0$ determines a measure $\mu$ on $S$ such that $\mu(f) = L(f)$ for $f \in \psi_0$, and (6) tells us that $\{\mu_n\}$ converges weakly to $\mu$. This measure is $\sigma$-smooth if the $\mu_n$ are $\sigma$-smooth (Theorem 4.1). If the $\pi_r$ are continuous a.s. $\mu$, then, by Theorem 2.2, the weak convergence of $\mu_n$ to $\mu$ implies that one of $\mu_n(\pi_r^{-1\prime})$ to $\mu(\pi_r^{-1\prime})$. The first part of the converse statement has just been proved. For continuous $\pi_r$ and $V_r$, let (v) be satisfied, and note that $\mu$ is $\sigma$-smooth as the weak limit of $\sigma$-smooth measures.

Since $f$ is continuous and $\pi_r$ is continuous a.s. $(\mu)$, we find by Theorem I.16.2 that $f(V_r\pi_r\cdot)$ is continuous a.s. $(\mu)$, and then, by Theorem I.16.3, that $|f - f(V_r\pi_r\cdot)|$ is continuous a.s. $(\mu)$. Hence, by Theorem 2.2.,

$$\mu_n[x:|f(x) - f(V_r\pi_r x)] \geqq \varepsilon] \to \mu[x:|f(x) - f(V_r\pi_r x)| \geqq \varepsilon]. \tag{8}$$

The last quantity tends to 0 as $r \to \infty$ by the dominated convergence theorem for integrals and by (v).

## 6. THE REDUCTION PROCEDURE FOR METRIC SPACES

In a metric space $S$, the real-valued, bounded, uniformly continuous functions form a Stone vector lattice ring, which we denote by $\psi_1$. Clearly it contains the function $x \to 1$ from $S$. Furthermore, $S$ is completely normal. However, any two disjoint closed sets $F_1$ and $F_2$ need not be connected by a function in $\psi_1$. According to Theorem I.4.3 they are connected if $S$ is pseudocompact, hence, especially if $S$ is compact. For pseudocompact metric spaces, we may thus apply Lemma 2.1 with $\psi_0 = \psi_1$, and then condition (iv) reduces to

(iv′) $$\lim_{r\to\infty} \limsup_{n\to\infty} \mu_n\{x:\rho(x, V_r\pi_r x) \geq \varepsilon\} = 0,$$

provided that $x \to \rho(x, V_r\pi_r x)$ is measurable. It is measurable if $S$ is separable (Lemma I.8.4).

For $f \in \psi_1$ the relations (5.1)–(5.7) remain true under the conditions in Lemma 5.1 even if (iv) is changed to (iv′). However, $L(f)$ is then only defined on $\psi_1$, and generally we cannot apply Alexandrov's second theorem (Theorem II.4.1). But we do not need this theorem if it can be shown that the sequence $\{\tilde{\mu}^{(r)}(V_r^{-1}\cdot)\}$ for the $\tilde{\mu}^{(r)}$ and $V_r$ given by Lemma 5.1 converges weakly to a $\sigma$-smooth measure $\mu$ on $S$. Indeed, it follows by (5.1)–(5.7) that

$$\mu_n(f) \to \mu(f) \qquad (n \to \infty) \qquad \text{for} \quad f \in \psi_1, \tag{1}$$

and by the corollary of Theorem 1.1 this relation implies the weak convergence of $\{\mu_n\}$ to $\mu$.

We shall consider a general situation when the weak convergence of a sequence $\{\mu_n\}$ of measures on a metric space is determined by the convergence of $\{\mu_n(f)\}$ for all $f \in \psi_1$.

**Lemma 1.** *Let the following conditions hold for a metric space $S$ with metric $\rho$:*

(i) *$S$ is mapped into a metric product space $\tilde{S}$ of metric spaces and onto a finite-dimensional product subspace $\tilde{S}^{(r)}$ of $\tilde{S}$ by a measurable mapping $\pi_r$ such that $\pi_r x = \tilde{\pi}_r \tilde{x} = \tilde{x}^{(r)}$ for $x \in S$, $\tilde{x} \in \tilde{S}$, $\tilde{x}^{(r)} \in \tilde{S}^{(r)}$, whenever $\tilde{x}^{(r)} = \pi_r x$ or $\tilde{x}^{(r)} = \tilde{\pi}_r \tilde{x}$ for the projection $\tilde{\pi}_r$ of $\tilde{S}$ onto $\tilde{S}^{(r)}$, where $\tilde{S}^{(r)}$ is a projection of $\tilde{S}^{(s)}$ for $r < s$. Furthermore, $\tilde{S}^{(r)}$ is mapped into $S$ by a continuous mapping $V_r$.*

(ii) *$\{\mu_n\}$ is a sequence of probability measures on $S$ and $\{\mu_n(\pi_r^{-1}\cdot)\}$ converges weakly to a probability measure $\tilde{\mu}^{(r)}$ on $\tilde{S}^{(r)}$ for $r = 1, 2, \ldots$ such that $\tilde{\mu}^{(r)}$ is tight on $\tilde{S}^{(r)}$.*

(iii) *$\rho(x, V_r \pi_r x) \to 0$, $r \to \infty$, and $x \to \rho(x, V_r \pi_r x)$ is measurable.*

(iv) *$\lim\limits_{r\to\infty} \limsup\limits_{n\to\infty} \mu_n\{x : \rho(x, V_r \pi_r x) \geq \varepsilon\} = 0$ for any $\varepsilon > 0$.*

*Then $\{\mu_n\}$ converges weakly to a probability measure $\mu$ on $S$. Conversely, let a sequence $\{\mu_n\}$ of probability measures converge weakly to a measure $\mu$ on $S$. If* (i) *holds with $\pi_r$ continuous* a.s. ($\mu$) *and $\rho(x, V_r \pi_r x) \to 0$* a.s. ($\mu$) *as $r \to \infty$, then* (ii) *holds with $\tilde{\mu}^{(r)} = \mu(\pi_r^{-1}\cdot)$ and* (iv) *is satisfied.*

*Remark:* The function $x \to \rho(x, V_r \pi_r x)$ is continuous if $\pi_r$ is continuous, and always measurable if $S$ is separable ($V_r$ being continuous). Note that any $\sigma$-smooth measure on $R^{(k)}$ is tight (see example in Section I.11).

*Proof:* The probability measures $\tilde{\mu}^{(r)}$ determined by (iii) in Lemma 1 determine a consistent class of probability measures $\tilde{\nu}^{(r)}$ on $\tilde{S}$:

$$\tilde{\nu}^{(r)}(\tilde{E}) = \tilde{\mu}^{(r)}[\tilde{E} \cap \tilde{S}^{(r)}].$$

Indeed, if $r < s$, then $\tilde{\pi}_{sr} \tilde{S}^{(s)} = \tilde{S}^{(r)}$ by a continuous mapping $\tilde{\pi}_{sr}$ and the weak convergence of $\mu_n(\pi_s^{-1}\cdot)$ to $\tilde{\mu}^{(s)}$ implies the weak convergence of $\mu_n(\tilde{\pi}_s^{-1}\tilde{\pi}_{sr}^{-1}\cdot)$ to $\tilde{\mu}^{(s)}(\pi_{sr}^{-1}\cdot)$. But

$$\tilde{\pi}_{sr}\pi_s x = \tilde{\pi}_{sr}\tilde{\pi}_s \tilde{x} = \tilde{\pi}_r \tilde{x} = \pi_r x.$$

Hence

$$\mu_n(\pi_s^{-1}\tilde{\pi}_{sr}^{-1}\cdot) = \mu_n(\pi_r^{-1}\cdot)$$

and

$$\tilde{\mu}^{(r)} = \tilde{\mu}^{(s)}(\tilde{\pi}_{sr}^{-1}\cdot).$$

The consistent class $\tilde{\mu}^{(r)}$ of probability measures determines a probability measure $\tilde{\mu}$ on $\tilde{S}$, according to Kolmogorov's theorem (Theorem I.15.2), such that

$$\tilde{\mu}^{(r)} = \tilde{\mu}(\tilde{\pi}_r^{-1}\cdot)$$

for all $r$.

Now let $\psi$ be the class of all bounded, real-valued, continuous functions from $S$. Then for any $f \in \psi$, the function $f(V_r\tilde{\pi}_r\tilde{x})$ is a continuous function from $\tilde{S}$ into $R$, and

$$f(V_r\tilde{\pi}_r\tilde{x}) = f(V_r\pi_r x).$$

By the assumption made,

$$\rho(V_r\pi_r x, x) \to 0 \qquad (r \to \infty)$$

for all $x$ and, hence,

$$f(V_r\pi_r x) \to f(x) \qquad (r \to \infty)$$

for all $x$ and $f(V_r\tilde{\pi}_r x)$ converges to a measurable function $\tilde{f}(\tilde{x})$. Using the relation $\tilde{\mu}^{(r)} = \tilde{\mu}(\tilde{\pi}_r^{-1}\cdot)$, we obtain from (5.5), by transformation of the integral,

$$L^{(r)}(f) = \int_{\tilde{S}^{(r)}} f(V_r\tilde{x}^{(r)})\,\tilde{\mu}(\tilde{\pi}_r^{-1}\,d\tilde{x}^{(r)}) = \int_{\tilde{S}} f(V_r\tilde{\pi}_r\tilde{x})\,\tilde{\mu}(d\tilde{x}).$$

Letting $r \to \infty$, we get, by the dominated convergence theorem for integrals,

$$\lim_{r\to\infty} L^{(r)}(f) = L(f) = \int_{\tilde{S}} \tilde{f}(\tilde{x})\,\tilde{\mu}(d\tilde{x}).$$

It is easily seen that $L$ so defined is a nonnegative, bounded, linear functional on $\psi$ and thus determines a measure $\mu$ on $S$, and then $\mu$ is the weak limit of $\tilde{\mu}^{(r)}(V_r^{-1}\cdot) = \tilde{\mu}(V_r^{-1}\tilde{\pi}_r^{-1}\cdot)$. As we have remarked at the beginning of this section, the sequence $\{\mu_n\}$ then converges weakly and to a probability measure, according to Alexandrov's third theorem [note that $\mu(S) = 1$ since $\mu_n(S) = 1$].

Suppose that $\{\mu_n\}$ converges weakly to a measure $\mu$, necessarily a probability measure since the $\mu_n$ are probability measures. Furthermore, let (i) hold with $\pi_r$ continuous a.s. $\mu$. Then $\{\mu_n(\pi_r^{-1}\cdot)\}$ converges weakly to $\mu(\pi_r^{-1}\cdot)$ according to Theorem 2.2. The mapping $x \to \rho(x, V_r\pi_r x)$ is continuous a.s. $(\mu)$ (by Theorem I.16.2). Hence, applying Theorem 1 and its corollary, we get

$$\limsup_{n\to\infty} \mu_n\{x: \rho(x, V_r\pi_r x) \geq \varepsilon\} \leq \mu\{x: \rho(x, V_r\pi_r x) \geq \varepsilon)\}.$$

The last quantity tends to 0 as $r \to \infty$, according to the dominated convergence theorem for integrals. Indeed, putting

$$E_r = \{x : \rho(x, V_r \pi_r) \geq 0\},$$

we get

$$\mu(E_r) = \int_S 1_{E_r}(x)\, \mu(dx) \to 0, \qquad r \to \infty$$

since $\mu$ is $\sigma$-smooth, and $\rho(x, V_r \pi_r x) \to 0$ $(r \to \infty)$. □

## 7. WEAK CONVERGENCE OF TIGHT SEQUENCES OF MEASURES ON METRIC SPACES

If a sequence $\{\mu_n\}$ of measures on a metric space $S$ is tight, then by the definition (4.3) there exists a sequence $\{K_m\}$ of compact sets $K_1 \subset K_2 \subset \cdots \subset K_m \subset S$ such that

$$\sup_n \mu_n(S \backslash K_m) < 1/m. \tag{1}$$

The measures $\mu_n$ determine measures on $K_m$ for all $m$.

**Theorem 1.** *If a sequence $\{\mu_n\}$ of measures on a metric space is tight, then the weak convergence of $\{\mu_n\}$ is determined by the weak convergence of $\mu_n$ on compact spaces, which approximate $S$ according to* (1).

*Proof:* Let $\psi$ be the class of real-valued, bounded, continuous functions from $S$:

$$L_n(f) = \int_S f(x)\, \mu_n(dx), \qquad L_n^{(m)}(f) = \int_{K_m} f(x)\, \mu_n(dx),$$

$$L^{(m)}(f) = \int_{K_m} f(x)\, \mu^{(m)}(dx),$$

where $\mu^{(m)}$ is the weak limit of $\mu_n$ on $K_m$. Then

$$|L^{(m_1)}(f) - L^{(m_2)}(f)| \leq |L^{(m_1)}(f) - L_n^{(m_1)}(f)| + |L_n^{(m_1)}(f) - L_n(f)| + |L_n(f) - L_n^{(m_2)}(f)| + |L_n^{(m_2)}(f) - L^{(m_2)}(f)|. \tag{2}$$

As $n \to \infty$, $L_n^{(m)}(f) \to L^{(m)}(f)$ according to the weak convergence of $\{\mu_n\}$ on $K_m$. By (1) we get

$$\sup_n |L_n(f) - L_n^{(m)}(f)| \leq \sup_n \left| \int_{S \backslash K_m} f(x)\, \mu_n(dx) \right| \leq \|f\| \cdot \frac{1}{m}.$$

Hence (2) implies

$$\lim_{m_1 \to \infty,\, m_2 \to \infty} |L^{(m_1)}(f) - L^{(m_2)}(f)| = 0,$$

and then

$$\lim_{m\to\infty} L^{(m)}(f) = L(f).$$

It is easy to show that $L$ is a nonnegative, bounded, linear functional on $\Psi$ and then, by Theorem I.4.1, this functional determines a measure $\mu$ on $S$. By the inequality

$$|L_n(f) - L(f)| \leq |L_n(f) - L_n^{(m)}(f)| + |L_n^{(m)}(f) - L(f)|$$

we then get

$$L_n(f) \to L(f) \qquad (n \to \infty) \qquad \text{for} \quad f \in \Psi.$$

Hence $\{\mu_n\}$ converges weakly to $\mu$. □

If a measure is regular and tight, then it is $\sigma$-smooth according to Theorem I.13.3. Hence, when we deal with tight measures, it is appropriate to assume that the measures are $\sigma$-smooth. If the $\mu_n$ are $\sigma$-smooth, then $\mu$ in the theorem is $\sigma$-smooth according to Theorem 4.1. The weak limit problem on a compact metric space is simpler than on general metric spaces, since we then only have to consider functionals of uniformly continuous functions. However, in order to reduce the limit problem for sequences $\{\mu_n\}$ of $\sigma$-smooth measures, according to Theorem 1, we have to deal with tight sequences of measures on $S$, and hence we can only deal with metric spaces with certain compactness properties.

A sequence $\{\mu_n\}$ of measures on an A-space is called weakly compact if it contains a subsequence that converges weakly.

**Theorem 2.** *If a sequence $\{\mu_n\}$ of measures on a metric space $S$ is tight and* $\sup_n \mu_n(S) < \infty$, *then $\{\mu_n\}$ is weakly compact and moreover it contains a subsequence which converges weakly to a $\sigma$-smooth measure.*

*Proof:* Since $\{\mu_n\}$ is tight, (1) holds for the sequence $\{K_m\}$ of compact sets considered above. As we have shown in Section II.6, there exists a measurable, disjoint, dense, and directed sequence of nets $N^{(m)}$ on $S$ such that the union of the meshes $E_j^{(m)}$ in $N^{(m)}$ is equal to $K_m$. Consider now the countable set of numbers $\{\mu_n(E_j^{(m)})\}$, $j = 1, 2, \ldots, r_m$, $m = 1, 2, \ldots$, where $n = 1, 2, \ldots$. By the selection principle explained in Section I.11 and used on the $l^2$-space, we can select a sequence $\bar{N}$ of positive integers such that $\{\mu_n(E_j^{(m)})\}$ converges to a number $\mu(E_j^{(m)})$ for all $j$ and $m$. Indeed, let the $E_i^{(m)}$ in suitable order be denoted by $\{A_j\}_{j=1}^{\infty}$.

As we have shown in Section II.6, there exists an infinite sequence $N_1$ of positive integers such that $\{\mu_n(A_2)\}_{n\in N_1}$ converges. The sequence $N_1$ contains a subsequence $N_2$ such that $\{\mu_n(A_1)\}_{n\in N_2}$ converges, and so on. Let $\bar{N}$ be the sequence whose first number is the first number in $N_1$, whose second number is the second number in $N_2$, and whose $i$th number is the $i$th number in $N_i$. Then clearly $\{\mu_n(A_i)\}_{n \in \bar{N}}$ converges to a finite value $\mu(A_i)$ for all $i$. It follows from the properties of limits that $\mu$ so defined on the meshes in the nets is nondecreasing and additive on the meshes. By the corollary of Theorem II.6.1, the set function $\mu$ on the meshes determines uniquely a $\sigma$-smooth measure $\nu$ on $S$. It follows, as in the proof of Theorem I, that $L_n(f)$ converges to a bounded linear functional $L(f)$ for $f \in \Psi$, where $L(f)$ is determined by the $\mu(A_i)$, and $L(f)$ determines the measure $\nu$, $L(f) = \nu(f)$. Hence $\{\mu_n\}$ converges weakly to $\nu$. $\square$

## 8. SEMINORMS ON AN ALGEBRA

We have seen in Section II.8 that there exists a commutative algebra over $R$ of signed measures under convolution as multiplicative composition and usual addition as additive composition. This algebra contains the semigroup of measures. An important part of probability theory is the theory of weak convergence of convolution products. It turns out that weak convergence of sequences of measures in general cases is equivalent to convergence in a certain class of seminorms. When we deal with convergence in seminorms, we can at once consider general commutative algebras. Hence, in the following, $\bar{\mathscr{M}}$ denotes a commutative algebra with unit element $e$ and zero element 0. We denote the elements in $\bar{\mathscr{M}}$ by $x, y, x_i, y_i$, etc., and the real numbers by $a, b, a_i, b_i$. By definition, $\bar{\mathscr{M}}$ has $R$ as operator field if $ax \in \bar{\mathscr{M}}$ for $x \in \bar{\mathscr{M}}$ and any real number $a$, and $a(x + y) = ax + ay$, $(ax)(by) = (ab)(xy)$.

A seminorm on $\bar{\mathscr{M}}$ is a nonnegative function $p$ from $\bar{\mathscr{M}}$ with the following properties:

$$\begin{aligned} p(x) \geq 0, \qquad p(x) = 0 \qquad &\text{for} \quad x = 0 \quad (0 = \text{zero element}), \\ p(x + y) \leq p(x) + p(y), & \\ p(ax) \leq |a| p(x). & \end{aligned} \tag{1}$$

Note that the second inequality implies

$$|p(x) - p(y)| \leq p(x - y), \tag{2}$$

since $x = y + (x - y)$, $y = x + (y - x)$. In the following, we also require that $p(e) = 1$ for $p \in \mathscr{P}$.

We consider classes $\mathscr{P}$ of seminorms on $\bar{\mathscr{M}}$. If, for a class $\mathscr{P}$, the relations $p(x) = 0$ for all $p \in \mathscr{P}$ implies $x = 0$, we call $\bar{\mathscr{M}}$ a seminormed algebra under $(\mathscr{P})$. By definition, $\{x_n\}$, $x_n \in \bar{\mathscr{M}}$ is bounded $(\mathscr{P})$ if $\sup_n p(x_n) < \infty$ for $p \in \mathscr{P}$; $\{x_n\}$ is convergent $(\mathscr{P})$ in $\bar{\mathscr{M}}$ to $x$ if $\lim_{n \to \infty} p(x_n - x) = 0$ for $p \in \mathscr{P}$, $x \in \bar{\mathscr{M}}$; $\{x_n\}$ is Cauchy convergent $(\mathscr{P})$ if $\lim_{n \to +\infty, m \to +\infty} p(x_n - x_m) = 0$ for $p \in \mathscr{P}$.

A subclass $\mathcal{M}'$ of $\bar{\mathcal{M}}$ is called complete ($\mathcal{P}$) if any Cauchy convergent ($\mathcal{P}$) sequence $\{x_n\}$ with $x_n \in \mathcal{M}'$ converges ($\mathcal{P}$) to a limit $x \in \mathcal{M}'$; $\mathcal{M}$ is *relatively compact* ($\mathcal{P}$) if any sequence $\{x_n\}$, $x_n \in \mathcal{M}'$, that is bounded ($\mathcal{P}$) contains a subsequence that converges ($\mathcal{P}$) to an element in $\mathcal{M}'$. A collection $\mathcal{C}$ of elements in $\bar{\mathcal{M}}$ is called *convex* if $x, y \in \mathcal{C}$ implies $ax + (1 - a)y \in \mathcal{C}$ for $0 \leq a \leq 1$. We call a semigroup $\mathcal{M}'$ of $\mathcal{M}$ a characteristic semigroup in $\bar{\mathcal{M}}$ if

$$p(xy) \leq p(x)p(y) \qquad \text{for} \quad p \in \mathcal{P}, \quad x \in \bar{\mathcal{M}}, \quad y \in \mathcal{M}'. \tag{3}$$

A characteristic semigroup $\mathcal{M}'$ is called distinguished if for any $x, y \in \mathcal{M}'$ we have

$$p[(x - e)(y - e)] \leq c(p)p(x - e)p(y - e), \qquad p \in \mathcal{P}$$

with a constant $c(p)$ only depending on $p$.

## 9. SOME FUNDAMENTAL IDENTITIES AND INEQUALITIES FOR PRODUCTS

For elements $x, y, \ldots$ in a commutative algebra with unit element $e$, we have the identity (easy to verify)

$$\prod_{i=1}^{n} x_i - \prod_{i=1}^{n} y_i = \sum_{j=1}^{n} \left( \prod_{i=j+1}^{n} x_i \right) \left( \prod_{i=1}^{j-1} y_i \right) (x_j - y_j), \tag{1}$$

where we put

$$\prod_{i=n+1}^{n} x_i = e, \qquad \prod_{i=1}^{0} x_i = e.$$

For $x_i = x$, $y_i = y$, $i = 1, \ldots, n$ it reduces to

$$x^n - y^n = (x - y) \sum_{i=1}^{n} x^{n-j} y^{j-1}. \tag{2}$$

A more complicated identity is the following one, which will be proved in the Appendix:

$$\begin{aligned} \prod_{i=1}^{n} x_i - \prod_{i=1}^{n} y_i = \bar{f}^{(n)} \sum_{i=1}^{n} (x_i - y_i) + \frac{1}{2n} \sum_{i=1}^{n} \sum_{j=1}^{n} \bar{u}_{ij}^{(n)}(x_i - y_i)(x_j - x_i) \\ + \bar{v}_{ij}^{(n)}(x_i - y_i)(y_j - y_i), \end{aligned} \tag{3}$$

where $\bar{f}^{(n)}$ is the mean value of products containing $n - 1$ factors of formally different $x_i$:*s* and $y_i$:*s*, and $\bar{\mu}_{ij}^{(n)}$ and $\bar{v}_{ij}^{(n)}$ are mean values of the same kind, but with $n - 2$ factors in the products (see the Appendix).

From these identities, we get inequalities in seminorms on the algebra. We deal with a class of seminorms and let $\bar{\mathcal{M}}$ be a seminormed algebra for $\mathcal{P}$. The products under consideration are of the form

$$\left\{\prod_{j=1}^{k_n} x_{nj}\right\}_{n=1}^{\infty}, \tag{4}$$

where $\{k_n\}$ is a nondecreasing sequence of positive integers tending to $\infty$ as $n \to \infty$. We say that the sequence (4) is weakly stable if

$$\sup_n \sum_{i=1}^{k_n} |p(x_{in}) - 1| < \infty \qquad \text{for} \quad p \in \mathcal{P}, \tag{5}$$

and it is stable if

$$\sup_n \sum_{i=1}^{n} p(x_{in} - e) < \infty \qquad \text{for} \quad p \in \mathcal{P}. \tag{6}$$

Since

$$|p(x) - 1| \leq p(x - e)$$

for any $x \in \bar{\mathcal{M}}$, the sequence (4) is weakly stable if it is stable. Note that for $x_{nj} = x_n$ for all $n$, (5) and (6) reduce.

**Lemma 1.** *Let $x_{in} \in \mathcal{M}$, $y_{in} \in \mathcal{M}$ where $\mathcal{M}$ is a characteristic semigroup in the seminormed algebra $\bar{\mathcal{M}}$ over R, and $e \in \mathcal{M}$. Furthermore, let $\{\prod_{i=1}^{k_n} x_{ni}\}$ and $\{\prod_{i=1}^{k_n} y_{ni}\}$ be weakly stable. Then*

$$p\left(\prod_{i=1}^{k_n} x_{in} - \prod_{i=1}^{k_n} y_{in}\right) \leq c_p \sum_{i=1}^{k_n} p(x_{in} - y_{in}), \qquad p \in \mathcal{P},$$

*where $c_p$ depends only on p.*

*Proof:* We have

$$p(x_{in}) \leq \exp|p(x_{in}) - 1|. \tag{7}$$

Thus, the products of any different factors $x_{in}$ are uniformly bounded. Since $\mathcal{M}$ is a characteristic semigroup in $\bar{\mathcal{M}}$, we obtain

$$p\left[\left(\prod_{i=j+1}^{k_n} x_{in}\right)\left(\prod_{i=1}^{j-1} y_{in}\right)(x_{jn} - y_{jn})\right] \leq \left(\prod_{i=j+1}^{k_n} p(x_{in})\right)\left(\prod_{i=1}^{j-1} p(y_{in})\right)p(x_{jn} - y_{jn})$$
$$\leq c_p p(x_{jn} - y_{jn}).$$

Hence, by identity (1), we get the inequality of Lemma 1. □

**Lemma 2.** *Let $x_{in} \in \mathcal{M}$, $y_{in} \in \mathcal{M}$, where $\mathcal{M}$ is a convex distinguished semigroup in the seminormed algebra $\bar{\mathcal{M}}$ over R and $e \in \mathcal{M}$. Furthermore, let*

*$\{\prod_{i=1}^{k_n} x_{in}\}$ and $\{\prod_{i=1}^{k'_n} y_{in}\}$ be weakly stable. Then*

$$\text{(i)}\quad p\left(\prod_{i=1}^{k_n} x_{in} - \prod_{i=1}^{k'_m} y_{im}\right) \le c'_p p\left[\sum_{i=1}^{k_n} (x_{in} - e) - \sum_{i=1}^{k'_m} (y_{im} - e)\right] + c''_p\left\{\sum_{i=1}^{k_n} [p(x_{jn} - e)]^2 + \sum_{i=1}^{k'_m} [p(y_{in} - e)]^2\right\}$$

*with constants $c'_p$ and $c''_p$ only dependent on $p$.*

*Proof:* Consider the case $k'_m \le k_n$ for some $m$ and $n$, and apply the identity (3) for products of $k_n$ factors $x_i$ changed to $x_{in}$ and $y_i$ changed to $y_{im}$, where we put $y_{im} = e$ for $i > k'_m$.

Note that $|p(y_{in}) - 1|$ is changed to 0 if $y_{in}$ is changed to $e$. Since $\mathscr{M}$ is convex and $\bar{u}_{ij}^{(n)}$ is a mean value of products of factors in $\mathscr{M}$, this mean value belongs to $\mathscr{M}$, and we get

$$p[\bar{u}_{ij}^{(n)}(x_{in} - y_{im})(x_{jn} - x_{in})] \le p(\bar{u}_{ij}^{(n)})p[(x_{in} - y_{im})(x_{jn} - x_{in})].$$

Since $\bar{u}_{ij}^{(n)}$ is the mean value of products of factors in $\mathscr{M}$ and the $p$-seminorm of any such factor is smaller than $\bar{c}_p$, we get by (5)

$$p(\bar{u}_{ij}^{(n)}) \le \bar{c}_p.$$

Furthermore,

$$p[(x_i - y_i)(x_j - x_i)] \le p[(x_i - e)(x_j - e)] + p[(x_i - e)^2] + p[(y_i - e)(x_j - e)] + p[(y_i - e)(x_i - e)].$$

Since $\mathscr{M}$ is distinguished,

$$p[(y_i - e)(x_j - e)] \le c(p)p(y_i - e)p(x_j - e) \le \tfrac{1}{2}c(p)\{[p(y_i - e)]^2 + [p(x_j - e)]^2\}.$$

By the same arguments, we get

$$p\left[\bar{f}\sum_{i=1}^{k_n} (x_{in} - y_{im})\right] \le c'_p p\left(\sum_{i=1}^{k_n} (x_{in} - y_{im})\right) \le c'_p p\left[\sum_{i=1}^{k_n} (x_{in} - e) - \sum_{i=1}^{k_n} (y_{im} - e)\right].$$

These inequalities imply the inequality of the lemma for $k_n \ge k'_m$. Corresponding statements may be made for $k_n \le k'_m$.

**Lemma 3.** *Let $\mathscr{M}$ contain the unit element $e$ and be a distinguished semigroup in $\mathscr{M}$ under $\mathscr{P}$. Furthermore, let $\{x_n^{k_n}\}_{n=1}^{\infty}$ be stable as $n \to \infty$. Then for any positive integer $r$ and*

$$l_n(r) = \left[\frac{k_n}{r}\right] = \text{largest integer} \le \frac{k_n}{r}$$

*we have*

(i) $$\lim_{r\to\infty}\limsup_{n\to\infty} p[r(x_n^{1_n(r)} - e) - k_n(x_n - e)] = 0$$

(*note that* $k_n \uparrow \infty$ *as* $n \uparrow \infty$).

*Proof:* For any positive integer $n$, we get, by the identity (1) applied twice,

$$x^n - e - n(x - e) = (x - e)^2 \sum_{i=1}^{n-1} \sum_{i=0}^{j-1} x^i.$$

Hence

$$\begin{aligned} p[r(x_n^{l_n(r)} - e) - r1_n(r)(x_n - e)] &\le rp(x_n - e)^2 \sum_{j=1}^{l_n(r)-1} \sum_{i=0}^{j-1} [p(x_n)]^i \\ &\le \tfrac{1}{2}\bar{c}_p r[l_n(r)]^2 p(x_n - e)^2 \\ &\le \tfrac{1}{2}\bar{\bar{c}}_p \frac{1}{r} [k_n p(x_n - e)]^2. \end{aligned}$$

The right-hand side tends to 0 as $r \to \infty$ uniformly with respect to $n$, since $\{x_n^{k_n}\}$ is stable, and hence

$$\sup_n k_n p(x_n - e) < \infty.$$

Furthermore,

$$p[k_n(x_n - e) - rl_n(r)(x_n - e)] \le (r/k_n)k_n p(x_n - e) \to 0 \qquad (n \to \infty)$$

for any $r$. □

## 10. CONVERGENCE IN SEMINORMS OF POWERS TO INFINITELY DIVISIBLE ELEMENTS

We use the same concepts and notations as in the two preceding sections.

**Theorem 1.** *Let* $\mathcal{M}$ *be a convex, complete, distinguished semigroup in a seminormed commutative algebra* $\bar{\mathcal{M}}$ *over* R, *the class of seminorms being* $\mathcal{P}$. *Let* $e \in \mathcal{M}$. *Furthermore, let* $\{x_n^{k_n}\}_{n=1}^{\infty}$ *be a sequence of powers of* $x_n \in \mathcal{M}$, $k_n \uparrow \infty$ *as* $n \uparrow \infty$. *If the Cauchy convergence*

(i) $$p[k_n(x_n - e) - k_m(x_m - e)] \to 0 \qquad \begin{pmatrix} n \to \infty \\ m \to \infty \end{pmatrix}$$

*is satisfied for any* $p \in \mathscr{P}$, *then for* $l_n(r) = [k_n/r]$, *r positive integer,*

$$\text{(ii)} \qquad p(x_n^{l_n(r)} - y_r) \to 0 \qquad (n \to \infty) \qquad \textit{for} \quad p \in \mathscr{P}$$

*with some* $y_r \in \mathscr{M}$ *such that* $y_r^r = y_1$, $r = 1, 2, \ldots$. *Hence* $y_1$ *is infinitely divisible. Conversely, if the sequence* $\{x_n^{k_n}\}$ *is stable and* (ii) *holds, then* (i) *is satisfied.*

*Remark:* Note that $\{x_n^{k_n}\}$ is stable when (i) holds.

**Corollary.** *If* $y = y_n^n$ *for all* $n = 1, 2, \ldots$, *and the sequence* $\{y_n^n\}$ *is stable, then* $\{n(y_n - e)\}$ *is Cauchy convergent* ($\mathscr{P}$) *provided that* $y_n$ *is uniquely determined by* $y$.

*Proof:* Let (i) hold. Then

$$\limsup_{n\to\infty} k_n p(x_n - e) < \infty \qquad \text{for} \quad p \in \mathscr{P}. \tag{1}$$

Thus $\{x_n^{k_n}\}$ is stable, and $\{x_n^{l_n(r)}\}$ is stable for any $r = 1, 2, \ldots$. Applying Lemma 9.2, we get

$$\begin{aligned} p[x_n^{l_n(r)} - x_m^{l_m(r)}] \le c_p' p[l_n(r)(x_n - e) - l_m(x_m - e)] \\ + c_p''\{l_n(r)p(x_m - e)^2 + l_m(r)p(x_m - e)^2\}. \end{aligned} \tag{2}$$

Here the two last terms tend to 0 as $n \to \infty$, $m \to \infty$ since (1) holds, and by (1) and (i) we also find that the first term on the right-hand side tends to 0. It then follows by (2) that $\{x_n^{l_n(r)}\}$ is Cauchy convergent in the class $\mathscr{P}$ of seminorms and thus convergent ($\mathscr{P}$) since $\mathscr{M}$ is complete. Denote the limit of $\{x_n^{l_n(r)}\}$ by $y_r$. To prove that $y_r^r = y$, consider

$$p(x_n^{rl}n^{(r)} - y) \le p(x_n^{l_n(r)} - y_r) \sum_{j=0}^{r} [p(x_n)]^{(r-j-1)l_n(r)}[p(y_r)]^j. \tag{3}$$

Since $[p(x_n)]^{l_n(r)}$ is uniformly bounded according to (1), and $p(y_r)$ is finite, and

$$p(x_n^{rl}n^{(r)} - y) \to 0 \qquad (n \to \infty),$$

according to (1) and (2), we get by (3)

$$p(y_1 - y_r^r) = 0.$$

Thus $y_1 = y_r^r$.

Suppose that (ii) holds and that $\{x_n^{k_n}\}$ is stable, i.e.,

$$\sup_n k_n p(\mu_n - e) < \infty.$$

Then by Lemma 9.3

$$\lim_{r\to\infty} \limsup_{n\to\infty} p[r(x_n^{l_n(r)} - e) - k_n(x_n - e)] = 0,$$

and since

$$p(x_n^{l_n(r)} - y_r) \to 0 \qquad (n \to \infty),$$

$$\lim_{r\to\infty} \limsup_{r\to\infty} p[r(y_r - e) - k_n(x_n - e)] = 0.$$

Hence

$$p[k_n(x_n - e) - k_m(x_m - e)] \to 0 \qquad \begin{pmatrix} n \to \infty \\ m \to \infty \end{pmatrix}. \quad \square$$

*Proof of the corollary:* If $y$ is uniquely infinitely divisible, $y = y_n^n$ for $n = 1, 2, \ldots$ and $\{y_n^n\}$ is stable, we have for positive integers $n$ and $r$

$$y_{nr}^{nr} = (y_{nr}^n)^r = y_r.$$

Applying Theorem 1 to $y_{nr}^{nr}$, $l_n(r) = nr/r = n$, proving the corollary. $\square$

## 11. CONVERGENCE IN SEMINORMS OF PRODUCTS

We say that the sequences $\{x_{nj}\}_{j=1}^{k_n}$, $n = 1, 2, \ldots, k_n \uparrow \infty$ as $n \uparrow \infty$, in a seminormed algebra over $R$ satisfy the u.a.n. condition (u.a.n. is the abbreviation for uniformly asymptotical negligibility), if

$$\lim_{n\to\infty} \sup_{1\le j\le k_n} p(x_{nj} - e) = 0 \qquad \text{for} \quad p \in \mathscr{P}. \tag{1}$$

Considering a sequence of products

$$\left\{\prod_{j=1}^{k_n} x_{nj}\right\}_{j=1}^{\infty}, \tag{2}$$

we say that it satisfies the u.a.n. condition if (1) holds. The convergence problem for products with respect to seminorms is reduced to a convergence problem for powers by the following theorem.

**Theorem 1.** *Let $\mathscr{M}$ be a convex distinguished semigroup in the seminormed commutative algebra $\bar{\mathscr{M}}$ with unit element $e \in \mathscr{M}$. If the sequence (2) with $x_{nj} \in \mathscr{M}$, $k_n \uparrow \infty$ as $n \uparrow \infty$, is weakly stable and satisfies the condition*

(i) $$\lim_{n\to\infty} \sum_{i=1}^{k_n} [p(x_{nj} - e)]^2 = 0 \qquad \textit{for} \quad p \in \mathscr{P},$$

*then for*

$$y_n = \frac{1}{k_n} \sum_{j=1}^{k_n} x_{nj},$$

(ii) *the sequence* $\{y_n^{k_n}\}$ *is weakly stable and*

$$\lim_{n\to\infty} p\left(y_n^{k_n} - \prod_{j=1}^{k_n} x_{nj}\right) = 0 \qquad \text{for} \quad p \in \mathscr{P}.$$

*Remark:* If the product (2) is stable and satisfies the u.a.n. condition, then (i) holds.

*Proof:* Let the convolution product (2) be weakly stable and let (i) hold. Since

$$p(y_n) \le \frac{1}{k_n} \sum_{j=1}^{k_n} p(x_{nj}),$$

we get

$$k_n[\max(p(y_n), 1) - 1] \le \sum_{j=1}^{k_n} \{\max[p(x_{nj}), 1] - 1\}.$$

Hence $\{y_n^{k_n}\}$ is weakly stable. Applying Lemma 9.2, we get

$$p\left(\prod_{j=1}^{k_n} x_{nj} - y_n^{k_n}\right) \le c'_p p\left[\sum_{i=1}^{k_n} (x_{nj} - e) - k_n(y_n - e)\right]$$
$$+ c''_p \sum_{i=1}^{k_n} \left[p(x_{nj} - e)\right]^2 + c''_p k_n [p(y_n - e)]^2. \qquad (3)$$

Here the first term on the right-hand side is equal to 0 by the definition of $y_n$. The second term tends to 0 as $n \to \infty$ by the condition (i). Furthermore, by Cauchy's inequality,

$$k_n[p(y_n - e)]^2 = \frac{1}{k_n}\left\{p\left[\sum_{j=1}^{k_n} (x_{nj} - e)\right]\right\}^2$$
$$\le \frac{1}{k_n}\left\{\sum_{j=1}^{k_n} p(x_{nj} - e)\right\}^2 \le \sum_{j=1}^{k_n} [p(x_{nj} - e)]^2.$$

The last sum tends to 0 as $n \to \infty$ according to (i). Thus also the third term on the right-hand side of (3) tends to 0 as $n \to \infty$. □

*Proof of the remark:* By definition the product (2) is stable if

$$\limsup_{n\to\infty} \sum_{j=1}^{k_n} p(x_{nj} - e) < \infty.$$

Then clearly (1) implies (i). □

CHAPTER IV

# WEAK CONVERGENCE ON $R^{(k)}$

## 1. $\sigma$-SMOOTH MEASURES ON $R^{(k)}$

The space $R^{(k)}$ was studied in Section I.11. Let $\mathscr{M}$ be the class of $\sigma$-smooth measures different from the zero measure on $R^{(k)}$, and $\bar{\mathscr{M}}$ the class of all signed $\sigma$-smooth measures on $R^{(k)}$. By the definition of convolution in Section II.8, the class $\bar{\mathscr{M}}$ is a commutative algebra over the real number field and $\mathscr{M}$ is a commutative semigroup. In $\bar{\mathscr{M}}$ and $\mathscr{M}$, the measure having all its mass in the zero point, and this mass being equal to 1, is unit element. We denote the unit measure by $e$.

The distribution function $\hat{\mu}$ of $\mu \in \bar{\mathscr{M}}$ is defined by

$$\hat{\mu}(x) = \int_{y \leq x} \mu(dy). \tag{1}$$

It is easy to show that $\hat{\mu}$ is continuous "from above" in the sense that $\hat{\mu}(x) \to \hat{\mu}(z)$ as $x \downarrow z$, i.e., $x^{(i)} \downarrow z^{(i)}$ for all $i = 1, 2, \ldots, k$. Hence $\hat{\mu}$ is uniquely determined by its values at a countable dense set of points (for instance, all vectors with rational coordinates).

The mapping $\pi^{(i)}$ defined by $\pi^{(i)}x = x^{(i)}$ is continuous, and more generally, the mapping $\pi(t)$ defined by

$$\pi(t)x = \sum_{i=e}^{k} t^{(i)}x^{(i)} = t \cdot x, \qquad \|t\| = 1 \tag{2}$$

is continuous. This follows by Lemma I.8.2, since

$$|t \cdot x - t \cdot y| = \|t \cdot (x - y)\| \leq \|t\| \cdot \|x - y\|.$$

The mapping $\pi^{(i)}$ determines the signed measure $\mu(\pi^{(i)-1} \cdot)$, called the $i$th margin of $\mu$ on $R$. By transforming the integral, we can write the distribution function of $\mu(\pi^{(i)-1}\cdot)$, called the $i$th margin distribution function of $\mu$,

$$\int_{y^{(i)}\le x^{(i)}} \mu(\pi^{(i)-1}\,dy^{(i)}) = \int_{y^{(i)}\le x^{(i)}} \mu(dy). \tag{3}$$

Since the distribution function of a measure on $R$ is nondecreasing, it has at most countably many discountinuity points.

**Theorem 1.** *A $\sigma$-smooth measure $\mu$ on $R^{(k)}$ is uniquely determined by its values on continuity intervals $(a,b]$, $a<b$ for $\mu$. Also $\mu$ is uniquely determined by the values of its distribution function at points belonging to a countable dense set on $R^{(k)}$. Even $\mu$ is uniquely determined by the values of its distribution function at points belonging to any countable dense set on $R^{(k)}$.*

*Proof:* Let $\mathscr{E}$ be the class of intervals $(a,b]$, $a<b$ such that the $a$ and $b$ belong to a dense set on $R^{(k)}$ and $(a_1,b_1]\cap(a_2,b_2]\in\mathscr{E}$ if $(a_1,b_1]\in\mathscr{E}$ and $(a_2,b_2]\in\mathscr{E}$. We get such a class if any coordinate $a^{(i)}$ and any coordinate $b^{(i)}$ belong to a dense set on $R$ for any $i$. The Borel $\sigma$-algebra on $R^{(k)}$ is generated by $\mathscr{E}$. A measure $\mu$ is uniquely determined by its values on $\mathscr{E}$ according to Theorem I.12.3. It is easy to show that $\mu$ on $\mathscr{E}$ is determined by the values of the distribution function at a dense set of points. Indeed, if $a<b$, then

$$E_1 = \{x{:}x\le b,\ x^{(1)}>a^{(1)}\} = \{x{:}x\le b\}\backslash\{x{:}x\le b_1\},$$

where $b_1^{(i)}=b^{(i)}$ for $i>1$, $b_1^{(i)}=a^{(1)}$ for $i=1$. Hence,

$$\mu(E_1)=\hat{\mu}(b)-\hat{\mu}(b_1).$$

In the same way,

$$\begin{aligned}&\{x{:}x\le b,\ x^{(1)}>a^{(1)},\ x^{(2)}>a^{(2)}\}\\ &\quad=\{x{:}x\le b;\ x^{(1)}>a^{(1)}\}-\{x\le b,\ x^{(1)}>a^{(1)},\ x^{(2)}\le a^{(2)}\}.\end{aligned}$$

Using induction, we conclude that $\mu\{x{:}a<x\le b\}$ is determined by the measures $\mu\{x{:}x\le c\}$, where $c$ has coordinates $a^{(i)}$ or $b^{(i)}$. □

## 2. GAUSSIAN MEASURES AND GAUSSIAN TRANSFORMS

The normalized Gaussian measure $\Phi$ on $R$ is given by its density function

$$\varphi(x)=(2\pi)^{-1/2}\exp(-\tfrac{1}{2}x^2),\qquad \Phi(E)=\int_E\varphi(x)\,dx \tag{1}$$

for any Borel set $E$. For the convolution of the measures $\Phi(\cdot/\alpha_1)$ and $\Phi(\cdot/\alpha_2)$, $\alpha_1>0$, $\alpha_2>0$ ($E/\alpha$ being defined as $\{x{:}\alpha x\in E\}$), we have the nice relation

$$\Phi\left(\frac{\cdot}{\alpha_1}\right)*\Phi\left(\frac{\cdot}{\alpha_2}\right)=\Phi\left(\frac{\cdot}{\alpha}\right),\qquad \alpha^2=\alpha_1^2+\alpha_2^2,\qquad \alpha>0. \tag{2}$$

To prove this, we consider the derivative of the distribution function of the convolution, which is equal to

$$\frac{1}{2\pi\alpha_1\alpha_2}\int \exp\left\{-\frac{1}{2}\left[\left(\frac{x-y}{\alpha_1}\right)^2+\left(\frac{y}{x_2}\right)^2\right]\right\}dy.$$

By the transformation

$$\frac{\alpha}{\alpha_1\alpha_2}\left(y-\frac{x\alpha_2^2}{\alpha^2}\right)=u,$$

it is transformed into

$$\left[\frac{1}{\sqrt{2\pi}}\exp\left(-\frac{1}{2}\frac{x^2}{\alpha^2}\right)\right]\frac{1}{\sqrt{2\pi}}\int \exp\left(-\frac{1}{2}u^2\right)du=\frac{1}{\sqrt{2\pi}}\exp\left(-\frac{1}{2}\frac{x^2}{\alpha^2}\right).$$

The right-hand side is the density function of $\Phi(\cdot/\alpha)$. From (1), we easily get the relation

$$\Phi\left(\frac{\cdot+m_1}{\alpha_1}\right)*\Phi\left(\frac{\cdot+m_2}{\alpha_2}\right)=\Phi\left(\frac{\cdot+m}{\alpha}\right),\quad \alpha^2=\alpha_1^2+\alpha_2^2,\quad m=m_1+m_2. \tag{3}$$

The Gaussian measure on $R$ determines the Gaussian measure on $R^{(k)}$ with distribution function

$$\Phi(x)=\prod_{i=1}^{k}\Phi(x^{(i)}) \tag{4}$$

and density function

$$\varphi(x)=(2\pi)^{-k/2}\exp(-\tfrac{1}{2}\|x\|^2),\qquad \|x\|^2=\sum_{i=1}^{k}[x^{(i)}]^2. \tag{5}$$

This is a special Gaussian measure on $R^{(k)}$. Generally, we call any measure on $R^{(k)}$ obtained by a linear transformation of $R^{(k)}$ a Gaussian measure. A linear transformation $T$ of $R^{(k)}$ is determined by a matrix

$$T=\begin{pmatrix} a_{11} & \cdots & a_{1k}\\ \vdots & & \vdots\\ a_{k1} & \cdots & a_{kk}\end{pmatrix}.$$

It is a continuous mapping of $R^{(k)}$ into $R^{(k)}$. For any Borel set $E$ in $R^{(k)}$, we define a measure $\mu$ on $R^{(k)}$ by

$$\mu(E)=\Phi(T^{-1}E)=\Phi(\{x:Tx\in E\}). \tag{6}$$

The set $T^{-1}E=\{x:Tx\in E\}$ is measurable, since $x\to Tx$ is a continuous function (by Lemma I.8.2). Note that $T^{-1}E$ is defined, but this does not mean that $T$ has an inverse. If it has, $R^{(k)}$ is mapped onto $R^{(k)}$ by $T$. Otherwise, it is mapped onto a subspace $TR^{(k)}$ of $R^{(k)}$, i.e., into $R^{(k)}$. The most general Gaussian measure on $R^{(k)}$ is obtained from $\Phi$ by the mapping $\upsilon : x \to \bar{\upsilon}x = Tx - m$, where $m$ is a constant vector in $R^{(k)}$. The corresponding Gaussian measure is then represented by

$$\Phi(\upsilon^{-1}E) = \Phi[\{x : \upsilon x \in E\}].$$

The measure $\Phi(T^{-1}\cdot)$ determines the nonnegative quadratic form

$$q(t) = \int_{R^{(k)}} (t'y)^2 \, \Phi(T^{-1}dy) = \int_{R^{(k)}} (t'Tz)^2 \, \Phi(dz).$$

(Note that we are using matrix notations for the vectors.) We have

$$(t'Tz)^2 = t'Tzz'T't,$$

and since

$$\int_{R^{(k)}} zz' \, \phi(dz)$$

is the diagonal matrix with all elements 1, we get

$$q(t) = t'TT't, \tag{7}$$

where $TT'$ is the covariance matrix

$$TT' = \begin{pmatrix} \sigma_{11} & \cdots & \sigma_{1k} \\ \vdots & & \vdots \\ \sigma_{k1} & \cdots & \sigma_{kk} \end{pmatrix},$$

which is symmetrical. Note that the nonnegative quadratic form $q(t)$ belongs to $TT'$. A matrix that has this property is called positive definite.

We also get

$$\int_{R^{(k)}} \|x\|^2 \, \Phi(T^{-1}dx) = \int_{R^{(k)}} \|Ty\|^2 \, \Phi(dy) = \sum_{i=1}^{k} \sigma_{ii}. \tag{8}$$

Indeed, if $T$ has the elements $a_{ij}$, we get

$$[x^{(i)}]^2 = \left[ \sum_{j=1}^{k} a_{ij} y_j \right]^2,$$

and

$$\int_{R^{(k)}} y_{j_1} v_{j_2} \, \Phi(dy) = \begin{cases} 0 & \text{for } j_1 \neq j_2, \\ 1 & \text{for } j_1 = j_2, \end{cases}$$

and

$$\sum_{j=1}^{k} a_{ij}^2 = \sigma_{ii}.$$

We now claim

**Theorem 1.** *For any given symmetrical and positive definite matrix B, there exists a real-valued linear transformation T such that $\Phi(T^{-1}\cdot)$ has the covariance matrix $B = TT'$ and satisfies the relations (8) and*

(i) $[\Phi(\sqrt{n}\,T^{-1}\cdot)]^{*n} = \Phi(T^{-1}\cdot)$.

*Proof:* Since $B$ is symmetrical, there exists an orthonormal matrix $\upsilon$ such that $\upsilon' B \upsilon$ is the diagonal matrix. It belongs to a nonnegative quadratic form since $B$ does. We may choose a diagonal matrix $S$ such that

$$\upsilon' B \upsilon = SJS',$$

where $J$ is the diagonal matrix with all elements equal to 1. Then

$$B = \upsilon SJS'\upsilon' = \upsilon SJ(\upsilon S)' = \upsilon S(\upsilon S)'.$$

Putting $T = \upsilon S$, we find that $\Phi(T^{-1}\cdot)$ determines the quadratic form $q(t) = t'TT't$ and, hence, the covariance matrix $TT'$. The relation (i) in Theorem 1 follows by (2.3) and (8) as above. □

For $\mu \in \bar{\mathscr{M}}$ on $R^{(k)}$, we consider the convolution $\Phi(\cdot/\alpha) * \mu$. Its distribution function

$$G_\alpha(x) = \int_{R^{(k)}} \Phi\left(\frac{x-y}{\alpha}\right)\mu(dy)$$

is called the Gaussian transform of $\mu$. It has continuous derivatives of all orders. In particular, its density function is

$$q_\alpha(x) = \alpha^{-k} \int_{R^{(k)}} \varphi\left(\frac{x-y}{\alpha}\right)\mu(dy). \tag{9}$$

**Theorem 2.** *$\mu \in \bar{\mathscr{M}}$ on $R^{(k)}$ is uniquely determined by its Gaussian transform.*

*Proof:* It is sufficient to prove the theorem for $\mu \in \mathscr{M}$. Indeed, if $\lambda_1 = \mu_1 - \nu_1, \lambda_2 = \mu_2 - \nu_2$, with $\mu_1, \nu_1, \mu_2, \nu_2 \in \mathscr{M}$, we have $\lambda_1 = \lambda_2$ if and only if $\mu_1 + \nu_2 = \mu_2 + \nu_1$. If $[y : y \leq x]$ is a continuity set for $\mu \in \mathscr{M}$, we get, by the dominated convergence theorem,

$$\lim_{\alpha \to 0} \int \phi\left(\frac{x-y}{\alpha}\right)\mu(dy) = \int_{y \leq x} \mu(dy) = \hat{\mu}(x),$$

since $\Phi(x - y/\alpha)$ tends to 1 for $y < x$ and to 0 for $y > x$. Thus the Gaussian transform determines the distribution function of $\mu$ and, hence, $\mu$, by Theorem 1.1. □

As $\alpha$ tends to 0, the Gaussian measure $\Phi(\cdot/\alpha)$ tends to the unit measure.

## 3. FOURIER TRANSFORMS AND THEIR RELATION TO GAUSSIAN TRANSFORMS

For $\lambda \in \bar{\mathscr{M}}$ on $R^{(k)}$, the characteristic function $\tilde{\lambda}$ of $\lambda$ is defined by

$$\tilde{\lambda}(t) = \int \exp(it \cdot x)\,\lambda(dx) = \int \cos(t \cdot x)\,\lambda(dx) + i \int \sin(t \cdot x)\,\lambda(dx), \tag{1}$$

where $t \cdot x$ is the scalar product of the vectors $t$ and $x$ in $R^{(k)}$. If $\lambda$ has a density function $g$, and thus

$$\lambda(E) = \int_E g(x)\,dx$$

for any set $E$ in the $\sigma$-algebra on $R^{(k)}$, we have

$$\tilde{\lambda}(t) = \int \exp(it \cdot x)\,g(x)\,dx.$$

Then we call $\tilde{\lambda}$ the Fourier transform of $g$ and use the notation $\tilde{g} = \tilde{\lambda}$.

**Theorem 1.** *The characteristic function of $\lambda * \mu$ for $\lambda$, $\mu \in \bar{\mathscr{M}}$ is equal to the product $\tilde{\lambda} \cdot \tilde{\mu}$.*

*Proof:* By the definition of convolutions and by the use of Fubini's theorem, we get

$$\iint \exp[it \cdot (x + y)]\,\lambda(dx)\,\mu(dy) = \int \exp(it \cdot x)\,\lambda(dx) \int \exp(it \cdot y)\,\mu(dy). \quad \square$$

**Theorem 2.** *The Fourier transform of $\varphi(x) = (2\pi)^{-k/2} \exp(-\frac{1}{2}\|x\|^2)$ on $R^{(k)}$ is $(2\pi)^{k/2}\varphi(t)$.*

*Proof:* Consider the case $k = 1$. According to the symmetry, we get

$$g(t) = \int \exp(it \cdot x)\exp(-\tfrac{1}{2}x^2)\,dx = \int \cos(t \cdot x)\exp(-\tfrac{1}{2}x^2)\,dx.$$

Taking the derivative with respect to $t$ and then integrating by parts, we obtain

$$g'(t) = -tg(t).$$

This differential equation only has the solution $g(t) = \exp(-\frac{1}{2}t^2)$, which, for $t = 0$, takes the value 0, which $g$ must take according to the definition. For

$k > 1$ the relation follows by (2.4) and Fubini's theorem

$$\int_{R^{(k)}} \exp(it \cdot x)\,\varphi(x)\,dx = \prod_{j=1}^{k} \int_R \exp(it^{(j)}x^{(j)})\,\varphi(x^{(j)})\,dx^{(j)}. \quad \square$$

**Theorem 3** (Inversion theorem). *Let $g(\alpha, \cdot)$ be the density function of the Gaussian transform of $\lambda \in \bar{\mathscr{M}}$, $\tilde{g}(\alpha, \cdot)$ the Fourier transform of $g(\alpha, \cdot)$, and $\tilde{\lambda}$ the characteristic function of $\lambda$. Then we have the relations*

(i) $$\tilde{g}(\alpha, t) = \tilde{\lambda}(t)\exp(-\tfrac{1}{2}\alpha^2\|t\|^2),$$

(ii) $$g(\alpha, x) = (2\pi)^{-k}\int \tilde{g}(\alpha, t)\exp(-it \cdot x)\,dt.$$

*Remark:* Note that $\tilde{g}(\alpha, t) = \int \exp(it \cdot x)g(\alpha, x)\,dx$, so that (ii) gives an inversion of this integral.

*Proof:* (i) follows by Theorems 1 and 2. By Fubini's theorem, we get

$$\int \tilde{g}(\alpha, t)\exp(-it \cdot x)\,dt = \int \exp(-\tfrac{1}{2}\alpha^2\|t\|^2)\left[\int \exp(it \cdot y)\,\lambda(dy)\right]\exp(-it \cdot x)\,dt$$

$$= \int\left\{\int \exp(-\tfrac{1}{2}\alpha^2\|t\|^2)\exp[it \cdot (y - x)]\,dt\right\}\lambda(dy).$$

By Theorem 2, the inner integral is equal to $(2\pi)^k\alpha^{-k}\varphi[(x - y)/\alpha]$. Hence, we obtain

$$\int \tilde{g}(\alpha, t)\exp(-it \cdot x)\,dt = (2\pi)^k\alpha^{-k}\int \varphi\left(\frac{x - y}{\alpha}\right)\lambda(dy) = (2\pi)^k g(\alpha, x).$$

As a corollary of Theorem 3 and Theorem 2.2, we get

**Theorem 4.** *$\lambda \in \bar{\mathscr{M}}$ is uniquely determined by its characteristic function.* $\square$

**Example.** Let $\Phi(T^{-1}\cdot)$ be the Gaussian measure on $R^{(k)}$ introduced at the end of the preceding section. Its characteristic function is

$$h(t) = \int \exp(it'x)\,\Phi(T^{-1}dx). \tag{2}$$

Here we deal with the vectors as matrices, observing that $t'x$ then is the scalar product $t \cdot x$ ($t'$ is the transpose of $t$). By the transformation $x = Ty$ we can write

$$h(t) = \int \exp(it'Ty)\,\Phi(dy). \tag{3}$$

Now if $T'$ denotes the transpose of $T$, we get

$$t'Ty = (T't)'y. \tag{4}$$

Remembering that $\Phi$ has the characteristic function $\exp(-\frac{1}{2}t't)$, we find by (3) that $\Phi(T^{-1}\cdot)$ has the characteristic function

$$\int \exp(it'x)\,\Phi(T^{-1}dx) = \exp(-\tfrac{1}{2}t'TT't) = \exp(-\tfrac{1}{2}\|T't\|^2). \tag{5}$$

The matrix $TT'$ is symmetrical, $(TT')' = TT'$, and the quadratic form $t'TT't = \|t'T\|^2$ is positive definite. We now state

**Theorem 5.** *A measure $\mu$ on $R^{(k)}$ is Gaussian if and only if its characteristic function has the form*

$$\exp\left[it'm - \tfrac{1}{2}t'Bt\right] \tag{6}$$

*with any constant vector $m$ and a symmetrical $k \times k$ matrix $B$, which belongs to a positive definite quadratic form. Here $B$ is the covariance matrix of the Gaussian measure.*

*Proof:* By definition, a Gaussian measure is equal to $\Phi(\upsilon^{-1}\cdot)$, where $\upsilon x = Tx + m$, and $\Phi$ is the special Gaussian measure introduced in Section 2. If $\mu$ has the characteristic function $\hat{\mu}$, then $\mu(\cdot - m)$ has the characteristic function

$$\exp(it'm)\hat{\mu}(t).$$

Indeed,

$$\int \exp(it'x)\,\mu(dx - m) = \exp(it'm)\int \exp(it'y)\,\mu(dy).$$

Thus by (5) the characteristic function of a Gaussian measure has the form (6), where $B$ fulfills the requirements in the theorem. Conversely, let $B$ fulfill these requirements. By the fundamental theorem for quadratic forms, for a symmetrical matrix there exists an orthonormal transformation $T$, i.e., $T'T = E$, where $E$ is the unit matrix, such that $T'T$ has the diagonal form $T'BT = \bar{B}$, $\bar{B} = (b_{ij})$, $b_{ij} = 0$ for $i \neq j$. Since the quadratic form is positive definite, $b_{ii} \geq 0$ for all $i$. Then defining $\upsilon_{ii} = |\sqrt{b_{ii}}|$, $\upsilon_{ij} = 0$ for $j \neq i$, the diagonal matrix $V$ with these elements satisfies the relation $V' = V$:

$$\bar{B} = VV', \qquad T'BT = VV', \qquad B = TV(TV)'.$$

The Gaussian measure $\Phi((TV')^{-1}\cdot)$ has the characteristic function

$$\exp[-\tfrac{1}{2}(t'TVV'Tt)] = \exp[-\tfrac{1}{2}(t'Bt)].$$

Hence (6) also is the characteristic function of a Gaussian measure.

Now let the Gaussian measure $\mu$ have the characteristic function (6), where $B$ is not a zero matrix. Then for any real number $\tau$, we have the identity

$$\int \exp(i\tau t'x)\,\mu(dx) = \exp(i\tau t'm_1 - \tfrac{1}{2}\tau^2 t'Bt).$$

Taking the derivative with respect to $\tau$ and then putting $\tau$ equal to 0, we get

$$\int t'x\,\mu(dx) = t'm. \tag{7}$$

Taking the derivative twice with respect to $\tau$ and putting $\tau = 0$, we obtain

$$\int (t'x)^2\,\mu(dx) = t'Bt + (t'm)^2. \tag{8}$$

It follows by (7) and (8) that $B$ and $m$ are uniquely determined by the mean values $\int x^{(i)}\,\mu(dx)$ and $\int x^{(i)}x^{(j)}\,\mu(dx)$, and conversely that these mean values determine $m$ and $B$ uniquely.

The mapping $\pi(t)$ defined by $\pi(t)x = t \cdot x$ for any $t \in R^{(k)}$ is a continuous mapping (it is called a projection).

**Theorem 6.** *Two $\sigma$-smooth measures $\mu$ and $\lambda$ are equal if and only if $\mu(\pi^{-1}(t)\cdot) = \lambda(\pi^{-1}(t)\cdot)$ for any vector $t \in R^{(k)}$.*

*Proof:* The characteristic function of $\mu(\pi^{-1}(t))$ can, by transformation of the integral, be written

$$\int_{R^{(k)}} \exp(i\alpha t \cdot x)\,\mu(dx) \qquad (\alpha \text{ is any number in } R),$$

which is the characteristic function of $\mu$ when it is considered as function of $\alpha t$. Thus, if $\mu(\pi^{-1}(t)\cdot)$ and $\lambda(\pi^{-1}(t)\cdot)$ are equal for any $t \in R^{(k)}$, they have the same characteristic function, and then $\mu$ and $\lambda$ have the same characteristic function. □

**Theorem 7.** *A $\sigma$-smooth measure $\mu$ in $R^{(k)}$ is Gaussian if and only if $\mu(\pi^{-1}(t)\cdot)$ is Gaussian for any $t \in R^{(k)}$.*

*Proof:* As above, the characteristic function of $\mu$ is determined by the characteristic functions of the $\mu[\pi^{-1}(t)\cdot]$, and these characteristic functions are determined by $\mu$. □

**Theorem 8.** *A sequence $\{\mu_n\}$ of Gaussian measures converges weakly and then to a Gaussian measure (eventually singular) if and only if the sequence of the mean-value vectors of $\mu_n$ and the sequence of moment matrices converge to a moment vector and a moment matrix belonging to a Gaussian measure.*

*Proof:* If $\{\mu_n\}$ converges weakly to $\mu$, then $\mu_n[\pi^{-1}(t)]$ converges weakly to $\mu[\pi^{-1}(t)]$. Applying Theorem 7, we conclude that the fact that $\mu$ is Gaussian follows for $k > 1$ if we prove it for $k = 1$. Now the characteristic function of $\mu_n$ for $k = 1$ has the form

$$\exp(im_n t - \tfrac{1}{2}\alpha_n^2 t^2),$$

and it converges if and only if $\{m_n\}$ and $\alpha_n^2$ converge to values $m$ and $\alpha^2$ such that $\exp[imt - \frac{1}{2}\alpha^2 t^2]$ is a characteristic function of a probability measure $\mu$, which is the case if and only if $m$ and $\alpha^2$ are finite. □

## 4. GAUSSIAN SEMINORMS

In Section III.8 we introduced the concept "class $\mathscr{P}$ of seminorms" on a commutative algebra $\bar{\mathscr{M}}$ with unit element $e$ over the real number field. Furthermore, we considered a semigroup $\mathscr{M}$ in $\bar{\mathscr{M}}$, which is characterized by the properties

$$e \in \mathscr{M}, \qquad p(e) = 1, \tag{1}$$

$$p(\mu\lambda) \leq p(\mu)p(\lambda) \qquad \text{for} \quad \mu \in \mathscr{M}, \quad \lambda \in \bar{\mathscr{M}}, \tag{2}$$

$$p[(\mu - e)(\nu - e)] \leq c(p)p(\mu - e)p(\nu - e) \qquad \text{for} \quad \mu, \nu \in \mathscr{M}, \tag{3}$$

where $c(p)$ is a constant depending only on $p$ and $\mathscr{M}$. Now the class $\bar{\mathscr{M}}$ of signed $\sigma$-smooth measures is a commutative algebra $\bar{\mathscr{M}}$ over the real number field with unit element $e$, when convolution is the multiplicative composition. We shall introduce a class $P_G$ of seminorms called Gaussian seminorms on the algebra of signed measures and show in Section 5 that the semigroup $\mathscr{M}$ of $\sigma$-smooth measures, different from the zero measure, is a convex distinguished semigroup in $\bar{\mathscr{M}}$ under the class of Gaussian seminorms.

We define the Gaussian seminorm $P_\alpha$ on $\bar{\mathscr{M}}$ for $\alpha > 0$ simply by

$$p_\alpha(\lambda) = \sup_x \left| \int \Phi\left(\frac{x-y}{\alpha}\right) \lambda(dy) \right|.$$

It follows at once that $G = \{p_\alpha : \alpha > 0\}$ has the properties required in Section III.8 for a class of seminorms. Indeed, if $p_\alpha(\lambda) = 0$ for all $\alpha > 0$, the Gaussian transform is zero for all $\alpha > 0$, and $\lambda$ is the zero measure by Theorem 2.2. The fact that $p_\alpha$ has the properties required for a seminorm is easily verified. Note that it is sufficient to consider a class $P_G$ with countably many $\alpha > 0$.

We use the same notation $P_\alpha$ for a Gaussian seminorm on $\bar{\mathscr{M}}$ on $R^{(k)}$ for any $k$, hence also on $\bar{\mathscr{M}}$ on $R$.

**Lemma 1.** *For $\lambda \in \bar{\mathscr{M}}$ on $R^{(k)}$, the* i*th margin $\lambda^{(i)} = \lambda(\pi^{(i)-1}\cdot)$ satisfies the inequality*

$$p_\alpha(\lambda^{(i)}) \leq p_\alpha(\lambda).$$

*Proof:* By the transformation $y^{(i)} = \pi^{(i)} y$ of the integral, we get

$$p_\alpha(\lambda^{(i)}) = \sup_{x^{(i)}} \left| \int_R \Phi\left(\frac{x^{(i)} - y^{(i)}}{\alpha}\right) \lambda(\pi^{(i)-1} dy^{(i)}) \right.$$

$$= \sup_{x^{(i)}} \left| \int_{R^k} \Phi\left(\frac{x^{(i)} - y^{(i)}}{\alpha}\right) \lambda(dy) \right| \le \sup_x \left| \int_{R^{(k)}} \Phi\left(\frac{x - y}{\alpha}\right) \lambda(dy) \right|,$$

since $\Phi[(x^{(i)} - y^{(i)})/\alpha] = \Phi[(x - y)/\alpha]$ if $x^{(j)} = \infty$ for $i \ne j$. □

**Lemma 2.** *For $\lambda = \mu - e$, $\mu \in \mathcal{M}$ on R, we have*

(i) $$\int_{|y| \ge \alpha} \mu(dy) \le C_1 p_\alpha(\lambda),$$

(ii) $$\frac{1}{\alpha} \left| \int_{|y| < \alpha} y\, \mu(dy) \right| \le C_2 p_\alpha(\lambda),$$

(iii) $$\frac{1}{\alpha^2} \int_{|y| < \alpha} y^2\, \mu(dy) \le C_3 p_\alpha(\lambda),$$

(iv) $$p_\alpha(\lambda) \le \int_{|y| \ge \alpha} \mu(dy) + |\lambda(R)| + \frac{1}{\alpha} \|\Phi'\| \left| \int_{|y| < \alpha} y\, \mu(dy) \right| + \frac{1}{2\alpha^2} \|\Phi''\| \int_{|y| < \alpha} y^2\, \mu(dy).$$

**Corollary.** *For $\lambda = \mu - e$, $\mu \in \mathcal{M}$ on $R^{(k)}$, and the* i*th margin $\lambda^{(i)}$ of $\lambda$, and*

$$D_\alpha = \{ y : |y^{(i)}| < \alpha \qquad (i = 1, \ldots, k),$$

*we have*

(i) $$\int_{D_\alpha^c} \mu(dx) \le C_1 \sum_{i=1}^{k} p_\alpha(\lambda^{(i)}) \le C_1 k p_\alpha(\lambda),$$

(ii) $$\frac{1}{\alpha} \left| \int_{D_\alpha} x^{(i)}\, \mu(dx) \right| \le C_2 p_\alpha(\lambda^{(i)}) + C_1 \sum_{i=1}^{k} p_\alpha(\lambda^{(i)}) \le (C_2 + C_1 k) p_\alpha(\lambda),$$

(iii) $$\frac{1}{\alpha^2} \int_{D_\alpha} [x^{(i)}]^2\, \mu(dx) \le C_3 p_\alpha(\lambda^{(i)}) \le C_3 p_\alpha(\lambda),$$

(iv) $$p_\alpha(\lambda) \le C_4 k \sum_{i=1}^{k} p_\alpha(\lambda^{(i)}),$$

*with constants $C_1, C_2, C_3, C_4$ (independent on k and x).*

*Proof:* Consider measures and points on $R$. Put

$$\Psi(x, y) = 2\Phi(x) - \Phi(x - y) - \Phi(x + y). \tag{4}$$

Clearly $\Psi(x, y) = \Psi(x - y)$, and the derivative with respect to $y$, $\varphi(x - y) - \varphi(x + y)$, is nonnegative for $x > 0$. Hence $\Psi(x, y)$ increases for $x \geq 0$ from 0 to $2\Phi(x) - 1$ as $y$ tends from 0 to $+\infty$. We get

$$\begin{aligned} p_\alpha(\lambda) &= \sup_x \left| \int_R \Phi\left(\frac{x - y}{\alpha}\right)[\mu(dy) - e(dy)] \right| \\ &= \sup_x \left| \int_R \left[ \Phi\left(\frac{x - y}{\alpha}\right) - \Phi\left(\frac{x}{\alpha}\right)\right] \mu(dy) - \Phi\left(\frac{x}{\alpha}\right)(\mu(R) - 1) \right|. \end{aligned} \tag{5}$$

Clearly $\mu(R) = p_\alpha(\mu)$ and

$$|p_\alpha(\mu) - 1| \leq p_\alpha(\mu - e) = p_\alpha(\lambda).$$

Hence by (5)

$$\left| \int_R \left[ \Phi\left(\frac{x - y}{\alpha}\right) - \Phi\left(\frac{x}{\alpha}\right)\right] \mu(dy) \right| \leq 2p_\alpha(\lambda). \tag{6}$$

This inequality remains true if we change $x$ to $-x$, and since

$$\Phi\left(\frac{-x - y}{\alpha}\right) - \Phi\left(\frac{-x}{\alpha}\right) = -\Phi\left(\frac{x + y}{\alpha}\right) + \Phi\left(\frac{x}{\alpha}\right),$$

the relation (6) also remains true if we change $y$ to $-y$. Thus we obtain from (6) and (4)

$$\int_R \Psi\left(\frac{x}{\alpha}, \frac{y}{\alpha}\right) \mu(dx) \leq 4p_\alpha(\lambda), \tag{7}$$

where $\Psi(x/\alpha, y/\alpha) \geq 0$ for $x \geq 0$ and all $y$. For $|y| \geq \alpha$, we choose $x$ so that $\Psi(x/\alpha, y/\alpha)$ has its largest value, which we denote by $4/C_1$. Then we get (i) from (4). By Taylor's formula we obtain

$$\Psi(x, y) = (-y^2/2)[\Phi''(x - \Theta y) + \Phi''(x + \Theta y)] = -y^2\Phi''(x - \Theta' y) \tag{8}$$

with $-1 < \Theta' < 1$. If $x \geq 1$ and $0 \leq y \leq x - 1$, we have $x - \Theta' y \geq x - y \geq 1$. Furthermore, $\Phi''(x)$ is negative and increasing for $x \geq 1$. Hence

$$\Psi(x, y) \geq y^2|\Phi''(x + |y|)| \qquad \text{for} \quad |y| \geq x - 1 \tag{9}$$

and

$$\Psi(x/\alpha, y/\alpha) \geq (y^2/\alpha^2)|\Phi''(3)|$$

for $x = 2\alpha$, $y \leq \alpha$. This inequality together with (7) implies (iii) with $C_3 = 4/|\Phi''(3)|$. To prove (ii), we use the approximation (6) and (iii) for

$|y| \leq \alpha$ and (i) for $|y| \geq \alpha$. We obtain (ii) for $x = 0$, with

$$C_2 = [2 + \tfrac{1}{2}C_3\|\Phi''\|]/\Phi'(0).$$

We get (iv) by (5) if we estimate the integral on the right-hand side directly for $|y| \geq \alpha$ and expand $\Phi[(x - y)/\alpha]$ by Taylor's formula for $|y| < \alpha$. □

*Proof of the corollary:* By Lemma 1, we have $p_\alpha(\lambda^{(i)}) \leq p_\alpha(\lambda)$, and then by Lemma 2,

$$\int_{D_\alpha^c} \mu(dy) \leq \sum_{i=1}^{k} \int_{|y^{(i)}| \geq \alpha} \mu(dy) \leq C_1 \sum_{i=1}^{k} p_\alpha(\lambda^{(i)}) \leq C_1 k p_\alpha(\lambda).$$

By (ii) in Lemma 2 we then get

$$\begin{aligned} \frac{1}{\alpha}\int_{D_\alpha} y^{(i)}\,\mu(dy) &= \frac{1}{\alpha}\left|\int_{D_\alpha \cap \{|y^{(i)}| < \alpha\}} y^{(i)}\,\mu(dy)\right| \\ &\leq \frac{1}{\alpha}\left|\int_{|y^{(i)}| < \alpha} y^{(i)}\,\mu(dy)\right| + \int_{D_\alpha^c} \mu(dy) \leq (C_2 + kC_1)p_\alpha(\lambda). \end{aligned}$$

The inequality (iii) in the corollary follows directly from (iii) in Lemma 2.

For the proof of (iv), we write

$$\int_{R^{(k)}} \Phi\left(\frac{x - y}{\alpha}\right)\lambda(dy) = \int_{D_\alpha^c} \Phi\left(\frac{x - y}{\alpha}\right)\mu(dy) + \int_{D_\alpha} \Phi\left(\frac{x - y}{\alpha}\right)\lambda(dy).$$

By (i) in the corollary, the first integral on the right-hand side is at most equal to

$$C_1 \sum_{i=1}^{k} p_\alpha(\lambda^{(i)}).$$

In the second integral, we expand $\Phi[(x - y)/\alpha]$ by Taylor's formula

$$\begin{aligned} \Phi\left(\frac{x - y}{\alpha}\right) &= \Phi\left(\frac{x}{\alpha}\right) - \sum_{i=1}^{k} y^{(i)} \frac{\partial}{\partial x^{(i)}} \Phi\left(\frac{x}{\alpha}\right) \\ &\quad + \sum_{i=1}^{k}\sum_{j=1}^{k} y^{(i)}y^{(j)}\left[\frac{\partial}{\partial x^{(i)}\,\partial x^{j}} \Phi\left(\frac{x}{\alpha}\right)\right]_{x = x_0} \end{aligned}$$

for some $x_0$. We have

$$\left|\int_{D_\alpha} \lambda(dy)\right| = \left|\int_{R^{(k)}} \lambda(dy) - \int_{D_\alpha^c} \mu(dy)\right| \leq |\lambda(R^{(k)})| + C_1 \sum_{i=1}^{k} p_\alpha(\lambda^{(i)}).$$

Here

$$|\lambda(R^{(k)})| = |\lambda^{(i)}(R^{(k)})| \leq p_\alpha(\lambda^{(i)}).$$

Hence

$$\left|\int_{D_\alpha} \lambda(dy)\right| \le C_1' \sum_{i=1}^{k} p_\alpha(\lambda^{(i)}).$$

By (ii) in the corollary, we get

$$\left|\int_{D_\alpha} y^{(i)}\,\lambda(dy)\right| = \left|\int_{D_\alpha} y^{(i)}\,\mu(dy)\right| \le \alpha C_2 p_\alpha(\lambda^{(i)}) + \alpha C_1 \sum_{i=1}^{k} p_\alpha(\lambda^{(i)}),$$

$$\int_{D_\alpha} |y^{(i)}y^{(j)}|\,\lambda(dy) = \int_{D_\alpha} |y^{(i)}y^{(j)}|\,\mu(dy)$$

$$\le \tfrac{1}{2}\int_{D_\alpha} ([y^{(i)}]^2 + [y^{(j)}]^2)\,\mu(dy) \le \tfrac{1}{2}\alpha^2[p_\alpha(\lambda^{(i)}) + p_\alpha\lambda^{(j)}].$$

Using these inequalities, we get

$$\left|\int_{R^{(k)}} \Phi\left(\frac{x-y}{\alpha}\right)\lambda(dy)\right| \le C_4 k \sum_{i=1}^{k} p_\alpha(\lambda^{(i)}).$$

**Lemma 3.** *If $\lambda_1 = \mu_1 - e$, $\lambda_2 = \mu_2 - e$, $\mu_1, \mu_2 \in \mathcal{M}$ on $R^{(k)}$, then*

$$p_\alpha[(\mu_1 - e) * (\mu_2 - e)] \le c(k) p_\alpha(\mu_1 - e) p_\alpha(\mu_2 - e)$$

*with a constant $c(k)$ dependent only on $k$.*

*Proof:* By Fubini's theorem, we get for $\lambda_1 = \mu_1 - e$, $\lambda_2 = \mu_2 - e$,

$$p_\alpha(\lambda_1 * \lambda_2) = \sup_x \left|\int_{R^{(k)}} \left[\int_{R^{(k)}} \Phi\left(\frac{x-y-z}{\alpha}\right)\lambda_1(dy)\right]\lambda_2(dz)\right|. \tag{10}$$

Put

$$g(z) = \int_{R^{(k)}} \Phi\left(\frac{x-y-z}{\alpha}\right)\lambda_1(dy).$$

Then $g(z)$ has bounded continuous derivatives of all orders. By Taylor's expansion we get

$$\begin{aligned}\left|\int_{R^{(k)}} g(z)\,\lambda_2(dz)\right| \le{}& \left|\int_{D_\alpha^c} g(z)\,\mu_2(dz)\right| \left|\int_{D_\alpha^c} g(0)\,\lambda_2(dz)\right| \\ &+ \max_{i=1,\dots,k} \left|\left[\frac{\partial g(z)}{\partial z^{(i)}}\right]_{z=0}\right| \sum_{i=1}^{k} \left|\int_{D_\alpha^c} z^{(i)}\,\mu_2(dz)\right| \\ &+ \max_{\substack{z,\, i=1,\dots,k \\ j=1,\dots,k}} \left|\frac{\partial^2 g(z)}{\partial z^{(i)}\,\partial z^{(j)}}\right| \sum_{i=1}^{k}\sum_{j=1}^{k} \left|\int_{D_\alpha^c} z^{(i)}z^{(j)}\,\mu_2(dz)\right|. \end{aligned} \tag{11}$$

Here

$$\left|\int_{D_\alpha} \lambda_2(dz)\right| \leq \left|\int_{R^{(k)}} \lambda_2(dz)\right| + \left|\int_{D_\alpha^c} \mu_2(dz)\right|.$$

The first integral on the right-hand side is at most equal to $p_\alpha(\lambda_2)$ by the definition of the Gaussian seminorm, and the second integral is at most equal to $C_1 k p_\alpha(\lambda_2)$ by (i) in the corollary of Lemma 2. Since

$$|z^{(i)}z^{(j)}| = \tfrac{1}{2}[(z_i)^2 + (z_j)^2],$$

we then obtain from (11), by the inequalities in this corollary,

$$\begin{aligned}\left|\int_{R^{(k)}} g(z)\,\lambda_2(dz)\right| &\leq (1 + 2C_1 k)p_\alpha(\lambda_2)\max_z |g(z)| \\ &\quad \times k(C_2 + C_1 k)\alpha p_\alpha(\lambda_2) \max_{i=1,\ldots,k}\left|\left[\frac{\partial g(z)}{\partial z^{(i)}}\right]_{z=0}\right| \\ &\quad + C_3 k^2 \alpha^2 p_\alpha(\lambda_2) \max_{z,\, i,\, j=1,\ldots,k}\left|\frac{\partial^2 g(z)}{\partial z^{(i)}\,\partial z^{(j)}}\right|. \qquad (12)\end{aligned}$$

Remembering now the definition of $g(z)$, we find that $g$ and its partial derivatives are of the same form as the integral on the left-hand side, but we must change $g(z)$ to $\Phi[(x - y)/\alpha]$ and its partial derivatives with respect to $y$. Observing that $\Phi$ and its partial derivatives are bounded, say numerically not larger than $c_0'$, we get by (11),

$$\begin{aligned}\max_z |g(z)| &\leq c_0'[(1 + 2C_1 k)p_\alpha(\lambda_1) + k(C_2 + C_1 k)p_\alpha(\lambda_1) + C_3 k^2 p_\alpha(\lambda_2)] \\ &= c'(k)p_\alpha(\lambda_1),\end{aligned}$$

and, in the same way,

$$\max_{z,\, i=1,\ldots,k}\left|\frac{\partial g(z)}{\partial z^{(i)}}\right| \leq c'(k)p_\alpha(\lambda_1),$$

$$\max_{z,\, i,\, j=1,\ldots,k}\left|\frac{\partial^2 g(z)}{\partial z^{(i)}\,\partial z^{(j)}}\right| \leq c'(k)p_\alpha(\lambda_2).$$

By (8) and (10), we then obtain

$$p_\alpha(\lambda_1 * \lambda_2) \leq c(k)p_\alpha(\lambda_1)p_\alpha(\lambda_2),$$

with $c(k) \leq c_0 k^2$ and an absolute constant $c_0$.

Clearly, $p_\alpha$ satisfies the relations (1) and (2), and by Lemma 3 it has property (3) also. Note that $c(k)$ in the inequality of the lemma does not depend on $\alpha$, i.e., it does not depend on the particular Gaussian seminorm. □

The following lemma is of importance for comparison of measures.

**Lemma 4.** *If $\mu$ and $\nu$ belong to $\mathcal{M}$ on $R$, then the distribution functions $\hat{\mu}$ and $\hat{\nu}$ of $\mu$ and $\nu$, respectively, satisfy the inequality:*

$$|\hat{\mu}(x) - \hat{\nu}(x)| \leq p_\alpha(\mu - \nu) + \hat{\nu}(x + 2\alpha\beta) - \hat{\nu}(x - 2\alpha\beta) + \Phi(-\beta)[\mu(R) + \nu(R)]$$

*for* $\alpha > 0$, $\beta \geq 1$.

*Proof:*

$$\begin{aligned} p_\alpha(\mu - \nu) &\geq \int_R \Phi\left(\frac{x + \alpha\beta - y}{\alpha}\right)[\mu(dy) - \nu(dy)] \\ &\geq \Phi(\beta)\int_{y \leq x} \mu(dy) - \int_{y \leq x + 2\alpha\beta} \nu(dy) - \Phi(-\beta)\int_{y > x + 2\alpha\beta} \nu(dy) \\ &\geq \hat{\mu}(x) - \hat{\nu}(x + 2\alpha\beta) - \Phi(-\beta)[\mu(R) + \nu(R)]. \end{aligned}$$

Here we may interchange $\mu$ and $\nu$ and at the same time change $x + 2\alpha\beta$ to $x$. Thus the inequality follows. □

## 5. THE SEMIGROUP OF $\sigma$-SMOOTH MEASURES

As we have pointed out in Section 4, the signed $\sigma$-smooth measures on $R^{(k)}$ form a commutative algebra $\bar{\mathcal{M}}$ over the real number field, and this algebra contains the commutative semigroup $\mathcal{M}$ of $\sigma$-smooth measures, different from the zero measure, the multiplicative composition in the semigroup being convolution. According to Lemma 4.3, $\mathcal{M}$ is a distinguished semigroup in $\bar{\mathcal{M}}$ for the Gaussian seminorms. It is convex, since for any $\mu_1$ and $\mu_2$ in $\mathcal{M}$ and $0 \leq \beta \leq 1$, $\beta\mu_1 + (1 - \beta)\mu_2 \in \mathcal{M}$. We now state

**Theorem 1.** *$\mathcal{M}$ on $R^{(k)}$ is a commutative convex semigroup under convolution. It is distinguished under the class $\mathcal{P}_G$ of Gaussian seminorms. Furthermore, it is complete under this class of seminorms. Convergence of a sequence $\{\mu_n\}_{n=1}^{+\infty}$ of $\sigma$-smooth measures under $\mathcal{P}_G$ is equivalent to weak convergence.*

*Proof:* It remains only to show the completeness and the stated equivalence. To begin with, we consider measures in $\mathcal{M}$ on $R$. By Lemma 4.3, we have

$$\begin{aligned} |\hat{\mu}_n(x) - \hat{\mu}_m(x)| \leq p_\alpha(\mu_n - \mu_m) + \hat{\mu}_m(x + 2\alpha\beta) - \mu_m(x - 2\alpha\beta) \\ + \Phi(-\beta)[\mu_m(R) + \mu_n(R)]. \end{aligned} \tag{1}$$

In (1) we let $n$, $m$, $\beta$, and $\alpha$ tend to limits successively as follows, assuming that

$$\begin{gathered} p_\alpha(\mu_n - \mu_m) \to 0 \qquad (n \to \infty, m \to \infty); \\ (1^\circ) \quad x \to \infty, \qquad (2^\circ) \quad \beta \to \infty, \qquad (3^\circ) \quad n, m \to \infty. \end{gathered} \tag{2}$$

We get

$$\hat{\mu}_n(\infty) - \hat{\mu}_m(\infty) \to 0 \qquad (n \to \infty, m \to \infty).$$

This Cauchy convergence implies the convergence of $\hat{\mu}_n(\infty)$ to a finite value as $n \to \infty$. Clearly, $\hat{\mu}_n(-\infty) = 0$ for all $n$.

Since $\hat{\mu}_n(x) \leq \hat{\mu}_n(\infty)$, we find by the selection principle, used in Section I.11 for the $l^2$-space, that for any infinite sequence on positive integers there exists a subsequence $\bar{N}$ such that $\{\hat{\mu}_n(x)\}_{n \in \bar{N}}$ converges at all rational points to a function $h$. Defining then

$$h(x) = \inf_{r \geq x,\, r \text{ rational number}} h(r)$$

at any point $x$, we conclude that $h$ is a nondecreasing bounded function. It is easy to show that $\{\mu_n(x)\}_{n \in \bar{N}}$ converges to $h$ at all continuity points of $h$ and that $h$ is right continuous.

In (1) we now let (1°) $m \to \infty$ such that $m \in \bar{N}$, and for continuity points $x$, $x + 2\alpha\beta$, and $x - 2\alpha\beta$, (2°) $\alpha \downarrow 0$ and $\beta \uparrow \infty$ such that $\alpha\beta \downarrow 0$, (3°) $n \to \infty$. We obtain

$$\lim_{n \to \infty} \hat{\mu}_n(x) - h(x) = 0 \tag{3}$$

at all continuity points of $h$ and at $x = -\infty$ and $x = \infty$, where $h(-\infty) = 0$. It could now easily be shown that this means that $\{\mu_n\}$ converges weakly to a measure $\mu$ with the distribution function $h$. However, we shall show this directly for measures on $R^{(k)}$ for any $k$.

Since $h(x) \to 0$ for $|x| \to \infty$, it follows that $\{\mu_n\}$ is tight.

Now for $\mu_n$ on $R^{(k)}$, the Cauchy convergence (2) implies this convergence for the marginal measures $\mu_n^{(i)}$ by Lemma 4.1. Hence, as we have shown above, $\{\mu_n^{(i)}\}$ is tight for any $i$ and

$$\limsup_{n \to \infty} \mu_n\{R^{(k)} \backslash [-a, a]\} \leq \limsup_{n \to \infty} \sum_{i=1}^{k} [\hat{\mu}_n^{(i)}(\infty) - \hat{\mu}_n^{(i)}(a^{(i)}) + \hat{\mu}_n^{(i)}(-a^{(i)})],$$

where the right-hand side is arbitrarily small for sufficiently large $a^{(i)}$, $i = 1, \ldots, k$. Hence, $\{\mu_n\}$ for $\mu_n$ on $R^{(k)}$ is tight. We apply Theorem III.7.2 and conclude that there exists a sequence $\bar{N}$ of positive integers such that $\{\mu_n\}_{n \in \bar{N}}$ converges weakly to a $\sigma$-smooth measure $\mu$ on $R^{(k)}$. It remains to show that (2) for measures on $R^{(k)}$ even implies the weak convergence of $\{\mu_n\}_{n=1}^{\infty}$.

Using the fact that $\Phi[(x - y)/\alpha]$ is uniformly continuous with respect to both $x$ and $y$ and is bounded, and that $\{\mu_n\}$ is tight, we shall show that

$$\lim_{\substack{n \to \infty \\ n \in \bar{N}}} \int_{R^{(k)}} \Phi\left(\frac{x - y}{\alpha}\right) \mu_n(dx) = \int_{R^{(k)}} \Phi\left(\frac{x - y}{\alpha}\right) \mu(dx) \tag{4}$$

uniformly with respect to $x \in R^{(k)}$. The convergence follows by the weak convergence of $\{\mu_n\}_{n \in \bar{N}}$ to $\mu$. According to the tightness of $\{\mu_n\}$, it is sufficient to prove (4), when $R^{(k)}$ is changed into arbitrarily large continuity intervals $[a, b]$, and such an interval can be decomposed into finitely many subintervals $\mathscr{T}_i$. According to the uniform continuity of $\Phi$, we can thus approximate the integrals in (4) uniformly with respect to $x$ and $n$ by sums

$$\sum f(x - g^{(i)})\mu_n(\mathscr{T}_i), \qquad \sum f(x - g^{(i)})\mu(\mathscr{T}_i),$$

where $\mu_n(\mathscr{T}_i)$ tends to $\mu(\mathscr{T}_i)$ as $n \to \infty$, since $\mathscr{T}_i$ is a continuity set for $\mu$. Thus the uniform convergence in (4) follows, and this means that

$$p_\alpha(\mu_n - \mu) \to 0 \qquad (n \to \infty, n \in \bar{N}) \quad (\alpha > 0). \tag{5}$$

Suppose that for some other subsequence $\bar{N}_1$ of positive integers

$$p_\alpha(\mu_n - \mu') \to 0 \qquad (n \to \infty, n \in \bar{N}_1)$$

with a $\sigma$-smooth measure $\mu'$. Then

$$p_\alpha(\mu - \mu') \leq p_\alpha(\mu_n - \mu') + p_\alpha(\mu_n - \mu_m) + p_\alpha(\mu_m - \mu'). \tag{6}$$

Letting $n \to \infty$, $n \in \bar{N}$, $m \to \infty$, $m \in \bar{N}_1$ and using the fact that $\mu_n$ is Cauchy convergent in the class of Gaussian seminorms, we get from (6) that $p_\alpha(\mu - \mu') = 0$ for $\alpha > 0$ and hence $\mu = \mu'$. Thus the Cauchy convergence of $\{\mu_n\}$ in the class of Gaussian seminorms implies the weak convergence of $\mu_n$. We have shown above in (4) that conversely, the weak convergence of $\mu_n$ to $\mu$ implies the convergence

$$p_\alpha(\mu_n - \mu) \to 0 \qquad (n \to \infty) \qquad \text{for} \quad \alpha > 0,$$

which obviously implies the Cauchy convergence

$$p_\alpha(\mu_n - \mu_m) \to 0 \qquad (n \to \infty, m \to \infty) \qquad \text{for} \quad x > 0. \quad \square$$

**Theorem 2.** *A sequence $\{\mu_n\}_{n=1}^{\infty}$ measures in $\mathscr{M}$ on $R^{(k)}$ converges weakly to a measure in $\mathscr{M}$ if and only if the sequences $\{\mu_n(\pi^{-1}(t))\}_{n=1}^{\infty}$ converge weakly for the projections $\pi(t) : \pi(t)x = t \cdot x$, $t$ being any vector in $R^{(k)}$.*

*Proof:* Let $\{\mu_n\}$ converge weakly to $\mu$. Then the sequences $\mu_n(\pi^{-1}(t)\cdot)$ converge weakly to $\mu(\pi^{-1}(t)\cdot)$ since $\pi(t)$ is a continuous mapping.

Conversely, let these sequences converge weakly for all $t$. In particular, the margins of $\mu_n$ converge. We then find as in the proof of Theorem 1 that $\{\mu_n\}$ is tight, and we conclude as in that proof that there exists a subsequence $\bar{N}$ of any sequence of positive integers such that $\{\mu_n\}_{n \in \bar{N}}$ converges weakly. By the same arguments as in the proof of Theorem 1, we conclude that $\mu_n$ converges weakly if the limit measure $\mu$ is the same for any subsequence $\bar{N}$

for which $\{\mu_n\}_{n\in\bar{N}}$ converges weakly. Now by Theorem 3.6, two $\sigma$-smooth measures $\lambda$ and $\mu$ are equal if $\lambda(\pi^{-1}(t)\cdot)$ and $\mu(\pi^{-1}(t)\cdot)$ are identical for all projections $\pi(t)$. □

## 6. STABILITY CONDITIONS FOR CONVOLUTION PRODUCTS THAT CONVERGE WEAKLY

In Section III.9 we introduced the concepts stable and weakly stable sequences.

**Lemma 1.** *Let $\{\mu_{nj}\}_{n=1}^{k_n}$ be sequences of probability measures on $R$, $k_n \uparrow \infty$ $(n \uparrow \infty)$, and suppose that $\{\prod_{i=1}^{k_n *} \mu_{nj}\}$ converges weakly. Then we have*

$$\text{(i)} \qquad \limsup_{n\to\infty} \sum_{j=1}^{k_n} \int_{|y|\geq a} \mu_{nj}(a_{nj}+dy) \begin{cases} < \infty & \text{for} \quad a>0, \\ \to 0 & \text{for} \quad a\to\infty, \end{cases}$$

$$\text{(ii)} \qquad \limsup_{n\to\infty} \sum_{j=1}^{k_n} \left| \int_{|y|<a} y\, \mu_{nj}(a_{nj}+dy) \right| < \infty,$$

$$\text{(iii)} \qquad \limsup_{n\to\infty} \sum_{j=1}^{k_n} \int_{|y|<a} y^2\, \mu_{nj}(a_{nj}+dy) < \infty,$$

*where the $a_{nj}$ are suitable constants. If the $\mu_{nj}$ are symmetrical,* (i)–(iii) *hold with $a_{nj}=0$. When* (i)–(iii) *hold, the sequence $\{\prod_{j=1}^{k_n *} \mu_{nj}\}$ is stable with respect to the Gaussian seminorms.*

*Proof:* First let the $\mu_{nj}$ be symmetrical, and consider the identity

$$e - \prod_{j=1}^{k_n *} \mu_{nj} = \sum_{j=1}^{k_n} \left( \prod_{i=0}^{j-1 *} \mu_{nj} \right) * (e-\mu_{nj}) = \sum_{j=1}^{k_n} \nu_{nj} * (e-\mu_{nj}) \tag{1}$$

with

$$\nu_{nj} = \prod_{j=0}^{j-1 *} \mu_{nj}, \qquad \nu_{n0} = e.$$

We form the Gaussian transform of this identity. Putting

$$h_{nj}(\alpha, x) = \int_{-\infty}^{\infty} \Phi\left(\frac{x-y}{\alpha}\right) \nu_{nj}(dy),$$

we obtain from (1)

$$2g_n(\alpha, x) = \sum_{j=1}^{k_n} \int_{-\infty}^{\infty} h_{nj}(\alpha, x-z)[e(dz) - \mu_{nj}(dz)]. \tag{2}$$

According to the symmetry, we can write this identity

$$2g_n(\alpha, x) = \sum_{j=1}^{k_n} \int_0^\infty [2h_{nj}(\alpha, x) - h_{nj}(\alpha, x - z) - h_{nj}(\alpha, x + z)]\, \mu_{nj}(dz). \quad (3)$$

By the definition of $h_{nj}$, we have

$$2h_{nj}(\alpha, x) - h_{nj}(\alpha, x - z) - h_{nj}(\alpha, x + z) = \int_{-\infty}^{\infty} \Psi\left(\frac{x-y}{\alpha}, \frac{z}{\alpha}\right) v_{nj}(dy), \quad (4)$$

where

$$\Psi(x, z) = 2\Phi(x) - \Phi(x - z) - \Phi(x + z)$$

is the function considered in (4.4).

Now we put $x = 2r\alpha$ and determine $r$ and $\alpha$ as suitable positive numbers. Put

$$\lambda_{nj} = \prod_{i=j+1}^{k_{n*}} \mu_{nj}, \qquad \mu_{n0} = e, \qquad \lambda_{n0} = \lambda_n.$$

Then $\lambda_n = \lambda_{nj} * v_{nj}$ for $j = 1, 2, \ldots$ . By assumption, $\lambda_n$ converges weakly to a probability measure $\lambda$. We choose $\alpha r$ so that $\{x : x \leq \alpha r\}$ is a continuity set for $\lambda$, and furthermore, so large that for given $\varepsilon > 0$,

$$\lambda[x : x \leq \alpha r] \geq 1 - \varepsilon/4.$$

Since $\lambda_n[x : x \leq \alpha r]$ converges to $\lambda[x : x \leq \alpha r]$, we may choose $n_0$ so large that $\lambda_n[x : x \leq \alpha r] \geq 1 - \varepsilon/2$ for $n > n_0$, and then

$$\frac{\varepsilon}{2} \geq \int_{x+y \geq \alpha r} \lambda_{nj}(dx)\, v_{nj}(dy) \geq \int_{x \geq 0} \lambda_{nj}(dx) \int_{y \geq \alpha r} v_{nj}(dy) \geq \frac{1}{2} \int_{y \geq \alpha r} v_{nj}(dy).$$

Hence

$$v_{nj}[x : x \leq \alpha r] \geq 1 - \varepsilon, \quad (5)$$

and by the definition of $h_{nj}(\alpha, x)$, we get

$$h_{nj}(\alpha, 2r\alpha) \geq \Phi(r) \cdot \int_{y \leq \alpha r} v_{nj}(dy) \geq (1 - \varepsilon)\Phi(r).$$

For $z \geq 2r\alpha$, $\Phi(r) \geq 1 - \varepsilon$, observing that

$$\begin{aligned} h_{nj}(\alpha, x) &\leq \tfrac{1}{2} \qquad \text{for } x \leq 0, \\ &\leq 1 \qquad \text{for all } \quad x, \end{aligned}$$

we obtain

$$2h_{nj}(\alpha, 2\alpha r) - h_{nj}(\alpha, 2r\alpha - z) - h_{nj}(\alpha, 2r\alpha + z) \geq 2(1 - \varepsilon)^2 - \tfrac{3}{2} = c_1(\varepsilon) \quad (6)$$

and $c_1(\varepsilon) > 0$ provided that $\varepsilon < 1 - \sqrt{3}/2$. Next we estimate the left-hand side of (6) for $z \leq 2r\alpha$, and then we use the integral representation (4). By (4.8) and (4.9), we have

$$\begin{aligned} \Psi(x, z) &\leq z^2|\Phi''(1)| && \text{for all} \quad x \text{ and } z, \\ \Psi(x, z) &\geq z^2|\Phi''(x + |z|)| && \text{for} \quad |z| < x - 1. \end{aligned} \tag{7}$$

Hence

$$\Psi\left(\frac{x}{\alpha}, \frac{z}{\alpha}\right) \geq \frac{z^2}{\alpha^2}\left|\Phi''\left(\frac{7r}{2}\right)\right| \qquad \text{for} \quad \alpha r \leq x \leq 3\alpha r, \quad 0 \leq z \leq \alpha r/2, \quad r \geq 2.$$

Furthermore, in the proof of (4.9) we observed that $\Psi(x, z)$ for $x > 0$ increases as $z$ passes from 0 to $+\infty$. Thus

$$\Psi\left(\frac{x}{\alpha}, \frac{z}{\alpha}\right) \geq \frac{z^2}{16\alpha^2}\left|\Phi''\left(\frac{7r}{2}\right)\right| \qquad \text{for} \quad \alpha r \leq x \leq 3\alpha r, \quad 0 \leq z \leq 2\alpha r, \quad r \geq 2. \tag{8}$$

Regarding (6), (7), and (8), we get for $0 \leq z \leq 2r\alpha$

$$\begin{aligned} \int_{-\infty}^{+\infty} \Psi\left(\frac{2r\alpha - y}{\alpha}, \frac{z}{\alpha}\right) v_{nj}(dy) &\geq \frac{z^2}{16\alpha^2}\left|\Phi''\left(\frac{7r}{2}\right)\right| \int_{|y| < r\alpha} v_{nj}(dy) \\ &\quad - \frac{z^2}{\alpha^2}|\Phi''(1)| \int_{|y| \geq r\alpha} v_{nj}(dy) \geq c_2(r, \varepsilon) \cdot \frac{z^2}{\alpha^2}, \end{aligned}$$

with

$$c_2(r, \varepsilon) = \tfrac{1}{16}|\Phi''(zr/2)(1 - 2\varepsilon) - 2\varepsilon|\Phi''(1)||.$$

Hence

$$h_{nj}(\alpha, 2r\alpha) - h_{nj}(\alpha, 2r\alpha - z) - h_{nj}(\alpha, 2r\alpha + z) \geq c_2(r, \varepsilon)(z^2/\alpha^2) \tag{9}$$

for $0 \leq z \leq 2r\alpha$. Clearly, we may choose $\varepsilon$ so small that $c_2(r, \varepsilon) > 0$. Now combining (3), (6), and (9) and observing that $g_n(\alpha, x) \leq 1$ for all $n$, $\alpha$, and $x$, we get the relations in the lemma. The fact that the sum in (i) tends to 0 as $a$ tends to $+\infty$ follows since $g_n(\alpha, x)$ tends uniformly on $-\infty < x + \infty$ to

$$\int \Phi\left(\frac{x - z}{\alpha}\right)[e(dz) - \lambda(dz)]$$

as $n \to \infty$ (compare IV.5.4). The last integral is arbitrarily small for sufficiently large $x$.

Let now $\mu_{nj}$ be arbitrary. If $\{\prod_{j=1}^{k_n *} \mu_{nj}\}$ converges weakly, this is also true if we change $\mu_{nj}$ to $\mu_{nj} * \bar{\mu}_{nj}$, where $\bar{\mu}_{nj}$ is the conjugate of $\mu_{nj}$ $[\bar{\mu}_{nj}(E) = \mu_{nj}(-E)$ for any Borel set $E]$. Hence (i)–(iii) in the lemma hold if we change

$\mu_{nj}(a_{nj} + \cdot)$ to the symmetrical measure $\mu_{nj} * \bar{\mu}_{nj}$. Using this fact and also the relation

$$\mu_{nj} * \bar{\mu}_{nj} = \mu_{nj}(a + \cdot) * \overline{\mu_{nj}(a + \cdot)} \tag{10}$$

for any real number $a$, we shall determine constants $a_{nj}$ successively so that (i)–(iii) hold.

To begin with, we choose $a_{nj}$ such that

$$\int_{y<0} \mu_{nj}(a_{nj} + dy) \le \tfrac{1}{2}, \qquad \int_{y>0} \mu_{nj}(a_{nj} + dy) \le \tfrac{1}{2}. \tag{11}$$

Changing the notation, we may then assume that this relation is already satisfied for $a_{nj} = 0$. Then we have

$$\begin{aligned} \int_{y \ge \beta/2} \mu_{nj} * \bar{\mu}_{nj}(dy) &\ge \int_{u \ge \beta} \int_{v \ge -\beta/2} \mu_{nj}(du)\, \bar{\mu}_{nj}(dv) \\ &\ge \tfrac{1}{2} \int_{y \ge \beta} \mu_{nj}(du). \end{aligned} \tag{12}$$

Hence the $\mu_{nj}$ satisfy (i) in the lemma. □

Put

$$m_{nj} = \int_{|y| < a_0} y\, \mu_{nj}(dy) \tag{13}$$

and choose $a$ so large that $a > 2a_0$. With $\nu_{nj} = \mu_{nj}(m_{nj} + \cdot)$, we get, since $|m_{nj}| < a_0$,

$$\int_{|y| \ge a} \nu_{nj}(dy) \le \int_{|y| \ge a/2} \mu_{nj}(dy).$$

Hence, the $\nu_{nj}$ satisfy (i) in the lemma for this $a$. Furthermore,

$$\begin{aligned} \left| \int_{|y|<a} y\, \nu_{nj}(dy) \right| &= \left| \int_{|z - m_{nj}| < a} (z - m_{nj})\, \mu_{nj}(dz) \right| \\ &\le \left| \int_{|z| < a_0} z\, \mu_{nj}(dz) - m_{nj} \right| \\ &\quad + \frac{3a}{2} \int_{|z| \ge a_0} \mu_{nj}(dz) + \frac{a}{2} \int_{|z| \ge a/2} \mu_{nj}(dz). \end{aligned} \tag{14}$$

Observing the relation (13) and the fact that the $\mu_{nj}$ satisfy (i) in the lemma with $a_{nj} = 0$ for any $a > 0$, we find by (14) that both (i) and (ii) are satisfied for $a > 2a_0$ with $a_{nj} = m_{nj}$. Moreover,

$$\sup_n \sum_{j=1}^{k_n} \left| \int_{|y|<a} y\, \nu_{nj}(dy) \right| < \infty. \tag{15}$$

Now

$$\nu_{nj} * \bar{\nu}_{nj} = \mu_{nj} * \bar{\mu}_{nj}$$

is symmetrical, and

$$\int_{|y| \leq 2a} y^2 \, \nu_{nj} * \bar{\nu}_{nj}(dy) \geq \int_{|u|<a} \int_{|v|<a} (u+v)^2 \, \nu_{nj}(du) \, \bar{\nu}_{nj}(dv)$$
$$= 2 \int_{|u|<a} u^2 \, \nu_{nj}(du) - 2 \left[ \int_{|u|<a} u \, \nu_{nj}(du) \right]^2,$$

where

$$\left[ \int_{|u|<a} u \, \nu_{nj}(du) \right]^2 \leq a \left| \int_{|u|<a} u \, \nu_{nj}(du) \right|.$$

Hence regarding (15) and using the fact that $\nu_{nj} * \bar{\nu}_{nj}$ being symmetric, satisfies (iii), we conclude that $\nu_{nj}$ satisfies (iii) for this $a$. But if (i)–(iii) hold for some $a > 0$, they clearly hold for any $a > 0$. □

## 7. THE UNIQUE DIVISIBILITY OF INFINITELY DIVISIBLE $\sigma$-SMOOTH MEASURES

The concept infinitely divisible element in a space was introduced in Section III.10. In accordance with this definition, a $\sigma$-smooth measure $\mu$ on $R^{(k)}$ is infinitely divisible if for any positive integer $n$ there exists a $\sigma$-smooth measure $\mu_n$ such that $\mu = \mu_n^{*n}$. An example of an infinitely divisible probability measure is the Gaussian measure. Indeed, it follows from (1.2), which also holds in $R^{(k)}$, that

$$[\Phi(\sqrt{n})]^{*n} = \Phi(\cdot).$$

The general Gaussian measure is also infinitely divisible. We find this easily by the help of the characteristic function as it is presented in Theorem 4.6, since by Theorem 4.1 the characteristic function of the convolution of $\sigma$-smooth measures is the product of the convolution factors, and since a $\sigma$-smooth measure is uniquely determined by its characteristic function. We now state

**Theorem 1.** *A $\sigma$-smooth infinitely divisible measure on $R^{(k)}$ is uniquely infinitely divisible.*

*Proof:* First we note that if $\mu$ is a $\sigma$-smooth measure on $R^{(k)}$ and $\mu = \mu_n^{*n}$ with a $\sigma$-smooth measure $\mu_n$ for any $n$, then $\mu = a\nu$, $\mu_n = \sqrt[n]{a}\,\nu_n$ with probability measures $\nu$ and $\nu_n$. Hence it is sufficient to prove the theorem for probability measures. Furthermore, if $\mu = \mu_n^{*n}$ on $R^{(k)}$, then for any projection $\pi(t)$ of $R^{(k)}$ on $R$ determined by $\pi(t)x = t \cdot x$ we have $\mu(\pi(t)^{-1}\cdot) = [\mu_n(\pi(t)^{-1}\cdot)]^{*n}$, according to the properties of convolutions, and a measure on $R^{(k)}$ is uniquely determined by its projections. Hence it is sufficient to prove the theorem for infinitely divisible probability measures on $R$. Thus consider probability measures $\mu$ and $\mu_n$ on $R$.

By Theorem 3.1, the characteristic functions $\tilde{\mu}$ of $\mu$ and $\tilde{\mu}_n$ of $\mu_n$ satisfy the relation $(\tilde{\mu}_n)^n = \tilde{\mu}$ if $\mu = \mu_n^{*n}$. Hence for any $t \in (-\infty, +\infty)$,

$$\tilde{\mu}_n(t) = |\tilde{\mu}(t)|^{1/n} \exp\{(i/n)[\arg \tilde{\mu}(t) + 2m(t)\pi]\}$$

with an integer $m(t)$. If follows from the definition of the characteristic function that it is a continuous function. Thus as $t$ tends from $-\infty$ to $+\infty$, $m(t)$ can change from one integer value to another integer value only at a point $t$, where $|\tilde{\mu}(t)| = 0$. We shall show that $|\tilde{\mu}(t)| > 0$ for any real number $t$. Then it follows that $m(t)$ is uniquely determined by its value at some point, say $t = 0$, and $\tilde{\mu}_n$ is uniquely determined by $\tilde{\mu}$. Consequently, $\mu_n$ is uniquely determined by $\mu$. It remains to prove that $|\tilde{\mu}(t)| > 0$. Now if $\mu = \mu_n^{*n}$, we have also $\bar{\mu} = (\bar{\mu}_n)^{*n}$, where $\bar{\mu}$ is the conjugate of $\mu$, and then $\mu * \bar{\mu} = (\mu_n * \bar{\mu}_n)^{*n}$, where $\mu * \bar{\mu}$ and $\mu_n * \bar{\mu}_n$ are symmetric. Hence it is sufficient to show that $|\tilde{\mu}(t)| > 0$ for a symmetric, infinitely divisible probability measure $\mu$ and $\mu = \mu_n^{*n}$ with symmetric $\mu_n$. Now by Lemma 6.1, this relation implies the inequality

$$\limsup_n n \int_{|x| \geq \varepsilon} \mu_n(dx) < +\infty \tag{1}$$

for any $\varepsilon > 0$. For a symmetric $\mu_n$, we get

$$\tilde{\mu}_n(t) = \int \exp(it \cdot x)\, \mu_n(dx) = \int \cos(t \cdot x)\, \mu_n(dx).$$

It follows by (1) that $\int_{|x| \geq \varepsilon} \mu_n(dx)$ is arbitrarily small for sufficiently large $n$, given any $\varepsilon > 0$. But $\int \mu_n(dx) = 1$. Thus $\tilde{\mu}_n(t)$ is arbitrarily close to 1 for sufficiently large $n$. Hence

$$|\tilde{\mu}(t)| = |\tilde{\mu}_n(t)|^n > 0. \quad \square$$

## 8. LÉVY MEASURES ON $R^{(k)}$; GAUSSIAN FUNCTIONALS

The weak convergence of convolution products

$$\prod_{j=1}^{k_{n*}} \mu_{nj}$$

of $\sigma$-smooth measures depends on the behavior of the measure

$$\lambda_n = \sum_{j=1}^{k_n} \mu_{nj}$$

on $R^{(k)}\backslash\{0\}$ (0 denoting the zero point) and on certain moments of $\lambda_n$ as $n \to \infty$. It turns out that the limit of $\lambda_n$ on $R^{(k)}\backslash\{0\}$ must exist and that this limit is a $\sigma$-smooth and $\sigma$-finite measure $\lambda$ on $R^{(k)}\backslash\{0\}$ such that

$$\int_{\|y\| \geq \beta} \lambda(dy) \begin{cases} < \infty & \text{for} \quad \beta > 0, \\ \to 0 & \text{for} \quad \beta \to \infty, \end{cases} \tag{1}$$

$$\int_{0 < \|y\| < \beta} \|y\|^2 \, \lambda(dy) < \infty. \tag{2}$$

We are going to show this in the following sections. A $\sigma$-smooth, $\sigma$-finite measure on $R^{(k)}\backslash\{0\}$ is called a Lévy measure when it satisfies conditions (1) and (2).

**Lemma 1.** *Let $\{\lambda_n\}$ be a sequence of $\sigma$-smooth measures on $R^{(k)}$ such that*

(i) $$\limsup_{n\to\infty} \int_{\|y\| \geq \beta} \lambda_n(dy) \begin{cases} < \infty & \text{for} \quad \beta > 0, \\ \to 0 & \text{as} \quad \beta \to \infty, \end{cases}$$

(ii) $$\limsup_{n\to\infty} \int_{\|y\| < \infty} \|y\|^2 \, \lambda_n(dy) < \infty.$$

*Then there exists a sequence $\hat{N}$ of positive integers such that $\{\lambda_n\}$ converges weakly to a Lévy measure $\lambda$ on $R^{(k)}\backslash\{0\}$ as $n \to \infty$, $n \in \hat{N}$.*

*Remark:* Note the definition of weak convergence to $\sigma$-finite measures.

**Corollary.** *If the sequence $\{\lambda_n\}$ of $\sigma$-smooth, $\sigma$-finite measures on $R^{(k)}\backslash\{0\}$ satisfies the conditions* (i) *and* (ii) *in the lemma, and furthermore*

(iii) $$\limsup_{n\to\infty} \left| \int_{\|y\| < \beta} y^{(i)} \, \lambda_n(dy) \right| < \infty \qquad \text{for} \quad i = 1, \ldots, k,$$

*then the sequence $\hat{N}$ may be chosen such that, as $n \to \infty$, $n \in \hat{N}$,*

(1°) *$\lambda_n$ converges weakly to a Lévy measure $\lambda$ on $R^{(k)}\backslash 0$,*

(2°) $$\lim_{n\to\infty} \int_{R^{(k)}} \frac{y^{(i)}}{1 + \|y\|^2} \, \lambda_n(dy) = m^{(i)},$$

$$|m^{(i)}| < \infty \qquad \textit{for} \quad i = 1, \ldots, k,$$

(3°) $$\lim_{\varepsilon \downarrow 0} \overline{\lim_{n\to\infty}} \int_{0 < \|y\| < \varepsilon} (y \cdot t)^2 \, \lambda_n(dy) = \sigma^2(t) < \infty \qquad \textit{for any } t \in R^{(k)}.$$

*Proof of Lemma 1:* For $\beta_0 > 0$, define $\lambda_n^{(0)}$ on $R^{(k)}$ by

$$\lambda_n^{(0)}(E) = \lambda_n(E \cap F_{\beta_0}), \qquad F_\beta = \{y:\|y\| \geq \beta\}. \tag{3}$$

Then

$$\sup_n \lambda_n^{(0)}(R^{(k)}) < \infty; \tag{4}$$

$\sup_n \lambda_n^{(0)}(F_\beta) \to 0\,(\beta \to \infty)$, i.e., $\{\lambda_n^{(0)}\}$, is tight on $R^{(k)}$. It follows by these relations and Theorem III.7.1 that $\{\lambda_n^{(0)}\}$ is weakly compact. Hence there exists a sequence $\hat{N}$ such that $\{\lambda_n^{(0)}\}$ converges weakly to a measure $\lambda^{(0)}$ which is $\sigma$-smooth according to (i). We may choose a sequence $\{\beta_i\}$, $\beta_{i-1} \supset \beta_i$, of positive numbers tending to 0 as $i \to \infty$, such that $\{\lambda_n^{(i)}\}$, defined by

$$\lambda_n^{(i)} = \lambda_n(E \cap F_{\beta_i})$$

for any Borel set $E$ on $R^{(k)}$, converges weakly to a measure $\lambda^{(i)}$ as $n \to \infty$, $n \in \hat{N}$ for a suitable sequence $\hat{N}$ of positive integers, and we may choose the $\beta_i$ such that $\lambda^{(i-1)}$ coincides with $\lambda^{(i)}$ on $F_{\beta_{i-1}}$. These $\lambda^{(i)}$ determine a $\sigma$-finite, $\sigma$-smooth measure $\lambda$ on $R^{(k)}\backslash\{0\}$. It follows by (i) and (ii) that $\lambda$ is a Lévy measure. □

*Proof of the corollary:* We must show that the sequence $\hat{N}$ may be chosen such that not only (1°), but also (2°) and (3°), hold. In (3°) we may let $\varepsilon \downarrow 0$ through a sequence $\varepsilon_j$ tending to 0. Then it follows that $\hat{N}$ may be chosen such that (3°) is satisfied. Indeed, we need only consider finitely many vectors $t$ in (3°) since the limit is a quadratic form. Clearly,

$$\sup_n \left| \int_{R^{(k)}} \frac{y^{(i)}}{1 + \|y\|^2} \lambda_n(dy) \right|$$

is bounded according to (ii) and (iii) since $\lambda_n$ is bounded on $\{y:\|y\| \geq \beta\}$, and

$$\left| \frac{y^{(i)}}{1 + \|y\|^2} - y^{(i)} \right| \leq \frac{|y^{(i)}|\,\|y\|^2}{1 + \|y\|^2} < \beta\|y\|^2$$

on $\{y:\|y\| < \beta\}$. Hence we may choose the sequence $\hat{N}$ such that (1°)–(3°) hold. □

**Lemma 2.** *Let $\{\lambda_n\}$ be a sequence of $\sigma$-smooth measures on $R^{(k)}$ such that*

(i) *$\lambda_n$ converges weakly to a Lévy measure $\lambda$ on $R^{(k)}\backslash 0$,*

(ii) $\displaystyle\lim_{n\to\infty} \int_{R^{(k)}} \frac{y^{(i)}}{1 + \|y\|^2} \lambda_n(dy) = m^{(i)}, \quad |m^{(i)}| < \infty \quad$ *for* $i = 1, \ldots, k$,

(iii) $$\lim_{\varepsilon\downarrow 0} \lim_{n\to\infty} \int_{\|y\|<\varepsilon} (t \cdot y)^2 \lambda_n(dy) = \sigma^2(t) = \sum_{i=1}^{k} \sum_{j=1}^{k} q_{ij} t_i t_j < \infty$$

*for* $t \in R^{(k)}$. *Then*

$$\Gamma(\alpha, x) = \lim_{n\to\infty} \int_{R^{(k)}} \left[\Phi\left(\frac{x-y}{\alpha}\right) - \Phi\left(\frac{x}{\alpha}\right)\right] \lambda_n(dy)$$

$$= -\sum_{i=1}^{k} m^{(i)} \frac{\partial}{\partial x^{(i)}} \Phi\left(\frac{x}{\sigma}\right) + \frac{1}{2}\sum_{i=1}^{k}\sum_{j=1}^{k} q_{ij} \frac{\partial^2}{\partial x^{(i)}\partial x^{(j)}} \Phi\left(\frac{x}{\alpha}\right)$$

$$+ \int_{R^{(k)}\setminus\{0\}} \left[\Phi\left(\frac{x-y}{\alpha}\right) - \Phi\left(\frac{x}{\alpha}\right) + \sum_{i=1}^{k} \frac{y^{(i)}}{1+\|y\|^2} \frac{\partial}{\partial x^{(i)}} \Phi\left(\frac{x}{\alpha}\right)\right] \lambda(dy).$$

*The convergence is uniform for* $x \in R^{(k)}$. *Furthermore,* $\lambda$, *the* $m^{(i)}$, *and the* $q_{ij}$ *are uniquely determined by* $\Gamma(\alpha, x)$ $(\alpha > 0, x \in R^{(k)})$. *The integral is uniformly convergent for* $x \in R^{(k)}$ *and given* $\alpha$.

*Remark 1:* We call $\Gamma(\alpha, x)$ a Gaussian functional. Note that $\Phi$ here is the Gaussian measure on $R^{(k)}$ with the density function $(2\pi)^{-k/2}\exp(-\frac{1}{2}\|y\|^2)$.

*Remark 2:* We may change $y^{(i)}/(1+\|y\|^2)$ in the integral to $y^{(i)}/[1+(y^{(i)})^2]$ if we at the same time change $m^{(i)}$ to (ii)

$$\lim_{n\to\infty} \int_{R^{(k)}} \frac{y^{(i)}}{1+(y^{(i)})^2} \lambda_n(dy).$$

*Proof:* We may write

$$\int_{R^{(k)}\setminus\{0\}} \left[\Phi\left(\frac{x-y}{\alpha}\right) - \Phi\left(\frac{x}{\alpha}\right)\right] \lambda_n(dy) = -\sum_{i=1}^{k} \frac{\partial}{\partial x^{(i)}} \Phi\left(\frac{x}{\alpha}\right) \int_{R^{(k)}} \frac{y^{(i)}}{1+\|y\|^2} \lambda_n(dy)$$

$$+ \int_{R^{(k)}\setminus\{0\}} \left[\Phi\left(\frac{x-y}{\alpha}\right) - \Phi\left(\frac{x}{\alpha}\right) + \sum_{i=1}^{k} \frac{y^{(i)}}{1+\|y\|^2} \frac{\partial}{\partial x^{(i)}} \Phi\left(\frac{x}{\alpha}\right)\right] \lambda_n(dy). \tag{5}$$

For any $\varepsilon > 0$ such that $\{y: \|y\| \geq \varepsilon\}$ is a continuity set for $\lambda$, the integral on the right-hand side of (5), considered on $\{y: \|y\| \geq \varepsilon\}$, converges to the corresponding integral for $\lambda$, since $\lambda_n$ converges weakly to $\lambda$. Hence, we have only to consider the integral on $\{y: \|y\| < \varepsilon\}$ for arbitrarily small $\varepsilon$. By the Taylor expansion, we get for $\|y\| < \varepsilon$,

$$\left|\Phi\left(\frac{x-y}{\alpha}\right) - \Phi\left(\frac{x}{\alpha}\right) + \sum_{i=1}^{k} \frac{y^{(i)}}{1+\|y\|^2} \frac{\partial}{\partial x^{(i)}} \Phi\left(\frac{x}{\alpha}\right) - \sum_{i=1}^{k}\sum_{j=1}^{k} y^{(i)}y^{(j)} \frac{\partial}{\partial x^{(i)}\partial x^{(j)}} \Phi\left(\frac{x}{\alpha}\right)\right| \leq \frac{c}{\alpha^3}\varepsilon\|y\|^2$$

with a constant $c$ (not dependent on $x$ and $\varepsilon$). Since

$$\limsup_{n\to\infty} \int_{\|y\|\le\varepsilon} \|y\|^2 \lambda_n(dy)$$

is finite, we find that $\Gamma(\alpha, x)$ has the given form.

We prove that $\lambda$, the $m^{(i)}$, and the $q_{ij}$ are uniquely determined by $\Gamma(\alpha, x)$. Then for some coordinate $x$ such that $x^{(i)} \ne 0$ for some $(i)$, let $\alpha \downarrow 0$. It follows that $\Gamma(\alpha, x)$ reduces to

$$\int_{R^{(k)}} [e(x-y) - e(x)]\, \lambda(dy) = \int_{y\le x} \lambda(dy) \tag{6}$$

if some coordinate $x^{(i)} < 0$ and $\{y : y \le x\}$ is a continuity set for $\lambda$, and

$$\int_{R^{(k)}} [e(x-y) - e(x)]\, \lambda(dy) = -\int_{y\ge x} \lambda(dy) \tag{7}$$

if all coordinates $x^{(i)} > 0$ and the set $\{y : y \ge x\}$ is a continuity set for $\lambda$. Clearly, $\lambda$ is uniquely determined by (6) and (7). Having proved this, we have only to show that the relation

$$\begin{aligned} &-\sum_{i=1}^{k} \tilde{m}^{(i)} \frac{\partial}{\partial x^{(i)}} \Phi\left(\frac{x}{\alpha}\right) + \frac{1}{2}\sum_{i=1}^{k}\sum_{j=1}^{k} \tilde{q}_{ij} \frac{\partial^2}{\partial x^{(i)}\partial x^{(j)}} \Phi\left(\frac{x}{\alpha}\right) \\ &\quad = -\sum_{i=1}^{k} m^{(i)} \frac{\partial}{\partial x^{(i)}} \Phi\left(\frac{x}{\alpha}\right) + \frac{1}{2}\sum_{i=1}^{k}\sum_{j=1}^{k} q_{ij} \frac{\partial^2}{\partial x^{(i)}\partial x^{(j)}} \Phi\left(\frac{x}{\alpha}\right) \end{aligned} \tag{8}$$

implies $\tilde{m}^{(i)} = m^{(i)}$, $\tilde{q}_{ij} = q_{ij}$ for all $i$ and $j$. Let $x^{(i)} \to \infty$, $x^{(j)} \to \infty$ for all $i$ and $j$ except $i = i_0, j = j_0$; multiply by $\alpha^2$; and let $\alpha \to 0$. Then we get $q_{ij} = \hat{q}_{ij}$ for $i = i_0, j = j_0$. In the same way we then find that $\tilde{m}_i = m_i$. □

## 9. WEAK CONVERGENCE OF CONVOLUTION POWERS OF $\sigma$-SMOOTH MEASURES

In Theorem III.10.1 we gave a general theorem on convergence of powers of elements in a semigroup under a class of seminorms. We can apply this theorem to a sequence of convolution powers of $\sigma$-smooth measures under the class of Gaussian seminorms. Indeed, the semigroup $\mathscr{M}$ of $\sigma$-smooth measures different from the zero measure fulfills the requirements in Theorem III.10.1, and the infinitely divisible measures in $\mathscr{M}$ are uniquely infinitely divisible according to Theorem 7.1.

**Theorem 1.**

(i) *A sequence $\{\mu_n^{*k_n}\}$ of convolution powers of $\sigma$-smooth measures on $R^{(k)}$, $k_n \uparrow \infty$ as $n \uparrow \infty$, converges weakly to a measure $\mu$ if and only if $\{k_n[\mu_n - e]\}$ is Cauchy convergent in the class of Gaussian seminorms.*

(ii) *The limit* $\nu$ *is a* $\sigma$*-smooth, infinitely divisible measure, and the weak convergence of* $\{\mu_n^{*k_n}\}$ *to* $\nu$ *implies the weak convergence of* $\{\mu_n^{*m_n(s)}\}$ *to* $\nu_s$ *with* $\nu_s^{*s} = \nu$ *for* $s = 1, 2, \ldots$ *where* $m_n(s) = [k_n/s]$.

(iii) *The Cauchy convergence of* $\{k_n[\mu_n - e]\}$ *in the class of Gaussian seminorms holds if and only if* $k_n\Phi(\cdot/\alpha) * [\mu_n - e]$ *converges uniformly on* $R^{(k)}$ *to the Gaussian functional* $\Gamma(\alpha, \nu, x)$ *with the representation*

$$\Gamma(\alpha, \nu, x) = c\Phi\left(\frac{x}{\alpha}\right) - \sum_{i=1}^{k} m^{(i)} \frac{\partial}{\partial x^{(i)}} \Phi\left(\frac{x}{\alpha}\right)$$

$$+ \frac{1}{2} \sum_{i=1}^{k} \sum_{j=1}^{k} q_{ij} \frac{\partial^2}{\partial x^{(i)} \partial x^{(j)}} \Phi\left(\frac{x}{\alpha}\right) + \int_{R^{(k)}\setminus\{0\}} \left[\Phi\left(\frac{x-y}{\alpha}\right) - \Phi\left(\frac{x}{\alpha}\right)\right.$$

$$\left. + \sum_{i=1}^{k} \frac{y^{(i)}}{1 + \|y\|^2} \frac{\partial}{\partial x^{(i)}} \Phi\left(\frac{x}{\alpha}\right)\right] \lambda(dy),$$

*where* $c$, $\lambda$, *the* $m^{(i)}$, *and the* $q_{ij}$ *are determined by the limits*

(1°) $$c = \lim_{n \to \infty} k_n[\mu_n(R^{(k)}) - 1],$$

(2°) $\lambda$ *is a Lévy measure and the weak limit of* $k_n\mu_n$ *on* $R^{(k)}\setminus\{0\}$, *as* $n \to \infty$,

(3°) $$m^{(i)} = \lim_{n \to \infty} k_n \int_{R^{(k)}} \frac{y^{(i)}}{1 + \|y\|^2} \mu_n(dy),$$

(4°) $$\lim_{\varepsilon \downarrow 0} \overline{\lim_{n \to \infty}} k_n \int_{\|y\| < \varepsilon} (t \cdot y)^2 \mu_n(dy) = \sum_{i=1}^{k} \sum_{j=1}^{k} q_{ij} t_i t_j = q(t)$$

*for any* $t \in R^{(k)}$, *i.e., this double sum is a nonnegative symmetrical quadratic form.*

*Note that these limits exist and are finite when* $k_n\Phi(\cdot/\alpha) * [\mu_n - e]$ *converges uniformly on* $R^{(k)}$ *for any* $\alpha > 0$.

*Remark 1:* We call $\Gamma(\alpha, \nu, \cdot)$ the Gaussian functional of the infinitely divisible measure $\nu$.

*Remark 2:* $m^{(i)}$ is uniquely determined by (2°) and the limit

$$\lim_{n \to \infty} k_n \int_{\|y\| < \beta} y^{(i)} \mu_n(dy)$$

and also by (2°) and the limit

$$\lim_{n \to \infty} k_n \int_{|y^{(i)}| < \beta} y^{(i)}(dy)$$

for any $\beta$ such that $\{y : \|y\| \geq \beta\}$, respectively, and $\{y : |y^{(i)}| \geq \beta\}$ are continuity sets for $\lambda$.

**Corollary.** *The sequence $\{\mu_n^{*k_n}\}$ in the theorem converges weakly if and only if $k_n\{(\mu_n[(\pi(t))^{-1}\cdot] - e)\}$ is Cauchy convergent in the class of Gaussian seminorms for all projections $\pi(t):\pi(t)x = t \cdot x$, and hence if and only if for any $t \in R^{(k)}$,*

$$\lim_{n\to\infty} k_n\left[\int_{R^{(k)}} \Phi\left(\frac{x\cdot t - y\cdot t}{\alpha}\right)\mu_n(dx) - \Phi\left(\frac{x\cdot t}{\alpha}\right)\right]$$

$$= \bar{\Gamma}(\alpha, v, x\cdot t)$$

$$= c\Phi\left(\frac{x\cdot t}{\alpha}\right) - \frac{\bar{m}\cdot t}{\alpha}\Phi'\left(\frac{x\cdot t}{\alpha}\right) + \frac{1}{2}q(t)\cdot\frac{1}{\alpha^2}\Phi''\left(\frac{t\cdot x}{\alpha}\right)$$

$$+ \int_{R^{(k)}}\left[\Phi\left(\frac{x\cdot t - y\cdot t}{\alpha}\right) - \Phi\left(\frac{x\cdot t}{\alpha}\right) + \frac{y\cdot t}{\alpha(1+\|y\|^2)}\Phi'\left(\frac{x\cdot t}{\alpha}\right)\right]\lambda(dy)$$

*uniformly for $x \in R^{(k)}$ with $c$, $\bar{m} = (m^{(1)}\cdots m^{(k)})$ and $q(t)$ as in the theorem ($\Phi$ is now the normalized Gaussian distribution on $R$).*

*Remark:* We call $\bar{\Gamma}(\alpha, v, x\cdot t)$ the projected Gaussian functional of $v$.

*Proof:* Let $k_n[\mu_n - e]$ be Cauchy convergent in the class of Gaussian seminorms. Then $k_n p_\alpha(\mu_n - e)$ is bounded,

$$\sup_n \quad k_n p_\alpha(\mu_n - e) < \infty, \tag{1}$$

and $\{\mu_n^{*k_n}\}$ is stable. It follows by Theorem III.10.1 that the sequence of these convolution powers converges weakly to a $\sigma$-smooth measure $v$ and that $v$ is infinitely divisible and $\{\mu_n^{*m_n(s)}\}$ converges to $v_s$ with $v = v_s^{*s}$ for $m_n(s) = [k_n/s]$, $s = 1, 2, \ldots$. Thus (ii) holds. We now show (iii). By (1) and the corollary of Lemma 4.2, we get

$$\limsup_{n\to\infty} k_n \int_{\|y\|\geq\beta}\mu_n(dy) < \infty \qquad \text{for} \qquad \beta > 0, \tag{2}$$

$$\limsup_{n\to\infty} k_n\left|\int_{\|y\|<\beta} y^{(i)}\,\mu_n(dy)\right| < \infty \qquad \text{for any} \quad \beta > 0, \tag{3}$$

$$\limsup_{n\to\infty} k_n\left|\int_{\|y\|<\beta}\|y\|^2\,\mu_n(dy)\right| < \infty \qquad \text{for any} \quad \beta > 0. \tag{4}$$

We claim that the limes supremum in (2) tends to 0 as $\beta \to \infty$. Indeed, if $\{\mu_n^{*k_n}\}$ converges weakly, then its margins converge weakly. Consider then first the case when $\{\mu_n^{*k_n}\}$ is a sequence on $R$ converging weakly. By Lemma 6.1(i), there exist constants $a_n$ such that

$$\limsup_{n\to\infty} k_n \int_{|y|\geq\beta}\mu_n(a_n + dy)\begin{cases} < \infty & \text{for} \quad \beta > 0, \\ \to 0 & \text{for} \quad \beta \to \infty. \end{cases} \tag{5}$$

Since $k_n \uparrow \infty$, $n \uparrow \infty$, both the inequalities (2) and (5) can hold for arbitrarily small $\beta > 0$ only if $a_n \to 0$ $(n \to \infty)$, but then (5) holds with $a_n = 0$. Since this is true for the margins of $\mu_n$, it is also true for $\mu_n$ on $R^{(k)}$. By Lemma 8.1 and its corollary, it follows that there exists a sequence $\tilde{N}$ of positive integers such that (2°)–(4°) hold, when $n \to \infty$, $n \in \tilde{N}$; and (1°) is obvious for probability measures $\mu_n$. Then by Lemma 8.2, $k_n\ \Phi(\cdot/\alpha) * (\mu_n - e)$ converges uniformly as $n \to \infty$, $n \in \tilde{N}$ to the Gaussian functional $\Gamma(\alpha, v\cdot)$ with the given representation, but this Gaussian functional is independent of the sequence $\tilde{N}$, and by Lemma 8.2, it determines $\lambda$, the $m^{(i)}$, and the $q_{ij}$ uniquely. Thus (iii) holds. Let $\{\mu_n^{*k_n}\}$ converge weakly. Then the Cauchy convergence of $k_n\Phi(\cdot/\alpha) * [\mu_n - e]$ in the class of Gaussian seminorms follows by Theorem III.10.1 if the sequence of convolution powers is stable. Hence we have to show this stability. At first, consider probability measures $\mu_n$. It follows by (iv) in the corollary of Lemma 4.2 that $\{\mu_n^{*k_n}\}$ is stable if its margins are stable. Hence it is sufficient to deal with probability measure on $R$. Applying Lemma 6.1, we can find that for suitable constants $a_n$, not only (5) holds, but also

$$\limsup_{n\to\infty} k_n \left| \int_{|y|<\beta} y\, \mu_n(a_n + dy) < \infty \right| \qquad \text{for} \quad \beta > 0, \tag{6}$$

$$\limsup_{n\to\infty} k_n \int_{|y|<\beta} y^2\, \mu_n(a_n + dy) < \infty. \tag{7}$$

We claim that

$$\limsup_{n\to\infty} k_n |a_n| < \infty. \tag{8}$$

On the contrary, suppose that $\sup_n k_n|a_n| = \infty$. We may then select a sequence $\tilde{N}$ of positive integers such that $k_n|a_n| \to \infty$, $n \to \infty$, $n \in \tilde{N}$. Applying Lemma 8.1 and its corollary and Lemma 8.2, we conclude that we may select a subsequence $\hat{N}$ of $\tilde{N}$ such that $k_n\Phi(\cdot/\alpha) * \{\mu_n(a_n + \cdot) - e\}$ converges uniformly on $R$ as $n \to \infty$, $n \in \tilde{N}$, but then, by that part of the theorem which has been proved, $[\mu_n(a_n + \cdot)]^{*k_n}$ converges weakly to a probability measure $v$ as $n \to \infty$. By assumption, $\mu_n^{*k_n}$ converges weakly to a probability measure $v_0$, and

$$[\mu_n(a_n + \cdot)]^{*k_n} = \mu_n^{*k_n}(k_n a_n + \cdot).$$

Clearly, the right-hand side cannot converge weakly to a probability measure as $n \to \infty$, $n \in \tilde{N}$, if $k_n|a_n|$ then tends to $\infty$. Hence (4) holds and $a_n \to 0$ $(n \to \infty)$. As above, we conclude that (5) holds with $a_n = 0$, and it easily follows that (6) and (7) also hold with $a_n = 0$ since (4) is satisfied; but (5)–(7) with $a_n = 0$ imply stability of $\mu_n^{*k_n}$ on $R$, according to Lemma 4.2(iv).

So far we have proved the stability of $\{\mu_n^{*k_n}\}$ for convolution powers of probability measures $\mu_n$, but if the $\mu_n$ are $\sigma$-smooth measures, then $\mu_n = a_n v_n$

with $a_n = \mu_n(R^{(k)})$ and probability measures $\nu_n$. Clearly $a_n^{k_n}$ converges to a number $a$ if and only if, $k_n[a_n - 1]$ converges to a number $c$. Then we easily find that $\{\mu_n^{*k_n}\}$ is stable when $\{\nu_n^{*k_n}\}$ is stable. □

*Proof of the corollary:* By Theorem 5.2, the sequence $\{\mu_n^{*k_n}\}$ converges weakly if and only if its projections converge weakly, and by Theorem II.8.3, we have

$$\mu_n^{*k_n}[(\pi(t))^{-1}\cdot] = \{\mu_n[(\pi(t))^{-1}\cdot]\}^{*k_n}$$

for any projection $\pi(t)$ (in fact, for any measurable mapping). The Gaussian transform of $k_n[\mu_n(\pi(t)^{-1}\cdot) - e]$ then converges uniformly to a Gaussian functional, which by transformation may be given as in the corollary. □

## 10. THE SEMIGROUP OF INFINITELY DIVISIBLE $\sigma$-SMOOTH MEASURES

**Theorem 1.** *The infinitely divisible $\sigma$-smooth measures on $R^{(k)}$ form a commutative semigroup $\mathscr{M}_0$ under convolution and $\mathscr{M}_0$ is* 1.1*-isomorph to the additive semigroup of Gaussian functionals $\Gamma(\alpha, \mu, x)$ and also to the additive semigroup of projected Gaussian functionals $\bar{F}(\alpha, \mu, x \cdot t)$, i.e., $\mu \Leftrightarrow \Gamma(\alpha, \mu, x)$ for $\mu \in \mathscr{M}_0 \Leftrightarrow \mu * \nu \Leftrightarrow \Gamma(\alpha, \mu, x) + \Gamma(\alpha, \nu, x)$ for $\mu, \nu \in \mathscr{M}_0$; correspondingly, for the $\bar{\Gamma}(\alpha, \mu, x \cdot t)$.*

*Remark:* Note that $\Gamma(\alpha, \mu, x)$ and $\bar{\Gamma}(\alpha, \mu, x \cdot t)$ are uniquely determined by $c$, the Lévy measure $\lambda$, the mean value vector $\bar{m}$, and the quadratic form $q(t)$, and that addition of Gaussian functionals of $\mu$ and $\nu$ corresponds to a simultaneous addition of the Lévy measures, the mean-value vectors, and the quadratic forms. (Compare Theorem 9.1.)

*Proof:* The infinitely divisible $\sigma$-smooth measure $\mu$ determines $\Gamma(\alpha, \nu, x)$ by the definition of the Gaussian functional. If $\mu = \mu_n^{*k_n} = \nu = \nu_n^{*k_n}$ we have $\mu_n = \nu_n$ since an infinitely divisible $\sigma$-smooth measure is uniquely infinitely divisible. Furthermore, if $\{\mu_n^{*n}\}$ and $\{\nu_n^{*n}\}$ converge weakly (which they do if $\mu_n^{*n} = \mu$, $\nu_n^{*n} = \nu$ for all $n$), these sequences are stable (compare the proof of Theorem 9.1). By Lemma III.9.1, we then get

$$\begin{aligned} p_\alpha(\mu - \nu) = p_\alpha(\mu_n^{*n} - \nu_n^{*n}) &\leq c \limsup_{n\to\infty} p_\alpha(n\mu_n - n\nu_n) \\ &= c \sup_x |\Gamma(\alpha, \mu, x) - \Gamma(\alpha, \nu, x)| \end{aligned}$$

with a positive constant $c$. Hence $\mu = \nu$ if $\Gamma(\cdot, \mu, \cdot) = \Gamma(\cdot, \nu, \cdot)$. For any $\mu$ and $\nu$ in $\mathscr{M}_0$, we get

$$\Gamma(\alpha, \mu * \nu, \cdot) = \lim_{n\to\infty} n\Phi(\cdot/\alpha) * [\mu_n * \nu_n - e],$$

where $\mu = \mu_n^{*n}$, $\nu = \nu_n^{*n}$, and the convergence holds in the uniform metric. Using the identity

$$n[\mu_n * \nu_n - e] = n(\mu_n - e) + n(\nu_n - e) + n(\mu_n - e) * (\nu_n - e),$$

we get, by Lemma 4.3,

$$np[(\mu_n - e) * (\nu_n - e)] \le \frac{1}{n} c(k) n^2 p_\alpha(\mu_n - e) p_\alpha(\nu_n - e) \to 0 \qquad (n \to \infty)$$

since the semigroup $\mathscr{M}$ is distinguished and $\{\mu_n^{*n}\}$ and $\{\nu_n^{*n}\}$ are stable. □

We shall now use the isomorphism between the two semigroups in order to find how they are generated, and then consider the following cases.

*Case 1:* $\lambda \in \mathscr{M}$ on $R^{(k)}$. Then $e + (\lambda/n) \in \mathscr{M}$, and

$$n\Phi(\cdot/\alpha) * [e + \lambda/n < e] \to \Phi(\cdot/\alpha) * \lambda \qquad (n \to \infty)$$

in the uniform metric. Hence $(e + \lambda/n)^{*n}$ converges weakly to an infinitely divisible measure with the Gaussian functional $\Phi(\cdot/\alpha) * \lambda$. Considering the analogy with the definition of $\exp \lambda$ for any number $\lambda$, we denote the infinitely divisible measure so defined by $\exp^* \lambda$. Note that it has the property

$$\exp^* \lambda_1 * \exp^* \lambda_2 = \exp^*(\lambda_1 + \lambda_2)$$

for $\lambda_1, \lambda_2 \in \mathscr{M}$. Also observe that $\lambda$ may be concentrated to a point.

*Case 2:* For any finite vector $\bar{m}$ in $R^{(k)}$, the measure $e(\cdot - \bar{m})$ is infinitely divisible:

$$e(\cdot - \bar{m}) = \left[e\left(\cdot - \frac{\bar{m}}{n}\right)\right]^{*n}$$

and has the Gaussian functional

$$\lim_{n \to \infty} \Phi\left[\frac{1}{\alpha}\left(x - \frac{\bar{m}}{n}\right)\right] - \Phi\left(\frac{x}{\alpha}\right) = -\sum_{i=1}^{k} m^{(i)} \frac{\partial}{\partial x^{(i)}} \Phi\left(\frac{x}{\alpha}\right).$$

*Case 3:* By Theorem 2.1, the general Gaussian measure with mean value 0 on $R^{(k)}$ has the representation $\Phi(T^{-1}\cdot)$ with a linear transformation $T$, which determines a nonnegative quadratic form

$$q(t) = t'TT't.$$

Any nonnegative symmetrical quadratic form has such a representation. Furthermore,

$$[\Phi(\sqrt{n}\,T^{-1}\cdot)]^{*n} = \Phi(T^{-1}\cdot).$$

The Gaussian functional of the infinitely divisible Gaussian measure $\Phi(T^{-1}\cdot)$ is then

$$\lim_{n\to\infty} \Phi\left(\frac{\cdot}{\alpha}\right) * [\Phi(\sqrt{n}\,T^{-1}\cdot) - e],$$

which at the point $x$ has the value

$$\lim_{n\to\infty} n \int_{R^{(k)}} \left[\Phi\left(\frac{x-y}{\alpha}\right) - \Phi\left(\frac{x}{\alpha}\right)\right] d\Phi(\sqrt{n}\,T^{-1}y)$$

$$= \lim_{n\to\infty} n \int_{R^{(k)}} \left[\Phi\left[\frac{1}{\alpha}\left(x - \frac{y}{\sqrt{n}}\right)\right] - \Phi\left(\frac{x}{\alpha}\right)\right] d\Phi(T^{-1}y).$$

By the help of the Taylor expansion, we find that this limit is equal to

$$\frac{1}{2}\sum_{i=1}^{k}\sum_{j=1}^{k} \frac{\partial}{\partial x^{(i)}\,\partial x^{(j)}} \int_{R^{(k)}} y^{(i)}y^{(j)}\, d\Phi(T^{-1}\,dy) = \frac{1}{2}\sum_{i=1}^{k}\sum_{j=1}^{k} q_{ij}\frac{\partial^2}{\partial x^{(i)}\,\partial x^{(j)}}\Phi\left(\frac{x}{\alpha}\right)$$

with

$$q_{ij} = \int_{R^{(k)}} y^{(i)}y^{(j)}\, d\Phi(T^{-1}y).$$

*Case 4:* For any positive number $a$, the measure $ae$ is infinitely divisible: $ae = [\sqrt[n]{a}\,e]^{*n}$, where

$$\lim_{n\to\infty} n\Phi\left(\frac{x}{\alpha}\right) * (\sqrt[n]{a} - 1)e = (\ln a)\Phi\left(\frac{x}{\alpha}\right).$$

By composition (convolution of infinitely divisible measures, addition of the corresponding Gaussian functionals), we get new infinitely divisible measures with their corresponding Gaussian functionals.

**Theorem 2.** *To a Lévy measure $\lambda$ on $R^{(k)}$, there belongs an infinitely divisible probability measure $\nu$ with the Gaussian functional*

(i) $$\Gamma(\alpha, \nu, x) = \int_{R^{(k)}} \left[\Phi\left(\frac{x-y}{\alpha}\right) - \Phi\left(\frac{x}{\alpha}\right) + \sum_{i=1}^{k} \frac{y^{(i)}}{1+\|y\|^2}\frac{\partial}{\partial x^{(i)}}\Phi\left(\frac{x}{\alpha}\right)\right]\lambda(dy)$$

*and the projected Gaussian functional*

(ii) $$\bar{\Gamma}(\alpha, \nu, x\cdot t) = \int_{R^{(k)}} \left[\Phi\left(\frac{x\cdot t - y\cdot t}{\alpha}\right) - \Phi\left(\frac{x\cdot t}{\alpha}\right) + \frac{y\cdot t}{\alpha(1+\|y\|^2)}\Phi'\left(\frac{x\cdot t}{\alpha}\right)\right]\lambda(dy).$$

**Corollary.** *To a given real number c, a given vector* $\bar{m} = (m^{(1)}, \ldots, m^{(k)})$, *and a given bounded symmetrical quadratic form*

$$q(t) = \sum_{i=1}^{k} \sum_{j=1}^{k} q_{ij} t_i t_j$$

*with* $\sum_{i=1}^{\infty} q_{ij}$, *there exists a σ-smooth, infinitely divisible measure μ with Gaussian invariant* $\Gamma(\alpha, v, x)$ *and projected Gaussian invariant* $\bar{\Gamma}(\alpha, v, x \cdot t)$:

$$\text{(iii)} \qquad \Gamma(\alpha, v, x) = c\Phi\left(\frac{x}{\alpha}\right) - \sum_{i=1}^{k} m^{(i)} \frac{\partial}{\partial x^{(i)}} \Phi\left(\frac{x}{\alpha}\right)$$

$$+ \frac{1}{2} \sum_{i=1}^{k} \sum_{j=1}^{k} q_{ij} \frac{\partial^2}{\partial x^{(i)} \partial x^{(j)}} \Phi\left(\frac{x}{\alpha}\right) + \int_{R^{(k)}\setminus\{0\}} \left[\Phi\left(\frac{x-y}{\alpha}\right) - \Phi\left(\frac{x}{\alpha}\right)\right.$$

$$\left. + \sum_{i=1}^{k} \frac{y^{(i)}}{1 + \|y\|^2} \frac{\partial}{\partial x^{(i)}} \Phi\left(\frac{x}{\alpha}\right)\right] \lambda(dy),$$

$$\text{(iv)} \quad \bar{\Gamma}(\alpha, v, x \cdot t) = c\Phi\left(\frac{x \cdot t}{\alpha}\right) - \frac{1}{\alpha} \bar{m} \cdot t\Phi'\left(\frac{x \cdot t}{\alpha}\right) + \frac{1}{2\alpha^2} q(t)\Phi''\left(\frac{x \cdot t}{\alpha}\right)$$

$$+ \int_{R^{(k)}\setminus\{0\}} \left[\Phi\left(\frac{x \cdot t - y \cdot t}{\alpha}\right) - \Phi\left(\frac{x \cdot t}{\alpha}\right)\right.$$

$$\left. + \frac{y \cdot t}{\alpha(1 + \|y\|^2)} \Phi'\left(\frac{t \cdot x}{\alpha}\right)\right] \lambda(dy),$$

*where v uniquely determines* $\Gamma(\alpha, v, x)$ *and* $\bar{\Gamma}(\alpha, v, x \cdot t)$, *and each of these Gaussian functionals determines v uniquely.*

*Proof of Theorem 2:* Let the Lévy measure $\lambda$ be given. For any $\beta > 0$, put $F_\beta = \{x : \|x\| \geq \beta\}$ and choose $\beta$ such that $F$ is a continuity set for $\lambda$. Consider a decreasing sequence $\{\beta_n\}$ of positive numbers chosen such that $F_{\beta_n}$ is a continuity set for $\lambda$ and

$$\frac{1}{n}\left[\int_{F_{\beta_n}} \lambda(dy)\right] \to 0, \qquad \beta_n \to 0, \qquad \text{as} \quad n \uparrow \infty. \tag{1}$$

Clearly the $\beta_n$ may be chosen in this way since

$$\int_{F_\beta} \lambda(dy) < \infty$$

for any $\beta > 0$. Now define the measure $\lambda_n$ on $R^{(k)}$ by

$$\lambda_n(E) = \lambda(E \cap F_{\beta_n})$$

for any Borel set $E$ in $R^{(k)}$, and put $\bar{m}_n = (m_n^{(1)}, \ldots, m_n^{(k)})$,

$$\bar{m}_n \cdot t = \int_{R_k} \frac{y \cdot t}{1 + \|y\|^2} \lambda_n(dy) = \int_{F_{\beta_n}} \frac{y \cdot t}{1 + \|y\|^2} \lambda(dy). \tag{2}$$

Now introduce the measure

$$\mu_n = \left[1 + \frac{1}{n}\lambda(F_{\beta_n})\right]^{-1}\left(e + \frac{1}{n}\lambda_n\right) * e(\cdot + \bar{m}_n/n)$$
$$= \left[1 + \frac{1}{n}\lambda(F_{\beta_n})\right]^{-1}\left[e(\cdot + \bar{m}_n/n) + \frac{1}{n}\lambda_n(\cdot - \bar{m}_n/n)\right]. \tag{3}$$

Then

$$\mu_n(R^{(k)}) = \left[1 + \frac{1}{n}\lambda(F_{\beta_n})\right]^{-1}\left[1 + \frac{1}{n}\lambda(F_{\beta_n})\right] = 1,$$

so that $\mu_n$ is a probability measure. We claim that $\{\mu_n^{*n}\}$ converges weakly to an infinitely divisible probability measure with the Gaussian functionals as in the theorem. Applying the corollary of Theorem 9.1, we have then to prove that

$$\lim_{n\to\infty} n \int_{R^{(k)}} \left[\Phi\left(\frac{x\cdot t - y\cdot t}{\alpha}\right) - \Phi\left(\frac{x\cdot t}{\alpha}\right)\right]\mu_n(dy)$$
$$= \int_{R^{(k)}} \left[\Phi\left(\frac{x\cdot t - y\cdot t}{\alpha}\right) - \Phi\left(\frac{x\cdot t}{\alpha}\right) + \frac{y\cdot t}{\alpha(1+\|y\|^2)}\Phi'\left(\frac{x\cdot t}{\alpha}\right)\right]\lambda(dy) \tag{4}$$

uniformly with respect to $x \in R^{(k)}$. Now

$$\lim_{n\to\infty} n \int_{R^{(k)}} \Phi\left(\frac{x\cdot t - y\cdot t}{\alpha}\right) - \Phi\left(\frac{x\cdot t}{\alpha}\right)\mu_n(dy)$$
$$= \lim_{n\to\infty}\left[1 + \frac{1}{n}\lambda(F_{\beta_n})\right]^{-1}\left\{n\left[\Phi\left(\frac{x\cdot t + (\bar{m}_n/n)\cdot t}{\alpha}\right) - \Phi\left(\frac{x\cdot t}{\alpha}\right)\right]\right.$$
$$\left. + \int_{F_{\beta_n}}\left[\Phi\left(\frac{x\cdot t + (\bar{m}_n\cdot t/n) - y\cdot t}{\alpha}\right) - \Phi\left(\frac{x\cdot t}{\alpha}\right)\right]\lambda(dy).\right. \tag{5}$$

Using the Taylor expansion and observing that

$$|\bar{m}_n \cdot t| \le \|t\|\lambda(F_{\beta_n}),$$

according to (2), we find by (1) and (2) that this limit, is uniformly equal with respect to $x \in R^{(k)}$ to

$$\lim_{n\to\infty}\left\{\bar{m}_n\cdot t + \int_{F_{\beta_n}}\left[\Phi\left(\frac{x\cdot t - y\cdot t}{\alpha}\right) - \Phi\left(\frac{x\cdot t}{\alpha}\right)\right]\lambda(dy)\right\}$$
$$= \lim_{n\to\infty}\int_{F_{\beta_n}}\left[\Phi\left(\frac{x\cdot t - y\cdot t}{\alpha}\right) - \Phi\left(\frac{x\cdot t}{\alpha}\right) + \frac{y\cdot t}{(1+\|y\|^2)}\Phi'\left(\frac{x\cdot t}{\alpha}\right)\right]\lambda(dy), \tag{6}$$

and thus (4) holds uniformly with respect to $x \in R^{(k)}$. Indeed, if we consider the integral on the right-hand side of (6) over $0 \leq \|y\| \leq \beta_n$ instead of $F_{\beta_n}$, we easily find by the help of the Taylor expansion that this integral is smaller than

$$\frac{1}{\alpha^2} c_0 \|t\|^2 \int_{0 < \|y\| \leq \beta_n} \|y\|^2 \lambda(dy)$$

with an absolute constant $c_0$, and this term tends to 0 as $n \to \infty$ since $\beta_n \to 0$. □

*Proof of the corollary:* The proof follows from the theorem and the results in cases (2) and (3) considered above when we use the isomorphism between $\mathscr{M}_0$ and the additive groups of Gaussian functionals. □

## 11. THE CHARACTERISTIC FUNCTION OF AN INFINITELY DIVISIBLE PROBABILITY MEASURE ON $R^{(k)}$ AND ITS CONNECTION WITH THE GAUSSIAN FUNCTIONALS

Using the fact that the characteristic function of a convolution product of $\sigma$-smooth measures is the product of the characteristic functions of the factors, and the characteristic function of the weak limit of a sequence $\sigma$-smooth measures is the limit of the characteristic functions of these measures, we get, by the construction in Section 10,

(a) The characteristic function of $e(\cdot - \bar{m})$ is $\exp(it \cdot \bar{m})$ for any vector $\bar{m}$.
(b) The characteristic function of $\Phi(T^{-1} \cdot)$ is, with matrix notations,

$$\int_{R^{(k)}} \exp(it'y)\, \Phi(T^{-1} dy) = \int_{R^{(k)}} \exp(it'Tz)\, \Phi(dz),$$

or, since the characteristic function of $\Phi$ is $\exp(-\frac{1}{2}t't)$ and $t'T = (T't)'$, we find that $\Phi(T^{-1} \cdot)$ has the characteristic function

$$\exp(-\tfrac{1}{2}t'TT't) = \exp[-\tfrac{1}{2}q(t)],$$

where $q(t)$ is the nonnegative symmetrical quadratic form

$$q(t) = \sum_{i=1}^{k} \sum_{j=1}^{k} q_{ij} t_i t_j$$

in Theorem 10.2(ii).

(c) Let $\lambda$ be a Lévy measure and put $\lambda_\beta = \lambda(F_\beta \cap E)$ for any Borel set $E \in R^{(k)}$. The characteristic function of $(e + \lambda_\beta/n)$ is

$$1 + \frac{1}{n} \int_{R^{(k)}} \exp(it \cdot y)\, \lambda_\beta(dy),$$

and thus the characteristic function of $\exp^* \lambda_\beta$ is

$$\lim_{n\to\infty}\left[1+\frac{1}{n}\int_{R^{(k)}}\exp(it\cdot y)\,\lambda_\beta(dy)\right]^n = \exp\int_{R^{(k)}}\exp(it\cdot y)\,\lambda_\beta(dy).$$

Since $\exp^* \lambda_\beta(R^{(k)}) = \exp \lambda_\beta(R^{(k)})$, we conclude that the characteristic function of the probability measure

$$\exp[-\lambda_\beta(R^{(k)})]\exp^* \lambda_\beta$$

is equal to

$$\exp\int_{R^{(k)}}[\exp(it\cdot y)-1]\,\lambda_\beta(dy).$$

For

$$m_\beta^{(i)} = \int_R \frac{y^{(i)}}{(1+\|y\|^2)}\,\lambda_\beta(dy), \qquad \bar{m}_\beta = (m_\beta^{(1)},\ldots,m_\beta^{(k)}),$$

$e(\cdot+\bar{m}_\beta)$ has the characteristic function $\exp(it\cdot\bar{m}_\beta)$ and hence

$$e(\cdot+\bar{m}_\beta) * \exp[-\lambda_\beta(R^{(k)})]\exp^* \lambda_\beta \tag{1}$$

has the characteristic function

$$\exp\int_{R^{(k)}}\left\{\exp[i(t\cdot y)]-1-\frac{it\cdot y}{1+\|y\|^2}\right\}\lambda_\beta(dy). \tag{2}$$

We have shown in Theorem 10.2 that the infinitely divisible probability measure (1) tends to a certain infinitely divisible probability measure as $\beta\downarrow 0$. The characteristic function of this probability measure is equal to the limit of (2) as $\beta\downarrow 0$. Also, regarding (a) and (b) above, we get

**Theorem 1.** *The characteristic function of an infinitely divisible probability measure on $R^{(k)}$ is equal to*

$$\text{(i)}\quad \exp\left\{im\cdot t-\tfrac{1}{2}q(t)+\int_{R^{(k)}}\left[\exp(it\cdot x)-1-\frac{it\cdot y}{1+\|y\|^2}\right]\lambda(dy)\right\},$$

*where $m$ is a real vector in $R^{(k)}$, $q(t)$ a definite nonnegative symmetrical quadratic form, and $\lambda$ a Lévy measure. For any given such vector $m$, quadratic form $q$, and Lévy measure $\lambda$, there exists an infinitely divisible probability measure with the characteristic function* (i).

There is a close relationship between the characteristic function (i) of an infinitely divisible probability measure $\nu$ and the Gaussian functional and the projected Gaussian functional. Indeed, the characteristic function of

$$k_n\Phi(\cdot/\alpha) * [\mu_n\cdot - e]$$

for probability measures $\mu_n$ is

$$k_n \exp(-\tfrac{1}{2}\alpha^2\|t\|^2) \int_{R^{(k)}} [\exp(it \cdot x) - 1]\, \mu_n(dx). \tag{3}$$

We can write the integral multiplied by $k_n$ as $n \to \infty$:

$$\lim_{n\to\infty} k_n \int_{R^{(k)}} \left[ \exp(it \cdot x) - 1 - \frac{it \cdot x}{1 + \|y\|^2} \right] \mu_n(dx) + i(t \cdot m), \tag{4}$$

with

$$t \cdot m = \lim_{n\to\infty} k_n \int_{R^{(k)}} \frac{t \cdot y}{1 + \|y\|^2}\, \mu_n(dy).$$

Proceeding as in the proof of Theorem 10.1, we then find that the quantity (4) tends to the logarithm of the characteristic function. In the same way, for the measure $\mu_n[\pi^{-1}(t)\cdot]$ on $R$, we find that

$$k_n \Phi(\cdot/\alpha) * \{\mu_n[\pi^{-1}(t)\cdot] - e\}$$

(now $e$ and $\Phi$ on $R$) has the characteristic function

$$k_n \exp(-\tfrac{1}{2}\alpha^2 T^2) \int_R [\exp(i\tau\tilde{x} - 1)]\, \mu_n[\pi^{-1}(t)\, d\tilde{x}]. \tag{5}$$

By the transformation $\tilde{x} = \pi(t)x$, the quantity (5) takes the form (3) for $\tau = 1$.

We may consider $\Gamma(\alpha, v, x)$ as a distribution function of a signed measure. Observing that $\Gamma(\alpha, v, x)$ has bounded, continuous derivatives of all orders, we find that $\Gamma(\alpha, v, x)$ has a density function $\gamma(\alpha, v, x)$. Its Fourier transform is then

$$\int_{R^{(k)}} \exp(it \cdot x)\gamma(\alpha, v, x) = \exp(-\tfrac{1}{2}\alpha^2\|t\|^2) f(t), \tag{6}$$

where $f(t)$ is the logarithm of the characteristic function of $v$. This Fourier transform can be inverted to $\gamma(\alpha, v, x)$ according to Theorem 3.3. Thus we have proved

**Theorem 2.** *The Gaussian functional $\Gamma(\alpha, v, x)$ of an infinitely divisible probability measure $v$ is the distribution of a signed measure. This distribution has a density function $\gamma(\alpha, v, x)$, whose Fourier transform is equal to*

$$\exp(-\alpha^2\|t\|^2) f(t),$$

*where $f(t)$ is the logarithm of the characteristic function of $v$. The inversion of this Fourier transform is defined and equal to $\gamma(\alpha, v, x)$.*

**Corollary.** *An infinitely divisible probability measure is uniquely determined by its characteristic function, by its Gaussian functional, and by its projected Gaussian transform.*

*Remark:* It is easy to compute the Fourier transform of $\gamma(\alpha, v, x)$ directly. Use the relation

$$\int_{R^{(k)}} \exp(it \cdot x)\, \varphi\left(\frac{x}{\alpha}\right) dx = \exp(-\tfrac{1}{2}\alpha^2 \|t\|^2),$$

where $\varphi$ is the density function of $\Phi$ on $R^{(k)}$ and the partial derivatives of this relation.

*Proof of the corollary:* The proof follows from Theorem 2 above and from Theorem 10.2. □

## 12. WEAK CONVERGENCE OF CONVOLUTION PRODUCTS

Consider the general convolution products

$$\prod_{i=1}^{k_n *} \mu_{nj} \tag{1}$$

of $\sigma$-smooth measure on $R^{(k)}$, $k_n \uparrow \infty$ as $n \uparrow \infty$. Convergence theorems for such products are given in Section III.10 and can be applied directly on $R^{(k)}$. These theorems essentially bring the limit theorems for convolution products (1) back to limit problems for convolution powers.

The u.a.n. condition introduced in Section III.10.1 obviously holds in the class of Gaussian seminorms on $R^{(k)}$ if and only if

$$\limsup_{n\to\infty,\ 1\le j\le k_n} \int_{\|y\|\ge\varepsilon} \mu_{nj}(dy) = 0 \qquad (\varepsilon > 0). \tag{2}$$

We now formulate Theorem III.10.1 for $R^{(k)}$.

**Theorem 1.** *If the sequence* (1) *is weakly stable and satisfies the condition*

(i) $$\lim_{n\to\infty} \sum_{j=1}^{k_n} p_\alpha^2(\mu_{nj} - e) = 0$$

*for the Gaussian seminorms* $p_\alpha$, $\alpha > 0$, *and*

(ii) $$\lambda_n = \frac{1}{k_n} \sum_{j=1}^{k_n} \mu_{nj},$$

*then*

(iii) $$\lim_{n\to\infty} p_\alpha\left(\lambda_n^{*k_n} - \prod_{j=1}^{k_n *} \mu_{nj}\right) = 0.$$

*Thus the convolution product in* (iii) *converges weakly if and only if the convolution power in* (iii) *converges weakly.*

*Remark:* $\{\prod_{j=1}^{k_{n*}} \mu_{nj}\}$ is weakly stable if the $\mu_{nj}$ are probability measures. If this sequence of convolution products is stable, the u.a.n. condition implies (i).

By Theorem 1 we get necessary and sufficient conditions for the weak convergence of the sequence of convolution products under the restrictions (i) and (ii), and by Remark 1 these restrictions are fulfilled if the sequence is stable and satisfies the u.a.n. condition. Theorem 2 shows that the convergence problem for sequences of convolution products satisfying the u.a.n. condition can be brought back to the corresponding problem for stable sequences, and this can be done by a suitable normalization.

**Theorem 2.** *Let the $\mu_{nj}$ in* (1) *be probability measures which satisfy the u.a.n. condition. Furthermore, suppose that the sequence* (1) *converges weakly. For some $\beta_0 > 0$, put*

$$a_{nj}^{(i)} = \int_{|y^{(i)}|<\beta_0} y^{(i)}\, \mu_{nj}(dy) - \frac{1}{k_n} \sum_{j=1}^{k_n} \int_{|y^{(i)}|<\beta_0} y^{(i)}\, \mu_{nj}(dy),$$

$$a_{nj} = (a_{nj}^{(1)}, \ldots, a_{nj}^{(k)}), \quad \nu_{nj} = \mu_{nj}(a_{nj} + \cdot).$$

*Then*

(i) $$\prod_{j=1}^{k_{n*}} \mu_{nj} = \prod_{j=1}^{k_{n*}} \nu_{nj},$$

(ii) $\left\{\prod_{j=1}^{k_{n*}} \nu_{nj}\right\}$ *is stable, and*

(iii) *the $\nu_{nj}$ satisfy the u.a.n. condition.*

*Proof:* (i) follows by the definition of convolutions. For the proof of (ii), we observe that by (iv) in the corollary of Lemma 4.2, a sequence of convolution products is stable if the sequences of its margins are stable. Thus it is sufficient to deal with probability measures on $R$. Then the vector $a_{nj}$ reduces to a number. We then use the same procedure as in the last part of the proof of Lemma 6.1.

The u.a.n. condition (2) implies

$$\limsup_{n\to\infty,\, 1\le j\le k_n} |a_{nj}| = 0. \tag{3}$$

By Lemma 6.1, we have

$$\limsup_{n\to\infty} \sum_{j=1}^{k_n} \int_{|y|\geq\beta} \mu_{nj}(a_{nj}+dy) \begin{cases} <\infty & \text{for} \quad \beta>0, \\ \to 0 & \text{for} \quad \beta\to\infty, \end{cases} \tag{4}$$

$$\limsup_{n\to\infty} \sum_{j=1}^{k_n} \left| \int_{|y|<\beta} y\mu_{nj}(a_{nj}+dy) \right| < \infty, \tag{5}$$

$$\limsup_{n\to\infty} \sum_{j=1}^{k_n} \int_{|y|<\beta} y^2\mu_{nj}(a_{nj}+dy) < \infty \tag{6}$$

for suitable constants. By the use of (3) and the estimations in the proof of Lemma 6.1, we find that we may choose the $a_{nj}$ as in the theorem and then (ii) follows. Furthermore, (2) and (3) imply (iii). □

## 13. STABLE PROBABILITY MEASURES

For normalized Gaussian probability measures on $R^{(k)}$, we have the nice relation 2.2. The question immediately arises: Do there exist other probability measures on $R^{(k)}$, satisfying the relation

$$\mu(\cdot/\alpha_1) * \mu(\cdot/\alpha_2) = \mu(\cdot/\alpha), \tag{1}$$

where $\alpha_1$ and $\alpha_2$ are positive numbers and $\alpha$ is determined by $\alpha_1$ and $\alpha_2$? If $\mu$ satisfies (1), we call it a stable measure in the restricted sense. If (1) should hold, then clearly we must have $\mu^{*n} = \mu(\cdot/\sigma(n))$, with a function $\sigma$ defined on the set of all positive integers. Furthermore,

$$\mu(\cdot/\sigma(nm)) = (\mu^{*n})^{*m} = [\mu(\cdot/\sigma(n))]^{*m} = \mu(\cdot/\sigma(n)\sigma(m)).$$

Hence, necessarily

$$\sigma(nm) = \sigma(n)\sigma(m), \qquad \sigma(1) = 1. \tag{2}$$

First we deal with measures on $R$.

A second necessary condition of the function $\sigma(n)$ we get by the help of the concept "concentration of a distribution function," introduced by Paul Lévy. For a distribution function $\hat{\mu}$ of a probability measure $\mu$ on $R$, the concentration $Q(\hat{\mu}, d)$ is defined by

$$Q(\hat{\mu}, d) = \sup_x [\hat{\mu}(x+d+0) - \hat{\mu}(x-0)].$$

Since for two probability measures $\mu$ and $\nu$

$$\widehat{\mu * \nu}(x) = \int \hat{\mu}(x-y)\,\nu(dy),$$

we obviously have

$$Q(\widehat{\mu * \nu}, d) \leq (\hat{\mu}, d).$$

Hence, $\mu^{*n} = \mu(\cdot/\sigma(n))$ implies

$$[\hat{\mu}(\cdot/\sigma(n)), d] \geq [\hat{\mu}(\cdot/\sigma(m)), d] \qquad \text{for} \quad n \leq m,$$

and thus

$$\sigma(n) \leq \sigma(m) \qquad \text{for} \quad m \geq n \quad \text{if} \quad \mu \neq e. \tag{3}$$

**Lemma 1.** *If $\sigma(nm) = \sigma(n)\sigma(m)$ for all n and m, $\sigma(n) \leq \sigma(m)$ for $n \leq m$, and $\sigma(1) = 1$, then $\sigma(n) = n^s$, where s is a positive number.*

*Proof:* Consider the mapping $n \to \sigma(n)$, where $\sigma(n)$ satisfies (2) and (3). The two inequalities

$$2^r < m^s, \qquad \sigma(2^r) < \sigma(m^s)$$

imply each other for any pairs $r, s$ of positive integers and any given positive integer $m$. Hence the inequalities

$$r/s < \frac{\ln m}{\ln 2}, \qquad r/s < \frac{\ln \sigma(m)}{\ln 2}$$

hold simultaneously, and consequently

$$\frac{\ln m}{\ln 2} = \frac{\ln \sigma(m)}{\ln 2}$$

for all positive integers $m$. Thus

$$\sigma(m) = m^s \qquad \text{with} \quad s = \frac{\ln \sigma(2)}{\ln 2}.$$

We have proved this relation for probability measures satisfying (1) on $R$, but then it also holds on $R^{(k)}$, since (1) for probability measures on $R^{(k)}$ implies this relation for the margins of $\mu$.

We now state

**Theorem 1.**

(i) *A probability $\mu \neq e$ on $R^{(k)}$ that is stable in a restricted sense belongs to a class $\mathcal{M}(\rho)$, $0 < \rho \leq 2$, of infinitely divisible probability measures $\mu$ such that*

$$\mu(\cdot/\sigma_1) * \mu(\cdot/\sigma_2) = \mu(\cdot/\sigma)$$

*for $\sigma_1^\rho + \sigma_2^\rho = \sigma^\rho$, $\sigma_1 > 0$, $\sigma_2 > 0$, $\sigma > 0$.*

(ii) *$\mu \in \mathcal{M}(2)$ if and only if it is Gaussian with mean-value vector zero.*

(iii) *For given* $\rho$, $0 < \rho < 2$, *we have* $\mu \in \mathcal{M}(\rho)$ *if and only if the Lévy measure of* $\mu$ *has the properties*

$$(1^\circ) \qquad \mu = r\mu(r^{1/\rho}\cdot)$$

*for any positive number r and*

$$(2^\circ) \qquad \lambda(dy) = c_1^{(i)}\beta^{-\rho}, \qquad \int_{y^{(i)} \leq -\beta} \lambda(dy) = c_2^{(i)}\beta^{-\rho}$$

*for* $\beta > 0$, *where* $c_1^{(i)} \geq 0$, $c_2^{(i)} \geq 0$, $c_1^{(i)} + c_2^{(i)} > 0$, *for some i, and if* $\rho = 1$, $c_1^{(i)} = c_2^{(i)}$ *for all* $i = 1, \ldots, k$. *The Gaussian functional of* $\mu$ *is then uniquely determined by* $\lambda$, *according to*

$$\Gamma(\alpha, \beta, \mu) = -\sum_{i=1}^{k} m^{(i)} \frac{\partial}{\partial x^{(i)}} \Phi\left(\frac{x}{\alpha}\right) + \int_{R^{(k)}} \left[\Phi\left(\frac{x-y}{\alpha}\right) - \Phi\left(\frac{x}{\alpha}\right) + \sum_{i=1}^{k} \frac{\partial}{\partial x^{(i)}} \Phi\left(\frac{x}{\alpha}\right) \frac{y^{(i)}}{1 + (y^{(i)})^2}\right] \lambda(dy),$$

*where*

$$m^{(i)} = [c_1^{(i)} - c_2^{(i)}]\left[\frac{\rho}{1-\rho}\beta^{1-\rho} - \rho\int_{0<t\leq\beta} \frac{t^{2-\rho}}{1+t^2}\,dt + \rho\int_{t\geq\beta} \frac{t^{-\rho}}{1+t^2}\,dt\right]$$

*for* $\rho \neq 1$, *and* $m^{(i)} = 0$ *for* $\rho = 1$.

*Remark 1:* Note the connection between the Gaussian functional and the characteristic function of an infinitely divisible probability measure (See Section 11).

*Remark 2:* Conditions for weak convergence of convolution powers and convolution products are given by the general limit theorems in Sections 9 and 12.

*Proof:* Let $\mu$ be a probability measure on $R^{(k)}$ such that it is stable in the restricted sense. By Lemma 1, we get, with $s = 1/\rho$,

$$\{\mu(n^{1/\rho}\cdot)\}^{*n} = \mu \tag{4}$$

for $n = 1, 2, \ldots$ . Hence $\mu$ is infinitely divisible, and applying Theorem 10.2, we conclude that $n\mu(n^{1/\rho}\cdot)$ converges weakly to a Lévy measure $\lambda$, but then also

$$\{\mu[(nr)^{1/\rho}\cdot]\}^{*nr} = \mu,$$

and thus

$$nr\mu[(nr)^{1/\rho}\cdot] \to r\lambda(r^{1/\rho}\cdot) \qquad (n \to \infty)$$

for any positive integer $r$. Hence

$$\lambda = r\lambda(r^{1/\rho}\cdot) \tag{5}$$

for any positive integer $r$, but it is easy to show that (5) must hold for any positive number $r$. Indeed, by (5), we get

$$s\lambda(s^{1/\rho}\cdot) = r\lambda(r^{1/\rho}\cdot), \qquad \lambda = (r/s)\lambda[(r/s)^{1/\rho}\cdot].$$

This is true for any positive numbers $r$ and $s$. Thus (5) holds for any positive rational number, and, passing to limits, we conclude that it holds for any positive number.

By (5), we obtain

$$\begin{aligned} g(\beta) &= \int_{y^{(i)} \geq \beta} \lambda(dy) = r\int_{y^{(i)} \geq \beta} \lambda(r^{1/\rho}dy) = r\int_{z^{(i)} \geq \beta r^{1/\rho}} \lambda(dz) \\ &= rg(\beta r^{1/\rho}) \end{aligned}$$

for any $\beta > 0$, $r > 0$. Putting $r^{1/\rho}\beta = u$, we can write this relation

$$g(u) = (u/\beta_0)^{-\rho} g(\beta_0), \qquad \beta_0 > 0.$$

Proceeding in the same way for $\lambda$ on $y^{(i)} \leq -\beta$, we find that the conditions on $\lambda$ stated in (2°) must be satisfied except the relations $c_1^{(i)} = c_2^{(i)}$ for $\rho = 1$ and $0 < \rho < 2$, which remain to be proved. Now $\lambda$ determines the margin $\lambda^{(i)}$, and this measure is determined by the integrals in (2°). Since $\lambda$ is a Lévy measure, we must have

$$\int_{0 < |y^{(i)}| < \beta} (y^{(i)})^2 \lambda(dy) = (c_1 + c_2)\rho \int_{0 < t < \beta} t^{-\rho+1}\, dt < \infty.$$

Hence, necessarily $0 < \rho < 2$ if $c_1 + c_2 \neq 0$, and clearly $\lambda$ is the zero measure for $c_1 + c_2 = 0$.

Since $\mu$ is infinitely divisible and satisfies (4), and hence also its margins $\mu^{(i)}$ are infinitely divisible and satisfy (4), it follows necessarily by the remark on Theorem 9.1 that

$$m^{(i)}(\beta) = \lim_{n\to\infty} n \int_{|y^{(i)}| < \beta} y^{(i)}\, \mu^{(i)}(n^{1/\rho}\, dy^{(i)})$$

exists and is finite. Hence, also for any positive integer $r$

$$m^{(i)}(\beta) = \lim_{n\to\infty} nr \int_{|y^{(i)}| < \beta} y^{(i)}\, \mu^{(i)}[(nr)^{1/\rho}\, dy^{(i)}].$$

By the substitution $y^{(i)} = r^{-1/\rho}t$, we can write the last relation

$$\begin{aligned} m^{(i)}(\beta) &= r^{(\rho-1)/\rho} \lim_{n\to\infty} n \int_{|t| < \beta} t\, \mu^{(i)}(n^{1/\rho}\, dt) \\ &\quad + r^{(\rho-1)/\rho} \lim_{n\to\infty} n \int_{\beta \leq |t| < \beta r^{1/\rho}} t\, \mu^{(i)}(n^{1/\rho}\, dt). \end{aligned} \tag{6}$$

The first limit on the right-hand side is equal to $m^{(i)}(\beta)$. Since $n\mu^{(i)}(n^{1/\rho}\cdot)$ tends weakly to the measure determined by the integrals in (2°), we find that the second limit on the right-hand side is equal to

$$(c_1^{(i)} - c_2^{(i)})\rho \int_{\beta \le t \le \beta r^{1/\rho}} t^{-\rho}\, dt$$
$$= \begin{cases} (c_1^{(i)} - c_2^{(i)}) \ln r & \text{for} \quad \rho = 1, \\ (c_1^{(i)} - c_2^{(i)}) \dfrac{\rho}{1-\rho} (r^{(1-\rho)/\rho} - 1)\beta^{1-\rho} & \text{for} \quad \rho \ne 1. \end{cases} \tag{7}$$

Since $r > 0$ is arbitrary, it necessarily follows by (6) that $c_1 = c_2$ for $\rho = 1$. For $\rho \ne 1$, the relations (6) and (7) imply

$$m^{(i)}(\beta) = r^{(\rho-1)/\rho} m^{(i)}(\beta) + (c_1^{(i)} - c_2^{(i)}) \frac{\rho}{1-\rho} (1 - r^{(\rho-1)/\rho})\beta^{1-\rho},$$

and thus

$$m^{(i)}(\beta) = (c_1^{(i)} - c_2^{(i)}) \frac{\rho}{1-\rho} \beta^{1-\rho}. \tag{8}$$

Note that (6) implies $m^{(i)}(\beta) = 0$ for $\rho = 2$. If $\lambda$ is the zero measure and $m^{(i)}(\beta) = 0$ for all (i), it follows by Theorem 10.2 that the Gaussian measure with the mean-value vector zero is the only possible probability measure $\ne e$, that is stable in the restricted sense. Returning to the case $0 < \rho < 2$, we compute the Gaussian functional of $\mu$. By the remarks on Theorem 10.2, we obtain

$$m^{(i)} = m^{(i)}(\beta) - \int_{0 < |y^{(i)}| < \beta} \frac{(y^{(i)})^3}{1 + (y^{(i)})^2} \lambda(dy) + \int_{|y^{(i)}| \ge \beta} \frac{y^{(i)}}{1 + (y^{(i)})^2} \lambda(dy),$$

and this relation reduces to the relation for $m^{(i)}$ given in (2°) since $\lambda$ has the special properties given there, and thus (8) holds. Thus we have proved that any probability measure, stable in the restricted sense satisfies the conditions in the theorem. Conversely, let $\mu$ be an infinitely divisible probability measure, which has the Lévy measure $\lambda$ with the properties stated in (2°). Then $\lambda$ determines the Gaussian functional in (2°) and, hence, an infinitely divisible probability measure $\mu$ belonging to this Gaussian functional. We have $\mu = \mu_n^{*n}$ with probability measures $\mu_n$ for all positive integers $n$, and hence

$$[\mu_n(r^{1/\rho}\cdot)]^{*nr} = [\mu(r^{1/\rho}\cdot)]^{*r}$$

for any positive integers $n$ and $r$. Hence $[\mu(r^{1/\rho}\cdot)]^{*r}$ is infinitely divisible. Furthermore, $n\mu_n$ converges weakly to $\lambda$, and thus $nr\mu_n(r^{1/\rho}\cdot)$ converges weakly to $r\lambda(r^{1/r}\cdot)$, but $r\lambda(r^{1/\rho}\cdot) = \lambda$ by assumption. Hence, the infinitely divisible measure $[\mu(r^{1/\rho}\cdot)]^{*r}$ also has the Lévy measure $\lambda$, which determines the Gaussian functional and thus the probability measure uniquely. Hence,

$$[\mu(r^{1/\rho}\cdot)]^{*r} = \mu.$$

This relation implies

$$\mu\left[\left(\frac{N}{r}\right)^{1/\rho}\cdot\right] * \mu\left[\left(\frac{N}{s}\right)^{1/\rho}\cdot\right] = [\mu(N^{1/\rho}\cdot)]^{*r} * [\mu(N^{1/\rho}\cdot)]^{*s}$$
$$= [\mu(N^{1/\rho}\cdot)]^{*r+s} = \mu\left[\left(\frac{N}{r+s}\right)^{1/\rho}\cdot\right]$$

for any positive integers $r$, $s$, and $N$, and thus the relation in (i) for positive rational numbers $\sigma_1$ and $\sigma_2$, but then this relation follows for any real numbers. [Note that $\mu(\cdot/\sigma_n)$ converges weakly to $\mu(\cdot/\sigma)$ if $\{\sigma_n\}$ converges to $\sigma$. □

A probability measure $\mu$ on $R^{(k)}$ is called stable if

$$u^{*n} = \mu\{\bar{c}(n) + [\cdot/\sigma(n)]\} \tag{9}$$

with a vector $\bar{c}(n)$ and a positive number $\sigma(n)$ for any positive integer $n$. Clearly $\mu * \bar{\mu}$ is stable in the restricted sense if $\mu$ is stable and, hence, $\sigma(n) = n^{1/\rho}$, $0 < \rho \leq 2$. For the positive integers $n$ and $m$ we obtain by (9)

$$\mu^{*nm} = \left[\mu\left(\bar{c}(m) + \frac{\cdot}{\sigma(m)}\right)\right]^{*n} = \mu^{*n}\left[nc(m) + \frac{\cdot}{\sigma(m)}\right]$$
$$= \mu\left\{\bar{c}(n) + \frac{\cdot}{\sigma(n)}\left[nc(m) + \frac{\cdot}{\sigma(m)}\right]\right\}.$$

Otherwise

$$\mu^{*nm} = \mu\left[\bar{c}(nm) + \frac{\cdot}{\sigma(nm)}\right].$$

Hence, since $\sigma(nm) = \sigma(n)\sigma(m)$,

$$\bar{c}(nm) = \bar{c}(n) + \frac{n}{\sigma(n)}\bar{c}(m). \tag{10}$$

For $\rho = 1$, we have $\sigma(n) = n$, and thus

$$\bar{c}(nm) = \bar{c}(n) + \bar{c}(m). \tag{11}$$

Otherwise, interchanging $m$ and $n$ in (10), we obtain

$$\bar{c}(n) + \frac{n}{\sigma(n)}\bar{c}(m) = \bar{c}(m) + \frac{m}{\sigma(m)}\bar{c}(n)$$

and

$$\frac{\bar{c}(n)\sigma(n)}{\sigma(n) - n} = \frac{\bar{c}(m)\sigma(m)}{\sigma(m) - m}$$

for all positive integers $n \geq 2$ and, hence,

$$\bar{c}(n) = \frac{\sigma(n) - n}{\sigma(n)}\bar{c} \tag{12}$$

with a constant vector $\bar{c}$. In this case where $\rho \neq 1$, putting $v = \mu(\bar{c} + \cdot)$, we get

$$v^{*n} = \mu^{*n}(n\bar{c} + \cdot) = \mu\left[\bar{c}(n) + \frac{n\bar{c} + \cdot}{\sigma(n)}\right] = \mu\left[\bar{c} + \frac{\cdot}{\sigma(n)}\right] = v\left[\frac{\cdot}{\sigma(n)}\right], \tag{13}$$

and, hence, $v$ is stable in the restricted sense. Conversely, it is easily seen that $\mu$ is stable if $v$ is stable in the restricted sense and $\sigma(n)$ and $c(n)$ are determined as above. In the case $\rho = 1$, the transformation (12) is not possible, and thus we get a special case for $\rho = 1$. We have then the relations (9) and (11), where $\sigma(n) = n$. These relations determine the $\bar{c}(n)$. To show this, it is sufficient to consider the case $k = 1$, since the coordinates of $\bar{c}(n)$ are determined by the margins of $\mu$. Now (9) on $R$ for $\rho = 1$ implies

$$[\mu(-c(n) + n\cdot)]^{*n} = \mu,$$

and then, by the remark on Theorem 9.1,

$$\lim_{n\to\infty} n\,\mu[-c(n) + n\cdot] = \lambda \tag{14}$$

with a Lévy measure $\lambda$, and

$$m(\beta) = \lim_{n\to\infty} n\int_{|x|<\beta} x\,\mu(-c(n) + n\,dx) \tag{15}$$

exists and is finite for any $\beta > 0$ such that $(-\infty, -\beta]$ and $[\beta, \infty]$ are continuity sets for $\lambda$. However, if (14) holds, necessarily $c(n)/n \to 0$ $(n \to \infty)$ and then (14) holds if we change $c(n)$ to 0. Then it follows, as in the proof of (2°) in the theorem, that our Lévy measure $\lambda$ has the properties (2°). Now (15) remains true if we change $n$ to $nr$ for a positive integer $r$. Observing (11), we then get, by transformation of the integral by $x \to x/r + c(r)/nr$,

$$m(\beta) = \lim_{n\to\infty} n \int_{|x+c(r)/n|<\beta r} \left(x + \frac{c(r)}{n}\right) \mu(-c(n)+n\,dx)$$

$$= \lim_{n\to\infty} \left\{ n \int_{|x|<\beta} x\mu[-c(n)+n(dx)] + n \int_{|x|\geq\beta,\, |x+c(r)/n|<\beta r} x\mu(-c(n)+n\,dx) \right.$$

$$\left. + c(r) \int_{|x+c(r)/n|<\beta r} \mu[-c(n)+n(dx)] \right\}. \tag{16}$$

On the right-hand side, the limit of the first integral multiplied by $n$ is $m(\beta)$. The second integral multiplied by $n$ tends to

$$\int_{\beta<|x|<r\beta} x\,\lambda(dx) = \int_{-r\beta}^{-\beta} x\,\lambda(dx) + \int_{\beta}^{r\beta} x\,\lambda(dx).$$

Using (2°), we find, as in (7), that this integral is equal to

$$(c_1 - c_2)\ln r.$$

The last integral on the right-hand side of (16) tends to 1. Hence, regarding (15), we get

$$c(r) = -(c_1 - c_2)\ln r. \tag{17}$$

By Theorem 1 and the remarks above on general stable measures, we have shown how these measures are determined by their Lévy measures.

## 14. GAUSSIAN TRANSFORMS AND GAUSSIAN SEMINORMS OF RANDOM VARIABLES: A COMPARISON METHOD

Let $\xi$ be a random variable from a probability space $(\Omega, \beta, P)$ into $R^{(k)}$, $k \geq 1$. Then $\xi$ and $P$ determine the probability measure $P(\xi^{-1}\cdot)$ on $R^{(k)}$, called the distribution of $\xi$, and

$$\boldsymbol{E}\left[\Phi\left(\frac{x-\xi}{\alpha}\right)\right] = \int_{R^k} \Phi\left(\frac{x-y}{\alpha}\right) P(\xi^{-1}\,dy) \tag{1}$$

is the Gaussian transform of $P(\xi^{-1}\cdot)$. We call $\boldsymbol{E}\{\Phi[(\cdot - \xi)/\alpha]\}$ the Gaussian transform of $\xi$. Different random variables $\xi, \eta, \ldots$ determine different Gaussian transforms. Then

$$\boldsymbol{E}\left[\Phi\left(\frac{x-\xi}{\alpha}\right) - \Phi\left(\frac{x-\eta}{\alpha}\right)\right]$$

is the Gaussian transform of the measure $P(\xi^{-1}\cdot) - P(\eta^{-1}\cdot)$ and

$$\sup_{x \in R^{(k)}} \left| \boldsymbol{E}\left[\Phi\left(\frac{x-\xi}{\alpha}\right) - \Phi\left(\frac{x-\zeta}{\alpha}\right)\right]\right| = p_\alpha[P(\xi^{-1}\cdot) - P(\eta^{-1}\cdot)] \tag{2}$$

is the Gaussian seminorm of $[P(\xi^{-1}\cdot) - P(\eta^{-1}\cdot)]$. We put

$$\overline{p}_\alpha(\xi, \eta) = p_\alpha[P(\xi^{-1}\cdot) - P(\eta^{-1}\cdot)].$$

Then the class $\{\overline{p}_\alpha, \alpha > 0\}$ is a semimetric on the class of random variables from $(\Omega, \beta, P)$ into $R^{(k)}$ in the following sense:

$$\begin{aligned} &\overline{p}_\alpha(\xi, \eta) \geq 0, \qquad \overline{p}_\alpha(\xi, \eta) \leq \overline{p}_\alpha(\xi, \zeta) + \overline{p}_\alpha(\zeta, \eta), \\ &\overline{p}_\alpha(\xi, \eta) = \overline{p}_\alpha(\eta, \xi). \end{aligned} \tag{3}$$

Furthermore, $\xi$ and $\eta$ have the same distribution if and only if $\overline{p}_\alpha(\xi, \eta) = 0$ for all $\alpha > 0$. Note that $\overline{p}_\alpha(\xi, \eta)$ is defined and also satisfies (3) if $\xi, \eta, \zeta$ are random variables into $R^{(k)}$ from different probability spaces. However, without loss of generality, we may consider them as random variables from the same probability space.

**Theorem 1.** *If $\{\xi_n\}$ is a sequence of random variables into $R^{(k)}$, then $\{\xi_n\}$ is Cauchy convergent in the semimetric $\{\overline{p}_\alpha\}$, i.e., $\overline{p}_\alpha(\xi_n, \xi_m) \to 0$ $(n \to +\infty, m \to +\infty)$ for all $\alpha > 0$ if and only if the sequence of distributions of the $\xi_n$ converges weakly.*

*Proof:* The proof follows from Theorem 5.1. □

**Corollary.** *If $\{\xi_n\}$ and $\{\eta_n\}$ are sequences of random variables into R, and the distribution of $\eta_n$ converges weakly to a probability measure $\mu$ on R, then the distribution of $\xi_n$ converges to $\mu$ if*

(i) $$\boldsymbol{E}[|\xi_n - \eta_n|] \to 0, \qquad n \to \infty,$$

*hence, certainly if*

(ii) $$\boldsymbol{E}[(\xi_n - \eta_n)^2] \to 0, \qquad n \to \infty.$$

*Proof:* By the use of the mean-value theorem for integrals, we get, by the definition of $\overline{p}_\alpha(\xi_n, \eta_n)$,

$$\overline{p}_\alpha(\xi_n, \eta_n) \leq (1/\alpha)\|\Phi'\| \boldsymbol{E}|\xi_n - \eta_n|.$$

Hence, (i) implies $\overline{p}_\alpha(\xi_n, \eta_n) \to 0$ $(n \to \infty)$, and by this relation it follows that $\{\xi_n\}$ is Cauchy convergent in the Gaussian semimetric since $\eta_n$ is Cauchy convergent in this metric. Furthermore, (ii) implies (i) since

$$\boldsymbol{E}|\xi_n - \eta_n| \leq \varepsilon + (1/\varepsilon)\boldsymbol{E}(\xi_n - \eta_n)^2$$

for any $\varepsilon > 0$. □

**Lemma 1.** *If* $\xi$, $\eta$, and $\zeta$ *are random variables into* $R^{(k)}$, *and* $\zeta$ *is independent of* $\xi$ *and* $\eta$, *then*

$$\bar{p}_\alpha(\xi + \zeta, \eta + \zeta) \leq \bar{p}(\xi, \eta).$$

*Proof:* Let $\mu_\xi$ denote the distribution of $\xi$. Then $\mu_{\xi+\zeta} = \mu_\xi * \mu_\zeta$ and $\mu_{\eta+\zeta} = \mu_\eta * \mu_\zeta$, since $\zeta$ is independent of $\xi$ and $\eta$, and thus

$$\left|\mathbf{E}\left[\Phi\left(\frac{x-\xi-\zeta}{\alpha}\right) - \Phi\left(\frac{x-\eta-\zeta}{\alpha}\right)\right]\right|$$

$$= \left|\int_{R^{(k)}} \Phi\left(\frac{x-y}{\alpha}\right)\mu_{\xi+\zeta}(dy) - \int_{R^{(k)}} \Phi\left(\frac{x-y}{\alpha}\right)\mu_{\eta+\zeta}\right|$$

$$= \left|\int_{R^{(k)}}\left[\int_{R^{(k)}} \Phi\left(\frac{x-y-z}{\alpha}\right)[\mu_\xi(dy) - \mu_\eta(dy)]\right]\mu_\zeta(dz)\right| \leq \bar{p}_\alpha(\xi, \eta).$$

In the preceding sections we considered convolution products of measures. The distribution of a sum of independent random variables is the convolution product of the distributions of their terms. We dealt with limits for such convolution products. The distribution of a sum of dependent random variables cannot be given by the distributions of the terms alone. In order to give weak limits of distributions of sums of dependent random variables, we compare these sums with corresponding sums of independent random variables and then use the following comparison method, which we give only for random variables into $R$. However, it is easy to see how it may be generalized to random variables into $R^{(k)}$.

**Lemma 2.** *Let* $\xi_i, \eta_i, i = 1, \ldots, n$ *be random variables into* $R$ *with* $\mathbf{E}(\xi_i^2) < +\infty$, $\mathbf{E}(\eta_i^2) < +\infty$, *and let the* $\eta_i$ *be mutually independent and independent of the* $\xi_i$. *Denote the σ-algebra generated by the* $\xi_i$, $i \leq j$, *by* $\mathscr{B}_j$. *Then*

(i) $$\bar{p}_\alpha\left(\sum_{i=1}^{n}\eta_i, \sum_{i=1}^{n}\xi_i\right) \leq \sum_{j=1}^{n}\bar{p}_\alpha\left(\sum_{i=1}^{j-1}\xi_i + \eta_j, \sum_{i=1}^{j}\xi_i\right),$$

(ii) $$\bar{p}_\alpha\left(\sum_{i=1}^{j-1}\xi_i + \eta_j, \sum_{i=1}^{j}\xi_i\right) \leq \frac{1}{\alpha}\mathbf{E}\{|\mathbf{E}(\eta_j) - \mathbf{E}(\xi_j|\mathscr{B}_{j-1})|\}$$

$$+ \frac{1}{2\alpha^2}\mathbf{E}\{|\mathbf{E}(\eta_j^2) - \mathbf{E}(\xi_j^2|\mathscr{B}_{j-1})|\} + \frac{1}{6\alpha^2}\varepsilon\mathbf{E}(\eta_j^2 + \xi_j^2)$$

$$+ \frac{1}{\alpha^2}\left[\int_{|\eta_j| \geq \alpha\varepsilon}\eta_j^2\,dP + \int_{|\xi_j| \geq \alpha\varepsilon}\xi_j^2\,dP\right].$$

*Remark 1:* (i) holds also for random variables into $R^{(k)}$, $k > 1$.

*Remark 2:* If $\{\xi_i\}$, $i = 0, \pm 1, \pm 2, \ldots$ is given and $\mathscr{B}_j$ denotes the σ-algebra generated by the $\xi_i$, $i \leq j$, then (ii) still holds.

*Remark 3:* Note that $\boldsymbol{E}[\xi_j|\mathscr{B}_{j-1}] = \boldsymbol{E}(\xi_j)$, $\boldsymbol{E}(\xi_j^2|\mathscr{B}_{j-1}) = \boldsymbol{E}(\xi_j^2)$ if the $\xi_i$ are mutually independent.

*Proof:* By the properties (3) of $\bar{p}_\alpha$ and the assumed independence, we get, applying Lemma 1,

$$\bar{p}_\alpha\left(\sum_{i=1}^n \eta_i, \sum_{i=1}^n \xi_i\right) \leq \sum_{j=1}^n \bar{p}_\alpha\left(\sum_{i=1}^{j-1} \xi_i + \sum_{i=j}^n \eta_i, \sum_{i=1}^j \xi_i + \sum_{i=j+1}^n \eta_i\right)$$
$$\leq \sum_{j=1}^n \bar{p}_\alpha\left(\sum_{i=1}^{j-1} \xi_i + \eta_j, \sum_{i=1}^j \xi_i\right). \tag{4}$$

We estimate the terms on the right-hand side by the help of the Taylor expansion

$$\Phi(x - y) = \Phi(x) - y\Phi'(x) + \tfrac{1}{2}\Phi''(x) + r(x, y) \tag{5}$$

with the remainder term

$$r(x, y) \begin{cases} \leq \frac{1}{6}\varepsilon y^2\|\Phi'''\| \leq \frac{1}{6}\varepsilon y^2 & \text{for} \quad |y| < \varepsilon, \\ \leq y^2\|\Phi''\| \leq y^2 & \text{for} \quad |y| \geq \varepsilon, \end{cases}$$

since $\|\Phi''\| < 1$, $\|\Phi'''\| < 1$. Observing that also $\|\Phi'\| < 1$, we get by (5)

$$\bar{p}_\alpha\left(\sum_{i=1}^{j-1} \xi_i + \eta_j, \sum_{i=1}^j \xi_i\right)$$
$$= \sup_x \left|\boldsymbol{E}\left\{\Phi\left[\frac{1}{\alpha}\left(x - \sum_{i=1}^{j-1}(\xi_i - \eta_j)\right)\right] - \Phi\left[\frac{1}{\alpha}\left(x - \sum_{i=1}^j \xi_i\right)\right]\right\}\right|$$
$$\leq \frac{1}{\alpha}\boldsymbol{E}\{|\boldsymbol{E}[(\eta_j - \xi_j)|\mathscr{B}_{j-1}]|\} + \frac{1}{2\alpha^2}\boldsymbol{E}\{|\boldsymbol{E}[(\eta_j^2 - \xi_j^2)|\mathscr{B}_{j-1}]|\}$$
$$+ \frac{1}{6\alpha^2}\varepsilon\boldsymbol{E}(\eta_j^2 + \xi_j^2) + \frac{1}{\alpha^2}\left[\int_{|\eta_i| \geq \alpha\varepsilon} \eta_j^2\, dP + \int_{|\xi_i| \geq \alpha\varepsilon} \xi_j^2\, dP\right]. \quad \square$$

The inequalities (i) and (ii) may be used when $\xi_j$ does not depend too strongly on the preceding random variables $\xi_i$, $i < j$. Clearly the summation may be made in such an order that this dependence is as weak as possible. In (ii) we require that the moments of the second order exist, and, hence, this inequality is useful when we consider weak convergence to the Gaussian measure. In the following sections we shall apply (i) and (ii) to random variables that are dependent in different ways. If the random variables $\xi_i$ are independent, with mean value zero and variance $\sigma^2$, and have the same distribution, we choose the $\eta_i$ in (ii) as independent Gaussian random variables with mean value 0 and variance $\sigma^2$. Comparing the sums

$$\frac{1}{\sqrt{n}}\sum_{i=1}^{n}\xi_i, \qquad \frac{1}{\sqrt{n}}\sum_{i=1}^{n}\eta_i,$$

we find directly by Lemma 2 that the first sum tends to the Gaussian distribution with mean value 0 and variance $\sigma^2$ since the second sum has this distribution.

We may also compare the distributions of $\sum_{i=1}^{n}\xi_i$ and $\sum_{i=1}^{n}\eta_i$ when both the $\xi_i$ and the $\eta_i$ are dependent, and then Lemma 3 is useful.

**Lemma 3.** *Let $\{\xi_i\}_{i=1}^{+\infty}$ and $\{\eta_i\}_{i=1}^{+\infty}$ be sequences of random variables from a probability space $(\Omega, \mathscr{B}, P)$ into $R$, and let the variables $\xi_i$ be independent of the random variables $\eta_i$. Then for $\alpha^2 = 2\beta^2$, $\alpha > 0$, $\beta > 0$,*

$$p_\alpha\left(\sum_{i=1}^{n}\eta_i, \sum_{i=1}^{n}\xi_i\right) \le \sum_{j=1}^{n}\left\| \boldsymbol{E}\left\{\Phi\left[\frac{1}{\beta}\left(\cdot - \sum_{i=1}^{j-1}\xi_i\right)\right] * \boldsymbol{E}\left\{\Phi\left[\frac{1}{\beta}\left(\cdot - \sum_{i=j}^{n}\eta_i\right)\right]\right\}\right.$$
$$\left. - \boldsymbol{E}\left\{\Phi\left[\frac{1}{\beta}\left(\cdot - \sum_{i=1}^{j}\xi_i\right)\right] * \boldsymbol{E}\Phi\left[\frac{1}{\beta}\left(\cdot - \sum_{i=j+1}^{n}\eta_i\right)\right]\right\}\right\|.$$

*Proof:* If $\xi$ and $\eta$ are independent random variables into $R$, observing that

$$\Phi\left(\frac{\cdot - \xi}{\beta}\right) * \Phi\left(\frac{\cdot - \eta}{\beta}\right) = \Phi\left(\frac{\cdot - \xi - \eta}{\alpha}\right),$$

we get for $\alpha^2 = 2\beta^2$ and using Fubini's theorem,

$$\boldsymbol{E}\left[\Phi\left(\frac{\cdot - \xi - \eta}{\alpha}\right)\right] = \boldsymbol{E}\left[\Phi\left(\frac{\cdot - \xi}{\beta}\right) * \boldsymbol{E}\Phi\left(\frac{\cdot - \eta}{\beta}\right)\right].$$

By the last relation and the first inequality of (4), we get the Lemma 3. □

## 15. WEAK LIMITS OF DISTRIBUTIONS OF SUMS OF MARTINGALE DIFFERENCES

A sequence $\{\xi_i\}_{i=1}^{+\infty}$ of real-valued random variables from a probability space $(\Omega, \mathscr{B}, P)$ is called a sequence of martingale differences if

$$\boldsymbol{E}[\xi_j | \mathscr{B}_{j-1}] = 0, \qquad j = 1, 2, \ldots, \tag{1}$$

where $\mathscr{B}_0 = \mathscr{B}$ and $\mathscr{B}_j$ for $j > 0$ is the $\sigma$-algebra generated by the variables $\xi_1, \ldots, \xi_j$. The sequence $\{\xi_i\}_{i=1}^{n}$ is called stationary if, for any positive integers $k$ and $j$, the random vectors $(\xi_1, \xi_2, \ldots, \xi_j)$ and $(\xi_k, \xi_{k+1}, \ldots, \xi_{k+j})$ have the same distribution. We call a stationary sequence $\{\xi_i\}_{i=1}^{+\infty}$ mean ergodic if

$$\lim_{n\to+\infty} \boldsymbol{E}\left[\left|\frac{1}{n}\sum_{i=1}^{n} \xi_i - \boldsymbol{E}(\xi_i)\right|\right] = 0. \tag{2}$$

Note that $\boldsymbol{E}[\xi_i\xi_j] = 0$ for $i < j$ when $\{\xi_i\}$ is a sequence of martingale differences. Indeed, $\boldsymbol{E}[\xi_i\xi_j] = \boldsymbol{E}\{\xi_i \boldsymbol{E}[\xi_j | \mathscr{B}_i]\} = 0$.

**Theorem 1.** *Let $\{\xi_i\}_{i=1}^{+\infty}$ be a stationary sequence of martingale differences with $\boldsymbol{E}(\xi_i) = 0$, $\boldsymbol{E}[\xi_j^2] = \sigma^2$, $\sigma > 0$, and suppose that $\{\xi_j^2\}_{j=1}^{+\infty}$ is mean ergodic. Furthermore, let $\{k_n\}_{n=1}^{+\infty}$ be a nondecreasing sequence of positive integers tending to $+\infty$ as $n \to +\infty$. Then the distribution of*

$$S_n = \frac{1}{\sqrt{k_n}} \sum_{i=1}^{k_n} \xi_i$$

*converges weakly to the Gaussian measure with mean value* 0 *and variance* $\sigma^2$.

**Corollary.** *If the $\xi_i$ satisfy the conditions in Theorem* 1 *and $\{k_{n,1}\}_{n=1}^{+\infty}, \ldots, \{k_{n,r}\}_{n=1}^{+\infty}$ are nondecreasing sequences such that $k_{n,1} < k_{n,2} < \cdots < k_{n,r} \leq k_n$, $k_{n,t}/k_n \to b_t$ $(n \to +\infty)$ and $\alpha_1, \ldots, \alpha_r$ are real numbers, then*

$$S_n = \frac{1}{\sqrt{k_n}} \sum_{t=1}^{r} \alpha_t \sum_{i=1}^{k_{n,t}} \xi_i$$

*converges weakly to the Gaussian measure with mean value* 0 *and variance*

$$\sum_{t=1}^{r} \alpha_t^2 b_t \sigma^2.$$

*Proof:* Let $\eta_i$, $i = 1, 2, \ldots$ be independent Gaussian random variables with mean value 0 and variance $\sigma^2$.

We form

$$S_n' = \frac{1}{\sqrt{k_n}} \sum_{i=1}^{k_n} \eta_j.$$

Note that we may consider the $\eta_j$ and $\xi_j$ as random variables from the same probability space (compare Section II.9). Clearly, $S_n'$ is Gaussian with mean value 0 and variance $\sigma^2$. For positive integers $l$, put $m_n(l) = [k_n/l]$. Using Lemma 14.2, we shall prove that the distributions of

$$\bar{S}_n = \frac{1}{\sqrt{k_n}} \sum_{i=1}^{lm_n(l)} \xi_i, \qquad \bar{S}_n' = \frac{1}{\sqrt{k_n}} \sum_{i=1}^{lm_n(l)} \eta_i$$

converge to the same distribution as first $n \to +\infty$ and then $l \to +\infty$, and hence to $\Phi(\cdot/\sigma)$ by the corollary of Theorem 14.1, since

$$\boldsymbol{E}[(S_n' - \bar{S}_n')^2] \le (1/k_n)l\sigma^2 \to 0 \qquad (n \to +\infty).$$

Then also the distribution of $S_n$ tends to $\Phi(\cdot/\sigma)$ by this corollary since

$$\boldsymbol{E}[(S_n - \bar{S}_n)^2] \le (1/k_n)l\sigma^2 \to 0 \qquad (n \to +\infty).$$

In order to compare the distributions of $\bar{S}_n$ and $\bar{S}_n'$, we put

$$\xi_{jn} = \frac{1}{\sqrt{k_n}} \sum_{i=(j-1)l+1}^{jl} \xi_i, \qquad \eta_{jn} = \frac{1}{\sqrt{k_n}} \sum_{i=(j-1)l+1}^{jl} \eta_i$$

for $j = 1, \dots, m_n(l)$ and write

$$\bar{S}_n = \sum_{j=1}^{m_n(l)} \xi_{jn}, \qquad \bar{S}_n' = \sum_{j=1}^{m_n(l)} \eta_{jn}.$$

Comparing these sums by the inequalities in Lemma 14.2, we observe that

$$\boldsymbol{E}[\xi_{jn} | \mathscr{B}_{(j-1)l}] = 0, \qquad \boldsymbol{E}[\eta_{jn}] = 0$$

and

$$\boldsymbol{E}[\xi_{jn}^2 | \mathscr{B}_{(j-1)l}] = \frac{1}{k_n} \boldsymbol{E}\left( \sum_{i=(j-1)l+1}^{jl} \xi_i^2 | \mathscr{B}_{(j-1)l} \right)$$

$$\boldsymbol{E}[\eta_{jn}^2] = \frac{l}{k_n} \sigma^2.$$

Then, using the Jensen inequality (corollary of Theorem II.11.1) and the stationarity of the sequence $\{\xi_i\}$, we get

$$\begin{aligned} \boldsymbol{E}\{|\boldsymbol{E}(\eta_{jn}^2) - \boldsymbol{E}(\xi_{jn}^2 | \mathscr{B}_{(j-1)l})|\} &= \frac{l}{k_n} \boldsymbol{E}\left\{ \left| \boldsymbol{E}\left( \sigma^2 - \frac{1}{l} \sum_{i=(j-1)l+1}^{jl} \xi_i^2 | \mathscr{B}_{(j-1)l} \right) \right| \right\} \\ &\le \frac{l}{k_n} \boldsymbol{E}\left[ \left| \sigma^2 - \frac{1}{l} \sum_{i=1}^{l} \xi_i^2 \right| \right]. \end{aligned}$$

By these estimations and using the stationarity of $\{\xi_i\}$, we obtain from Lemma 14.2

$$\begin{aligned} \bar{p}_\alpha(\bar{S}_n', \bar{S}_n) \le{}& \frac{1}{2\alpha^2} \frac{l \cdot m_n(l)}{k_n} \boldsymbol{E}\left[ \left| \sigma^2 - \frac{1}{l} \sum_{i=1}^{l} \xi_i \right| \right] + \frac{\varepsilon}{6\alpha^2} m_n(l) [\boldsymbol{E}(\eta_{1n}^2) + \boldsymbol{E}(\xi_{1n}^2)] \\ &+ \frac{1}{\alpha^2} m_n(l) \left[ \int_{|\xi_{1n}| \ge \varepsilon\alpha} \xi_{1n}^2 \, P(d\omega) + \int_{|\eta_{1n}| \ge \varepsilon\alpha} \eta_{1n}^2 \, P(d\omega) \right]. \end{aligned} \tag{3}$$

As $n \to +\infty$, $lm_n(l)/k_n$ tends to 1. Hence, by (2) the first term on the right-hand side of (3) tends to 0 as first $n \to +\infty$, and then $l \to +\infty$. The second term is equal to

$$\frac{2\varepsilon}{6\alpha^2} \frac{lm_n(l)}{k_n} \sigma^2 \to \frac{2\varepsilon}{6\alpha^2} \qquad (n \to +\infty).$$

For the third term we get

$$\begin{aligned}
&\frac{1}{\alpha^2} m_n(l) \left[ \int_{|\xi_{1n}| \geq \varepsilon} \xi_{1n}^2 \, P(d\omega) + \int_{|\eta_{1n}| \geq \varepsilon} \eta_{1n}^2 \, P(d\omega) \right] \\
&\quad = \frac{1}{\sigma^2} \frac{m_n(l)}{k_n} \left\{ \int_{|\sum_{i=1}^{l} \xi_i| \geq \varepsilon\alpha\sqrt{k_n}} \left( \sum_{i=1}^{l} \xi_i \right)^2 P(d\omega) \right. \\
&\qquad \left. + \int_{|\sum_{i=1}^{l} \eta_i| \geq \varepsilon\alpha\sqrt{k_n}} \left( \sum_{i=1}^{l} \eta_i \right)^2 P(d\omega) \right\}.
\end{aligned}$$

This quantity tends to 0 as $n \to \infty$ for fixed $l$, since $\sum_{i=1}^{l} \xi_i$ and $\sum_{i=1}^{l} \eta_i$ are random variables with finite variance $l\sigma^2$. It now follows by (3) that $\overline{p}_\alpha(\overline{S}'_n, \overline{S}_n)$, which depends also on $l$, tends to 0 as first $n \to \infty$ and then $l \to \infty$. □

*Proof of the corollary:* Let $\xi_i$, $\eta_i$, $\xi_{in}$, and $\eta_{in}$ be given as above, and put

$$\begin{aligned}
S_n &= \frac{1}{\sqrt{k_n}} \sum_{t=1}^{r} \alpha_t \sum_{i=1}^{k_{nt}} \xi_i, \\
S'_n &= \frac{1}{\sqrt{k_n}} \sum_{t=1}^{r} \alpha_t \sum_{i=1}^{k_{nt}} \eta_i, \\
m_{nt}(l) &= [k_{nt}/l], \\
\overline{S}_n &= \sum_{t=1}^{r} \alpha_t \sum_{j=1}^{m_{nt}(l)} \xi_{jn}, \\
\overline{S}'_n &= \sum_{t=1}^{r} \alpha_t \sum_{j=1}^{m_{nt}(l)} \eta_{jn}.
\end{aligned}$$

It then follows as above that

$$\boldsymbol{E}[S_n - \overline{S}_n)^2] \to 0 \qquad (n \to \infty), \qquad \boldsymbol{E}[(S'_n - \overline{S}'_n)^2] \to 0 \qquad (n \to \infty).$$

Hence it is sufficient to show that

$$\overline{p}_\alpha(\overline{S}_n, \overline{S}'_n) \to 0 \qquad (n \to \infty, l \to \infty)$$

and this follows as in the proof of the theorem. □

## 16. WEAK LIMITS OF DISTRIBUTIONS OF SUMS OF RANDOM VARIABLES UNDER INDEPENDENCE AND $\varphi$-MIXING

Consider random variables $\xi_i$, $i = 1, 2, \ldots$ from a probability space $(\Omega, \mathscr{B}, P)$ into $R$, and let $\mathscr{B}_a^b$ denote the sub–$\sigma$-algebra of $\mathscr{B}$ generated by the $\xi_i$ for $a \le i < b$. The sequence $\{\xi_i\}_{i=1}^{\infty}$ is said to be $\varphi$-mixing if

$$|P(E \cap E') - P(E)P(E')| \le \varphi(k)P(E) \tag{1}$$

with a function $\varphi$ that tends to 0 as $k \to +\infty$ for any $E$ and $E'$ such that $E \in \mathscr{B}_1^a$, $E' \in \mathscr{B}_{a+k}^{\infty}$. Note that (1) can also be written

$$|P(E' \mid E) - P(E')| \le \varphi(k). \tag{1'}$$

The concept of $\varphi$-mixing was introduced by J. A. Ibragimov [21] who also gave the following relation:

**Lemma 1.** *Let $\{\xi_i\}_{i=1}^{\infty}$ be a $\varphi$-mixing sequence from a probability space $(\Omega, \mathscr{B}, P)$ into $R$, and let $\mathscr{B}_a^b$ denote the sub–$\sigma$-algebra of $\mathscr{B}$ generated by the $\xi_i$ for $a \le i < b$. If $\xi \in \mathscr{B}_1^a$, $\eta \in \mathscr{B}_{a+k}^{\infty}$ and $p \ge 1$, $q \ge 1$, $1/p + 1/q = 1$, then*

(i) $$|\mathbf{E}\xi\eta - \mathbf{E}\xi\mathbf{E}\eta| \le 2\varphi^{1/p}(k)\mathbf{E}^{1/p}|\xi|^p\mathbf{E}^{1/q}|\eta|^q.$$

**Corollary.** *If $\{\xi_i\}$ is $\varphi$-mixing and $\mathbf{E}(\xi_i) = 0$ for all $i$, then*

$$\mathbf{E}\left(\sum_{i=1}^{m} \xi_i\right)^2 \le \left[1 + 2\sum_{k=1}^{m-1} \varphi^{1/2}(k)\right]\sum_{i=1}^{m} \mathbf{E}(\xi_i^2).$$

*Proof of Lemma 1:* It is sufficient to prove Lemma 1 for simple random variables since any random variable is the limit of a sequence of such random variables. Hence, consider simple random variables

$$\xi = \sum_i a_i 1_{E_i}, \qquad \eta = \sum_j b_j 1_{E_j'},$$

where the $E_i$ are pairwise disjoint and the $E_j'$ are pairwise disjoint. Then

$$|\mathbf{E}\xi\eta - \mathbf{E}\xi\mathbf{E}\eta| = \left|\sum_{ij} a_i b_j [P(E_i E_j') - P(E_i)P(E_j')]\right|. \tag{2}$$

Since $1 = 1/p + 1/q$, we can write (2) as

$$|\mathbf{E}\xi\eta - \mathbf{E}\xi\mathbf{E}\eta| = \left|\sum_i a_i P^{1/p}(E_i) g_i\right|, \tag{3}$$

$$\begin{aligned} |g_i| &\le \sum_j |b_j| [P(E_i) | P(E_j' \mid E_i) - P(E_j')|]^{1/q} [|P(E_j' \mid E_i) - P(E_j')|]^{1/p} \\ &\le \sum_j 2^{1/q} |b_j| [P(E_j')]^{1/q} |P(E_j' \mid E_i) - P(E_j')|^{1/p}. \end{aligned} \tag{4}$$

Applying Hölders inequality to the sum in (3) and the last sum in (4), we get

$$|\mathbf{E}\xi\eta - \mathbf{E}\xi\mathbf{E}\eta| \leq \left[\sum_i |a_i|^p P(E_i)\right]^{1/p} \left[\sum_i |g_i|^q\right]^{1/q}, \tag{5}$$

$$|g_i| \leq \left[2\sum_j |b_j|^q P(E'_j)\right]^{1/q} \left\{\sum_j |P(E'_j|E_i) - P(E'_j)|\right\}^{1/p}. \tag{6}$$

We now estimate the last sum in (6). Let $\hat{E}'$ be the union of those $E'_j$ for which $P(E'_j|E_i) - P(E'_j) > 0$ and $\tilde{E}'$ be the union of those $E'_j$ for which $P(E'_j|E_i) - P(E'_j) < 0$. Then, regarding (1), we get

$$\sum_j \left|P(E'_j|E_i) - P(E'_j)\right| = P(\hat{E}|E_i) - P(\hat{E}') + P(\tilde{E}') - P(\tilde{E}'|E_i) \leq 2\varphi(k).$$

Using this estimation and combining (5) and (6), we get the inequality of Lemma 1.

*Proof of the corollary:* By the use of Lemma 1, we get

$$\mathbf{E}\left(\sum_{i=1}^m \xi_i\right)^2 = \sum_{i=1}^m \mathbf{E}(\xi_i^2) + 2\sum_{i=1}^m \sum_{k=1}^{m-i} \mathbf{E}\xi_i\xi_{i+k}$$
$$\leq \sum_{i=1}^m \mathbf{E}(\xi_i^2)\left[1 + 2\sum_{k=1}^{m-1} \varphi^{1/2}(k)\right]. \quad \square$$

A $\varphi$-mixing sequence of random variables is rather close to an independent sequence of random variables and, of course, an independent sequence is $\varphi$-mixing with $\varphi(k) = 0$ for all $k$.

In the following, our limit theorems for sums of random variables in an independent sequence are obtained as special cases of limit theorems for sums of random variables in $\varphi$-mixing sequences.

We say that a $\varphi$-mixing sequence $\{\xi_i\}$ of random variables satisfies the Ibragimov (I) condition if

$$\sum_{i=1}^{\infty} \varphi^{1/2}(k) < \infty. \tag{7}$$

The famous Lindeberg condition (L) for a sequences $\{\xi_i^{(n)}\}_{i=1}^{\infty}$ of random variables $\xi_i^{(n)}$ from a probability space $(\Omega, \mathscr{B}, P)$ and a nondecreasing sequence $\{k_n\}$ of positive integers is given by

$$\lim_{n\to\infty} \frac{1}{s_n^2} \sum_{i=1}^{k_n} \int_{|\xi_i^{(n)}| \geq \varepsilon s_n} (\xi_i^{(n)})^2 \, P(dw) = 0, \tag{8}$$

where

$$s_n^2 = \sum_{i=1}^{k_n} \mathbf{E}[\xi_i^{(n)}]^2.$$

For any $\xi_i^{(n)}$, define $\xi_i^{(n)}(\varepsilon)$ by

$$\xi_i^{(n)}(\varepsilon) = \begin{cases} \xi_i^{(n)} & \text{for} \quad |\xi_i^{(n)}| < \varepsilon s_n, \\ 0 & \text{for} \quad |\xi_i^{(n)}| \geq \varepsilon s_n. \end{cases}$$

**Lemma 2.** *If the sequence $\{\xi_i^{(n)}\}_{i=1}^{\infty}$ for the nondecreasing sequence $\{k_n\}$ of positive integers satisfies the Lindeberg condition, then*

$$\lim_{n\to\infty} \sum_{i=1}^{k_n} P\{|\xi_i^{(n)} - \xi_i^{(n)}(\varepsilon)| \geq \varepsilon s_n\} = 0$$

*for any $\varepsilon > 0$.*

*Proof:* By Chebyshev's inequality, we get

$$P[|\xi_i^{(n)} - \xi_i^{(n)}(\varepsilon)| \geq \varepsilon s_n] \leq \frac{1}{\varepsilon^2 s_n^2} \int_{|\xi_i^{(n)}| \geq \varepsilon s_n} [\xi_i^{(n)}]^2 \, P(dw).$$

Hence, the sum considered above tends to 0 for any $\varepsilon > 0$ when (8) holds. □

**Theorem 1.** *Let $\{\xi_i^{(n)}\}_{i=1}^{\infty}$ be $\varphi$-mixing sequences of random variables with $E[\xi_i^{(n)}] = 0$, and $\{k_n\}$ a nondecreasing sequence of positive integers. Put*

$$s_n^2 = \sum_{i=1}^{k_n} \boldsymbol{E}[\xi_i^{(n)}]^2, \qquad s_n > 0.$$

*If $\{\xi_i^{(n)}\}$ satisfies the I condition and $\varphi$ is independent of n and the L condition, and if*

$$\lim_{n\to\infty} \frac{1}{s_n^2} \boldsymbol{E}\left(\sum_{i=1}^{k_n} \xi_i^{(n)}\right)^2 = \sigma^2, \qquad 0 < \sigma < \infty$$

*exists, then the distribution function of $(1/s_n)\sum_{i=1}^{k_n} \xi_i^{(n)}$ converges weakly to the Gaussian measure with mean value 0 and variance $\sigma^2$.*

**Corollary.** *If the sequence $\{\xi_i^{(n)}\}_{i=1}^{\infty}$ is independent for all n, then $\sigma = 1$.*

*Proof:* Changing $\xi_i^{(n)}/s_n$ to $\xi_i^{(n)}$ in our notation, we may consider the case $s_n = 1$. We introduce the Gaussian random vectors

$$(\zeta_i^{(n)}, \ldots, \zeta_{k_n}^{(n)})$$

for $n = 1, 2, \ldots$ such that

$$\boldsymbol{E}\zeta_i^{(n)} = 0,$$

$$\boldsymbol{E}\zeta_j^{(n)}\zeta_{j-v}^{(n)} = \begin{cases} \boldsymbol{E}\xi_j^{(n)}\xi_{j+v}^{(n)} & \text{for} \quad 0 \leq v < j, \quad j + v \leq k_n, \\ \boldsymbol{E}\xi_j^{(n)}\xi_{j-v}^{(n)} & \text{otherwise.} \end{cases}$$

Note that $(\zeta_1^{(n)} + \cdots + \zeta^{(n)})$ is then Gaussian with mean value 0 and variance

$$E(\zeta_1^{(n)} + \cdots + \zeta^{(n)}]^2.$$

Furthermore, by Theorem 3.8 a sequence of Gaussian measures $\mu_n$ with mean values 0 converges weakly if and only if the sequence of variances of the $\mu_n$ converges to a finite value.

In the following, we omit $n$ in our notations for the random variables; hence, we write $\zeta$ instead of $\zeta^{(n)}$, remembering that they depend on $n$. We now compare the distributions of $\sum_{i=1}^{k_n} \xi_i$ and $\sum_{i=1}^{k_n} \zeta_i$ by the help of Lemma 14.3. We have to prove that

$$\lim_{n\to\infty} \Bigg\| \sum_{j=1}^{k_n} \Bigg\{ E\Phi\left[\frac{1}{\beta}\left(\cdot - \sum_{i=1}^{j-1} \zeta_i\right)\right] * E\Phi\left[\frac{1}{\beta}\left(\cdot - \sum_{i=j}^{k_n} \xi_i\right)\right] - E\Phi\left[\frac{1}{\beta}\left(\cdot - \sum_{i=1}^{j} \zeta_i\right)\right] * E\Phi\left[\frac{1}{\beta}\left(\cdot - \sum_{i=j+1}^{k_n} \xi_i\right)\right]\Bigg\}\Bigg\| = 0. \tag{9}$$

If we change $\xi_i$ to $\xi_i\beta$ and $\zeta_i$ to $\zeta_i\beta$, all our assumptions for Theorem 1 remain unchanged except $s_n$, which changes to $s_n\beta$. Hence, it is sufficient to consider the case $\beta = 1$ in order to prove (9). By the L condition, $E[\xi_i^2]$ and, hence, $E[\zeta_i^2]$ tend to 0 uniformly for $1 \le i \le k_n$ as $n \to \infty$. Using this fact and Lemma 2, we shall reduce the left-hand side of (9). To begin with, we conclude that (9) holds, if

$$\lim_{k\to\infty} \limsup_{n\to\infty} \Bigg\| \sum_{j=k+1}^{k_n-k-1} \Bigg\{ E\Phi\left(\cdot - \sum_{i=1}^{j-1} \zeta_i\right) * E\Phi\left(\cdot - \sum_{i=j}^{k_n} \xi_i\right) - E\Phi\left(\cdot - \sum_{i=1}^{j} \zeta_i\right) * E\Phi\left(\cdot - \sum_{i=j+1}^{k_n} \xi_i\right)\Bigg\}\Bigg\| = 0. \tag{10}$$

Now we observe that $(\zeta_1, \ldots, \zeta_{k_n})$ is a Gaussian vector with $E(\zeta_i) = 0$. Hence $\sum_{i=1}^{j} \zeta_i$ is Gaussian with mean value 0 and variance

$$\hat{\sigma}_j^2 = E\left(\sum_{i=1}^{j} \zeta_i\right)^2.$$

Then, since

$$\Phi\left(\frac{\cdot}{\sigma_j}\right) = \Phi(\cdot) * \Phi\left(\frac{\cdot}{\hat{\sigma}_j}\right) = E\left[\Phi\left(\cdot - \sum_{i=1}^{j} \zeta_i\right)\right]$$

with

$$\sigma_j^2 = E\left(\sum_{i=1}^{j} \zeta_i\right)^2 + 1,$$

the relation (10) holds if

$$\Delta = \lim_{k\to\infty} \limsup_{n\to\infty} \left\| \sum_{j=k+1}^{k_n-k-1} \left\{ \Phi(\cdot) * \boldsymbol{E}\Phi\left(\cdot - \sum_{i=j}^{k_n} \xi_i\right) - \Phi\left(\frac{\cdot}{\delta_j}\right) * \boldsymbol{E}\Phi\left(\cdot - \sum_{i=j+1}^{k_n} \xi_i\right)\right\}\right\| = 0 \tag{11}$$

with $\delta_j^2 = 1 + \sigma_j^2 - \sigma_{j-1}^2 = 1 + 2\boldsymbol{E}(\zeta_j \sum_{i=1}^{j-1} \zeta_i) + \zeta_j^2$. In order to show that $\Delta = 0$, we use

**Lemma 3.**

(i) $$\left|\Phi\left(\frac{x}{\delta_j}\right) - \Phi(x) - \tfrac{1}{2}(\sigma_j^2 - \sigma_{j-1}^2)\Phi''(x)\right| \le c[\sigma_j^2 - \sigma_{j-1}^2]^2$$

*with an absolute constant c for* $|\sigma_j^2 - \sigma_{j-1}^2| < \frac{1}{2}$ *where*

(ii) $\sigma_j^2 - \sigma_{j-1}^2$ *tends to* 0 *uniformly with respect to j, as* $n \to \infty$, *and*

(iii) $\sum_{j=k+1}^{k_n-k-1} |\sigma_j^2 - \sigma_{j-1}^2| \le$ *constant (independent of k and n). Furthermore,*

(iv) $$\lim_{k\to\infty} \limsup_{n\to\infty} \sum_{j=k+1}^{k_n-k-1} \left|\sigma_j^2 - \sigma_{j-1}^2 - \boldsymbol{E}\left[\xi_j^2 + 2\xi_j \sum_{i=j+1}^{j+k} \xi_i\right]\right| = 0.$$

*Proof:* Expanding $\Phi(x/\sqrt{y})$ for constant $x$ and $y$ in the interval $y \ge \frac{1}{2}$ by Taylor's formula, we get

$$\frac{d}{dy}\Phi\left(\frac{x}{\sqrt{y}}\right) = \frac{1}{2y}\Phi''\left(\frac{x}{\sqrt{y}}\right), \qquad \frac{d^2}{dy^2}\Phi\left(\frac{x}{\sqrt{y}}\right) = \frac{1}{4y^2}\Phi^{(IV)}\left(\frac{x}{\sqrt{y}}\right)$$

and, hence,

$$\left|\Phi\left(\frac{x}{\delta_j}\right) - \Phi(x) - \tfrac{1}{2}(\sigma_j^2 - \sigma_{j-1}^2)\Phi''(x)\right| \le \tfrac{1}{8}(\sigma_j^2 - \sigma_{j-1}^2)^2 \max_x |\Phi^{(IV)}(x)|.$$

The relations (ii)–(iv) are corollaries of Lemma 4. Note that

$$\sigma_j^2 - \sigma_{j-1}^2 = \boldsymbol{E}\left[\zeta_j^2 + 2\zeta_j \sum_{i=1}^{j-1} \zeta_i\right].$$

**Lemma 4.**

(i) $$\left|\boldsymbol{E}\left[\xi_j \sum_{\substack{v>k>1\\ \text{all } j+v \text{ different}}} \xi_{j+v}\right]\right| \le \boldsymbol{E}^{1/2}(\xi_j^2)\left[\sum_{i=1}^{k_n} \boldsymbol{E}(\xi_i^2)\right]^{1/2}\left[\sum_{v=k}^{\infty} \varphi(v)\right]^{1/2},$$

(ii) $$\sum_{j=k+1}^{k_n-k-1} \left|\boldsymbol{E}\left(\xi_j \sum_{\substack{v>k>1\\ \text{all } j+v \text{ different}}} \xi_{j+v}\right)\right| \le \sum_{i=1}^{k_n} \boldsymbol{E}(\xi_i^2) \sum_{v=k}^{\infty} \varphi^{1/2}(v).$$

*Proof:* Since $\boldsymbol{E}(\xi_j) = 0$, we get by Lemma 1

$$|\boldsymbol{E}(\xi_j \xi_{j+v})| \leq \boldsymbol{E}^{1/2}(\xi_j^2)\boldsymbol{E}^{1/2}(\xi_{j+v}^2)\varphi^{1/2}(v).$$

By Cauchy's inequality we then get (i). Furthermore, by this inequality we obtain

$$\sum_{j=k+1}^{k_n-k-1} \sum_{\substack{v \geq k \\ \text{all } j+v \text{ different}}} |\boldsymbol{E}(\xi_j \xi_{j+v})| \leq \sum_{j=k+1}^{k_n-k-1} \sum_{v \geq k} \boldsymbol{E}^{1/2}(\xi_j^2)\boldsymbol{E}^{1/2}(\xi_{j+v}^2)\varphi^{1/2}(v)$$

$$\leq \sum_{i=1}^{k_n} \boldsymbol{E}[\xi_i^2] \sum_{v=k}^{\infty} \varphi^{1/2}(v).$$

We now return to the estimation of (11). Observing that

$$\Phi''(\cdot) * \Phi(-m) = \Phi(\cdot) * \Phi''(\cdot - m)$$

for any real number $m$, we find by the use of Lemma 3 that (11) and, hence, (10) are satisfied if

$$\Delta = \lim_{k\to\infty} \limsup_{n\to\infty} \left\| \sum_{j=k+1}^{k_n-k-1} \left\{ \boldsymbol{E}\left[ \Phi\left(\cdot - \sum_{i=j}^{k_n} \xi_i\right) - \Phi\left(\cdot - \sum_{i=j+1}^{k_n} \xi_i\right)\right] \right.\right.$$
$$\left.\left. -\tfrac{1}{2}\boldsymbol{E}\left[\xi_j^2 + 2\xi \sum_{i=j+1}^{j+k} \xi_i\right] \boldsymbol{E}\Phi''\left(\cdot - \sum_{i=j+1}^{k_n} \xi_i\right)\right\}\right\| = 0. \tag{12}$$

We perform different changes in the terms of the sums on the right-hand side of (12). Such a change is permitted if it does not change $\Delta$. It follows by Lemma 2 that the change of $\xi_i$ to $\xi_i(\varepsilon)$ for any $\xi_i$, $i = j, j+1, \ldots, j+2k$ appearing in the term for the index $j$ of the outer sum is permitted. Then, using the Taylor expansion, we change

$$\Phi\left(x - \sum_{i=j}^{k_n} \xi_i\right) - \Phi\left(x - \sum_{i=j+1}^{k_n} \xi_i\right) \tag{13}$$

to

$$-\xi_j(\varepsilon)\Phi'\left(x - \sum_{i=j+1}^{k_n} \xi_i\right) + \tfrac{1}{2}[\xi_j(\varepsilon)]^2\Phi''\left(x - \sum_{i=j+1}^{k_n} \xi_i\right). \tag{14}$$

This change for any term (13) is permitted since it only introduces an error at most equal to $\frac{1}{6}\varepsilon\|\Phi'''\|\boldsymbol{E}[\xi_j^2]$, and the sum of these errors for the different terms (13) is at most equal to

$$\tfrac{1}{6}\varepsilon\|\Phi'''\| \sum_{i=1}^{k_n} \boldsymbol{E}(\xi_i^2) = \tfrac{1}{6}\varepsilon\|\Phi\|'''\beta \qquad \left(\beta = \sum_{i=1}^{k_n} \boldsymbol{E}\xi_i^2\right)$$

with $\varepsilon$ arbitrarily small. Again using the Taylor expansion we change (14) to

$$\xi_j(\varepsilon) \sum_{i=j+1}^{j+k} \xi_i(\varepsilon)\Phi''\left(x - \sum_{i=j+k+1}^{k_n} \xi_i\right) + \tfrac{1}{2}\xi_j^2(\varepsilon)\Phi''\left(x - \sum_{i=j+k+1}^{k_n} \xi_i\right). \tag{15}$$

In the Taylor expansion, we have canceled the first term

$$\xi_j(\varepsilon)\Phi'\left(\cdot - \sum_{i=j+k+1}^{k_n} \xi_i\right).$$

Indeed, by Lemma 2 we may as well consider the term

$$\xi_j\Phi'\left(x - \sum_{i=j+k+1}^{k_n} \xi_i\right). \tag{16}$$

By Lemma 1, we get

$$\begin{aligned}\left|\boldsymbol{E}\xi_j\Phi'\left(x - \sum_{i=j+k+1}^{k_n} \xi_i\right)\right| &= \left|\boldsymbol{E}\xi_j\left[\Phi'\left(x - \sum_{j=k+1}^{k_n} \xi_i\right) - \Phi'(x)\right]\right| \\ &\le \sum_{v=k+1}^{k_n-j-1} \left|\boldsymbol{E}\xi_j\left[\Phi'\left(x - \sum_{i=j+v}^{k_n} \xi_i\right)\right.\right. \\ &\qquad \left.\left. - \Phi'\left(x - \sum_{i=j+v+1}^{k_n} \xi_i\right)\right]\right| \\ &\le \|\Phi''\| \sum_{v=k}^{k_n-j-1} \boldsymbol{E}^{1/2}\xi_j^2 \boldsymbol{E}^{1/2}\xi_{j+v}^2 \varphi^{1/2}(v).\end{aligned}$$

By Lemma 4, the sum over these terms corresponding to the terms (16) is at most equal to

$$\|\Phi''\|\beta \sum_{v=k}^{\infty} \varphi^{1/2}(v).$$

By the $I$ condition, this term tends to 0 as $k \to \infty$. The remainder term in the Taylor expansion, when we change (14) to (15), introduces an error at most equal to

$$\tfrac{1}{2}\varepsilon\|\Phi'''\|\boldsymbol{E}\left(\sum_{i=j+1}^{j+k} \xi_i\right)^2 + \tfrac{1}{2}k\varepsilon\|\Phi'''\|\boldsymbol{E}\xi_j^2(\varepsilon) \le \tfrac{1}{2}k\varepsilon\|\Phi'''\|\boldsymbol{E}\sum_{i=j}^{j+k} \xi_j^2,$$

and the sum over these terms for different $j$ is at most equal to $\frac{1}{2}k^2\varepsilon\|\Phi'''\|$ ($\varepsilon \to 0$ as $n \to \infty$). Thus the change (14) to (15) in the terms is permitted. By the procedure used above, we find that in (15) we may cancel $\xi_i$ for $j + k + 1 < i \le j + 2k + 1$ and then change the $\xi_i(\varepsilon)$ in (15) so obtained back to $\xi_i$. Hence, it is permitted to change all terms (13) in (12) to

$$\xi_j \sum_{i=j+1}^{j+k} \xi_i \Phi''\left(x - \sum_{i>j+2k} \xi_i\right) + \tfrac{1}{2}\xi_j^2 \Phi''\left(x - \sum_{i>j+2k} \xi_i\right).$$

Applying Lemma 1, we obtain

$$\left| \boldsymbol{E}\xi_j \sum_{i=j+1}^{j+k} \xi_i \Phi''\left(x - \sum_{i>j+2k} \xi_i\right) - \boldsymbol{E}\left(\xi_j \sum_{i=j+1}^{j+k} \xi_i\right) \boldsymbol{E}\Phi''\left(x - \sum_{i>j+2k} \xi_i\right)\right.$$
$$\leq 2\|\Phi''\| \left| \boldsymbol{E}\left(\xi_j \sum_{i=j+1}^{j+k} \xi_i\right) \varphi(k)\right|,$$

$$\left| \boldsymbol{E}\xi_j^2 \Phi''\left(x - \sum_{i>j+2k} \xi_i\right) - \boldsymbol{E}(\xi_i^2) \boldsymbol{E}\Phi''\left(x - \sum_{i>j+2k} \xi_i\right)\right|$$
$$\leq 2\|\Phi''\| \boldsymbol{E}[\xi_j^2] \varphi(k),$$

and we conclude as that it is permitted to change (15) and, hence, (13) into

$$\tfrac{1}{2}\boldsymbol{E}\left[\xi_j^2 + 2\xi_j \sum_{i=j+1}^{j+k} \xi_i\right] \Phi''\left(x - \sum_{i>j+2k} \xi_i\right).$$

It is now easy to show that it is permitted to change this term to

$$\tfrac{1}{2}\boldsymbol{E}\left[\xi_j^2 + 2\xi_j \sum_{i=j+1}^{j+k} \xi_i\right] \boldsymbol{E}\left[\Phi''\left(x - \sum_{i=j+1}^{k_n} \xi_i\right)\right]$$

so that (12) is satisfied. Indeed, since by Lemma 3

$$\limsup_{n\to\infty} \sum_{j=k+1}^{k_n-k-1} \left| \boldsymbol{E}\left[\xi_j^2 + 2\xi_j \sum_{i=j+1}^{j+k} \xi_i\right]\right| < \infty$$

for fixed $k$, we have only to prove that

$$\limsup_{n\to\infty} \left\| \boldsymbol{E}\left[\Phi''\left(\cdot - \sum_{i=j+1}^{k_n} \xi_i\right) - \Phi''\left(\cdot - \sum_{i-j+2k+1}^{k_n} \xi_i\right)\right]\right\| = 0,$$

and this relation holds since by the Lindeberg condition

$$\lim_{n\to\infty} \boldsymbol{E}\left\{\sum_{i=j+1}^{j+2k} |\xi_i|\right\} = 0$$

uniformly with respect to $j$. □

CHAPTER V

# WEAK CONVERGENCE ON THE $C$- AND $D$-SPACES

## 1. THE $C$- AND $D$-SPACES

The space $D[\alpha, \beta]$ on a finite interval $[\alpha, \beta]$ is the class of all real-valued, bounded functions $x$ on $[\alpha, \beta]$ such that $x$ has a right limit $x(t+)$ for any $t \in [\alpha, \beta)$ and a left limit $x(t-)$ for any $t \in (\alpha, \beta]$. Furthermore, it is required that $x$ is right continuous on $[\alpha, \beta)$, i.e., $x(t) = x(t+)$ for $t \in [\alpha, \beta)$. The last requirement, if not satisfied, can be obtained by a normalization such that $x(t)$ is changed into $x(t+)$. The space $C(\alpha, \beta)$ is the class of all continuous functions contained in $D(\alpha, \beta)$.

In order to analyze the $D$-space, we introduce the finite partition property for the oscillation of a function from these spaces. We say that a real-valued function on $[\alpha, \beta]$ has this property if for any given $\varepsilon > 0$ there exists an integer $n$ and points $t_i$, $\alpha = t_0 < t_1 < t_2 < \cdots < t_n = \beta$, such that the oscillation of $x$ on $(t_{i-1}, t_i)$ is smaller than $\varepsilon$. Note that we consider open intervals.

**Theorem 1.** *A real-valued function $x$ on $[\alpha, \beta]$ is bounded and has right limits at $t \in [\alpha, \beta)$ and left limits at $t \in (\alpha, \beta]$ if and only if $x$ has the finite partition property for the oscillation on $[\alpha, \beta]$. Furthermore, this property admits at most countably many discontinuity points for $x$ on $[\alpha, \beta]$.*

*Proof:* Suppose that $x$ is bounded and has right and left limits as required in the theorem, but does not have the finite partition property for the oscillation on $[\alpha, \beta]$. Then it does not have this property on one of the intervals $[\alpha, \alpha_1]$, $[\alpha_1, \beta]$, where $\alpha_1$ is the middle point of $[\alpha, \beta]$. Repeating this argument, we conclude that for some $\varepsilon > 0$ there exists a sequence $[\alpha_n, \beta_n]$, $[\alpha_n, \beta_n] \subset [\alpha_{n-1}, \beta_{n-1}]$, $\beta_n - \alpha_n = 2^{-n}(\beta - \alpha)$ of subintervals of $[\alpha, \beta]$ such that $x$ does not have this property on any interval $[\alpha_n, \beta_n]$. These intervals have a limit point $\tau$, which may be $\alpha$ or $\beta$; but certainly $x$ cannot have both right and left limits at $\tau$ if $\tau \in (\alpha, \beta)$, not a right limit if $\tau = \alpha$, and not a left limit if $\tau = \beta$. This contradiction proves that $x$ has the finite partition property for the oscillation on $[\alpha, \beta]$ under the assumption made. Conversely, let $x$ have this property. Clearly $x$ is then bounded. If $t \in [\alpha, \beta)$, there exists a decreasing sequence $\{\delta_n\}$ of positive numbers tending to 0 as $n \to +\infty$ such that the oscillation of $x$ on $(t, t + \delta_n) \subset [\alpha, \beta)$ is smaller than $1/n$. Thus $x$ has a right limit at $t$. In the same way, we conclude that $x$ has a left limit at any point $t \in (\alpha, \beta]$. Furthermore, there can be at most finitely many discontinuity points $t \in (\alpha, \beta)$ with jumps $|x(t+) - x(t-)| \geq 1/m$ for a given positive integer $m$. Thus the discontinuity points on $[\alpha, \beta]$ are countable. □

The spaces $D[\alpha, \beta]$ and $C(\alpha, \beta)$ can be transformed into the spaces $D[0, 1]$ and $C[0, 1]$ by the transformation

$$y(t) = [x(t) - x(\alpha)]/(\beta - \alpha).$$

Hence, in the following we only deal with $D[0, 1]$ and $C[0, 1]$ and use the notations $D$ and $C$ for these spaces. They become metric spaces under the uniform metric

$$(x, y) \to \rho_0(x, y) = \sup_{t \in [0,1]} |x(t) - y(t)|. \tag{1}$$

On $D$ we use the Skorokhod metric $\rho$, defined by

$$(x, y) \to \rho(x, y) = \inf_{\lambda \in \Lambda} \max \left\{ \sup_{t \in [0,1]} |x \circ \lambda(t) - y(t)|, \sup_{t \in [0,1]} |\lambda(t) - t| \right\}, \tag{2}$$

where $\Lambda$ is the class of strictly increasing continuous functions $\lambda$ on $[0, 1]$ with $\lambda(0) = 0, \lambda(1) = 1$.

**Theorem 2.** *$(x, y) \to \rho(x, y)$ for $x, y \in D$ is a metric.*

*Proof:* Since $\lambda^{-1}$ belongs to $\Lambda$, we have $\rho(x, y) = \rho(y, x)$. Clearly $\rho(x, y) \geq 0$ and $\rho(x, y) = 0$ if $x(t) = y(t)$ for all $t$ since $\lambda \in \Lambda$ if $\lambda(t) = t$ for $t \in [0, 1]$. To prove the triangle inequality for $\rho$, we consider arbitrary functions $x, y, z$ from $D$ into $[0, 1]$ and choose $\lambda_1, \lambda_2 \in \Lambda$ for given $\varepsilon > 0$ such that

$$\rho(x, z) \geq \max \left\{ \sup_{t \in [0,1]} |x \circ \lambda_1(t) - z(t)|, \sup_{t \in [0,1]} |\lambda_1(t) - t| \right\} - \varepsilon, \tag{3}$$

$$\rho(z, y) \geq \max \left\{ \sup_{t \in [0,1]} |z \circ \lambda_2(t) - y(t)|, \sup_{t \in [0,1]} |\lambda_2(t) - t| \right\} - \varepsilon. \tag{4}$$

Observing that $\lambda_1 \circ \lambda_2 \in \Lambda$, we get

$$\rho(x, y) \leq \max\left\{ \sup_{t \in [0,1]} \left|x \circ \lambda_1 \circ \lambda_2(t) - y(t)\right|, \sup_{t \in [0,1]} \left|\lambda_1 \circ \lambda_2(t) - t\right| \right\},$$

$$\begin{aligned} \sup_{t \in [0,1]} \left|x \circ \lambda_1 \circ \lambda_2(t) - y(t)\right| &\leq \sup_{t \in [0,1]} \left|x \circ \lambda_1 \circ \lambda_2(t) - z \circ \lambda_2(t)\right| \\ &\quad + \sup_{t \in [0,1]} \left|z \circ \lambda_2(t) - y(t)\right| \\ &\leq \rho(x, z) + \rho(z, y) + 2\varepsilon, \end{aligned} \tag{5}$$

and

$$\begin{aligned} \sup_{t \in [0,1]} \left|\lambda_1 \circ \lambda_2(t) - t\right| &\leq \sup_{t \in [0,1]} \left|\lambda_1 \circ \lambda_2(t) - \lambda_2(t)\right| + \sup_{t \in [0,1]} \left|\lambda_2(t) - t\right| \\ &\leq \rho(x, z) + \rho(z, y) + 2\varepsilon. \end{aligned}$$

Hence, by (5),

$$\rho(x, y) \leq \rho(x, z) + \rho(z, y) + 2\varepsilon.$$

Since $\varepsilon > 0$ is arbitrary, we have

$$\rho(x, y) \leq \rho(x, z) + \rho(z, y). \quad \square$$

The Skorokhod metric is more suitable than the uniform metric in the $D$-space. Indeed, the $\rho_0$-distance $\rho_0(x, y)$ between two functions $x$ and $y$ in $D$ may be large even if they differ only in an arbitrarily small neighborhood of a certain point $t \in [0, 1]$, but the $\rho$-distance may be small.

The $\rho_0$-metric determines metrices on $C$ and $D$, and $\rho$ determines a metric on $D$ and also a metric on $C$. These metrics determine topologies on $C$ and $D$. However, we have

**Theorem 3.** *The metrics $\rho_0$ and $\rho$ determine the same topology on $C$. On $D$ we have $\rho(x, y) \leq \rho_0(x, y)$.*

*Proof:* By Lemma I.8.2, a real-valued function $f$ from a metric space with metric $\rho$ is continuous if and only if for any $x$ and for any sequence $\{x_n\}$ in the metric space with $\rho(x_n, x) \to 0$ $(n \to +\infty)$, we have $f(x_n) - f(x) \to 0$ $(n \to +\infty)$. Clearly, $\rho(x, y) \leq \rho_0(x, y)$ for $x, y \in D$ or $C$ and the metrics (1) and (2). Hence, if $f$ is a continuous function from $C$ with respect to the topology determined by $\rho_0$, it is a continuous function with respect to the topology $\rho$; but if $\rho(x_n, x) \to 0, n \to +\infty$, for $x_n, x \in C$, then clearly $\rho_0(x_n, x) \to 0, n \to +\infty$. Thus $\rho$ and $\rho_0$ determine the same class of continuous functions from $C$, and the class of continuous functions from a metric space determines the topology (Theorem I.7.3.). $\square$

## 2. PROJECTIONS

A set $\Gamma_r = (t_1, t_2, \ldots, t_r)$ of distinct points on $[0, 1]$ determines a mapping $\pi_r$:

$$\pi_r x = [x(t_1), \ldots, x(t_r)] = \tilde{x}^{(r)} \tag{1}$$

of $D$ onto an $r$-dimensional vector space $\pi_r D$. We call $\pi_r$ a projection of $D$ onto $\pi_r D$. In order to mark the points $t_i$ on which $\pi_r$ depends, we write $\pi(t_1, \ldots, t_r)$ instead of $\pi_r$. In particular, $\pi(t)x = x(t)$ denotes a projection of $D$ into $R$. In the same way, the projection of $C$ onto $\pi_r C$ is defined.

**Theorem 1.** *The mapping $\pi_r$ is a continuous mapping of $C$ onto $\pi_r C$ for the topology $\rho_0$ and a measurable mapping of $D$ onto $\pi_r D$ for the topology $\rho$. The projection $\pi(t)$ of $D$ for $t \in (0, 1)$ is continuous at $x$ for the topology $\rho$ if and only if $x$ is continuous at $t$. The projections $\pi(0)$ and $\pi(1)$ of $D$ are continuous.*

*Proof:* If $x$ is any element in $C$ and $\{x_n\}$ a sequence in $C$ with $\rho_0(x_n, x) \to 0$ as $n \to +\infty$, then clearly $\{\tilde{x}_n^{(r)}\}$ converges to $\{\tilde{x}^{(r)}\}$ in the metric on the vector space $\pi_r C$. Hence $\pi_r$ is a continuous mapping by Lemma I.8.2.

Consider the projection $\pi_r$ of $D$. For $\varepsilon > 0$, put

$$x(\varepsilon, t) = \frac{1}{\varepsilon} \int_t^{t+\varepsilon} x(s)\, ds. \tag{2}$$

Since $x \in D$ is bounded and has at most countably many discontinuity points, the function $x(\varepsilon, \cdot)$ is continuous and thus belongs to $C$. The mapping $x \to x(\varepsilon, t)$ from $D$ into $R$ is continuous. Indeed, if $x, x_n \in D$, $n = 1, 2, \ldots$, and $\rho(x, x_n) \to 0$ $(n \to \infty)$, then by the definition of $\rho$ there exists a sequence $\{\lambda_n\}$, $\lambda_n \in \Lambda$ such that

$$\sup_{t \in (0,1]} \|\lambda_n(t) - t\| \to 0, \qquad \sup_{t \in [0,1]} |x \circ \lambda_n(t) - x_n(t)| \to 0$$

as $n \to +\infty$. These relations imply the convergence of $x_n(t)$ to $x(t)$ at all continuity points $t$ of $x$, and hence by the definition of $x(\varepsilon, \cdot)$, the sequence $\{x_n(\varepsilon, t)\}$ converges to $x(\varepsilon, t)$ for all $t$ if $\rho(x_n, x) \to 0$ $(n \to \infty)$. Thus by Lemma I.8.2, the mapping $x \to x(\varepsilon, t)$ is continuous. Now let $\varepsilon \to 0$ through a decreasing sequence of positive numbers. Since $x$ is continuous to the right, $x(\varepsilon, t)$ tends to $x(t)$ as $\varepsilon \to 0$. Thus the mapping $x \to x(t)$ from $D$ into $R$ as limit of the mappings $x \to x(\varepsilon, t)$ of $D$ into $R$ is measurable. Now put $\{x : x(t_i) \in B_i\} = E_i$, where $B_i$ is a measurable set on $R$. Then

$$\{x : [x(t_1), \ldots, x(t_r)] \in B_1 \otimes \cdots \otimes B_r\} = \bigcap_{i=1}^{r} E_i$$

is measurable, and the $\sigma$-algebra on the product space is generated by the measurable rectangles. Hence, $\pi_r$ is measurable. Consider the projection $\pi(t)$ of $D$. By Lemma I.8.2, $\pi(t)$ is continuous if and only if for any given $x \in D$ and any sequence $\{x_n\}$, $x_n \in D$ with $\rho(x_n, x) \to 0$, $n \to +\infty$, we have $|\pi(t)x_n - \pi(t)x| \to 0$. Clearly, the first relation implies the second one if $t$ is a continuity point of $x$. The first relation does not imply the second one if $t \in (0, 1)$ is a discontinuity point of $x$. Indeed, for $x \in D$, with a discontinuity point $t_0 \in (0, 1)$, let the function $\lambda_n$ in $\Lambda$ satisfy the conditions $\lambda_n(0) = 0$, $\lambda_n(1) = 1$, $\lambda_n(t_0) = t_0 - 1/n$, $\lambda_n$ linear on $[0, t_0 - 1/n]$ and $[t_0 - 1/n, 1]$. Put $x_n = x \circ \lambda_n$. Then $x_n(t) = x(t - 1/n)$ and $|x_n(t - x(t)| > 0$, though $\rho(x_n, x) \to 0$, $n \to +\infty$.

The fact that the projections $\pi(0)$ and $\pi(1)$ of $D$ are continuous follows since $\rho(x, x_n) \to 0$ $(n \to +\infty)$ implies $x_n(0) \to x(0)$, $x_n(1) \to x(1)$ [any function $\lambda$ in $\Lambda$ satisfies the relations $\lambda(0) = 0$, $\lambda(1) = 1$].

If $\mu$ is a $\sigma$-smooth measure on $D$, then by the definition of continuity almost everywhere $(\mu)$, the projection $\pi(t)$ is continuous almost everywhere $(\mu)$, if $\pi(t)$ is continuous at all points $x \in D$ except points belonging to a null set $(\mu)$. □

**Theorem 2.** *Let $\mu$ be a $\sigma$-smooth measure on D and put*

$$E_t = \{x \in D : |x(t) - x(t-)| > 0\}.$$

*Then $\mu(E_t) > 0$ for at most countably many t. If $\mu(E_t) = 0$, then the projection $\pi(t)$ is continuous almost everywhere $(\mu)$. The projections $\pi(0)$ and $\pi(1)$ are continuous.*

**Corollary.** *To a given $\sigma$-smooth measure $\mu$ on D, there exists a countable, dense set $\Gamma_\mu$ such that the projection $\pi(t_1, \ldots, t_r)$ is continuous almost everywhere $(\mu)$ for any points $t_1, \ldots, t_r$ in $\Gamma_\mu$.*

*Remark:* If $\mu(E_t) > 0$, we call $t$ an essential discontinuity point $(\mu)$ of $\pi(t)$.

*Proof:* Put

$$E_t(\varepsilon) = \{x \in D : |x(t) - x(t-)| > \varepsilon$$

for $t \in (0, 1)$. Clearly $E_t(\varepsilon)$ and, hence, $E_t$ are measurable. There can be at most finitely many values $t$ for which $\mu(E_t(\varepsilon)) \geq \delta$ for given $\delta > 0$. Indeed, assuming that this inequality holds for an infinite sequence $\{t_j\}$ of distinct points and using the fact that $\mu$ is $\sigma$-smooth, we should get

$$\mu\left[\limsup_{n \to +\infty} E_{t_n}(\varepsilon)\right] = \mu\left[\bigcap_{n=1}^{+\infty} \bigcup_{k=n}^{+\infty} E_{t_k}(\varepsilon)\right] = \lim_{m \to +\infty} \mu\left[\bigcap_{n=1}^{m} \bigcup_{k=n}^{+\infty} E_{t_k}(\varepsilon)\right] \geq \delta.$$

It means that the set of functions $x$, which have jumps $\geq \varepsilon$ at infinitely many points $t_k$, is not empty. This, however, is not possible for functions in $D$.

Thus there can be at most finitely many points $t$ for which $\mu(E_t(\varepsilon)) \geq \delta$. Letting $\delta$ take the values $1, \frac{1}{2}, \frac{1}{3}, \ldots, 1/n, \ldots$, we conclude that there can be at most countably many points $t$ for which $\mu(E_t(\varepsilon)) > 0$. Letting $\varepsilon$ take the values $1, \frac{1}{2}, \ldots, 1/n, \ldots$, and observing that $E_t(1/n) \downarrow E_t$ and, hence, $\mu[E_t(1/n)]$ tends to $\mu(E_t)$ as $n \uparrow \infty$, we conclude that there can be at most countably many distinct points $t$ for which $\mu(E_t) > 0$. If $x \notin E_t$, then $x$ is a continuity point of $x$, and by Theorem 1 the projection $\pi(t)$ is continuous at $x$. Hence, $\pi(t)$ is continuous almost everywhere ($\mu$). The corollary is an immediate consequence of the theorem. □

## 3. APPROXIMATIONS OF FUNCTIONS BY SCHAUDER SEQUENCES

Let $\Gamma_r$ be finite sets of distinct points on $[0, 1]$ for $r = 1, 2, \ldots$, such that $\Gamma_r \subset \Gamma_{r+1}$. If $\Gamma_r$ contains the points $0 < t_0^{(r)} < t_1^{(r)} < \cdots < t_{n_r}^{(r)} = 1$, any real-valued continuous function $x$ on $[0, 1]$ can be approximated by a polygon $x^{(r)}$ belonging to $\Gamma_r$ such that its corner points are the points $[t_i^{(r)}, x(t_i^{(r)})]$. We call $\Gamma_r$ a Schauder net on $[0, 1]$. Of course, any real-valued function $x$ on $[0, 1]$ can be approximated by polygons $x^{(r)}$, but the approximations may not be good. For functions in $D$, step functions $\hat{x}^{(r)}$, defined as follows for the net above, are more suitable. Put $\hat{x}^{(r)}(t) = x(t_i)$ for $t_i \leq t < t_{i+1}$, $i = 0, \ldots, i-1$, $\hat{x}(1) = x(1)$. This approximation is not always good in the uniform metric. However, we have

**Theorem 1.** *Let the set $\Gamma_r$ consist of the points $t_j^{(r)}, j = 1, \ldots, n_r$, satisfying the conditions*

$$\text{(i)} \qquad \tfrac{1}{2}\delta_r \leq t_{j+1}^{(r)} - t_j^{(r)} \leq \delta_r, \qquad \Gamma_r \subset \Gamma_{r+1}, \qquad r = 1, 2, \ldots,$$
$$\delta_r \downarrow 0 \qquad \textit{as} \quad r \uparrow \infty.$$

*Then for the Skorokhod metric $\rho$,*

$$\text{(ii)} \qquad \rho(x, \hat{x}^{(r)}) \to 0 \qquad (r \to \infty) \qquad \textit{for} \quad x \in D,$$

*and for the uniform metric*

$$\text{(iii)} \qquad \rho_0(x, x^{(r)}) \to 0 \qquad (r \to \infty) \qquad \textit{for} \quad x \in C.$$

**Corollary:** *Any $x \in D$ is uniquely determined with respect to the Skorokhod metric by the values $x(t)$, $t \in \bigcup_{r=1}^{\infty} \Gamma_r$, hence, by the values $x(t)$ belonging to any dense set of points on $[0, 1]$. For $x \in C$, this is true even with respect to the uniform metric.*

*Proof:* Since $x \in D$ has at most countably many discontinuity points in $[0,1]$, for given $\varepsilon > 0$ there are finitely many points $\tau$ with jumps larger than $\varepsilon$. Denote the set of these points by $\Gamma(\varepsilon)$ and choose $r$ so large that there are at least three points $t_j^{(r)}$ in $\Gamma_r$ between any two adjacent points in $\Gamma(\varepsilon)$, and furthermore, so that the oscillation of $x$ is smaller than $2\varepsilon$ on $[t_i, t_{i+1}]$ if there is no point in $\Gamma(\varepsilon)$ that belongs to this interval, and the oscillation of $x$ is smaller than $2\varepsilon$ on $[t_i, \tau)$ and $[\tau, t_{i+1}]$ if $[t_i, t_{i+1}]$ contains a point $\tau \in \Gamma(\varepsilon)$. Then, if $\tau_1 < \tau_2 < \tau_3$ are three adjacent points in $\Gamma(\varepsilon)$, we have the situation

$$\tau_1 \leq t_{i-2} < t_{i-1} < t_i < \tau_2 \leq t_{i+1} < t_{i+2} < t_{i+3} < \tau_3,$$

where $t_{i-2} \cdots t_{i+3}$ are the points in $\Gamma_r$ that are closest to $\tau_2$ and situated as above. Let the function $\lambda_r$ be defined on $[t_{i-2}, t_{i+3}]$ by $\lambda_r(t_{i-1}) = t_{i-1}$, $\lambda_r(t_{i+1}) = \tau_2$, $\lambda_r(t_{i+2}) = t_{i+2}$, and so that it is linear on $[t_{i-2}, t_{i+1}]$ and $[t_{i+1}, t_{i+2}]$. In this way, $\lambda_r$ is defined on certain intervals. At the points not belonging to these intervals, we define $\lambda_r(t) = t$. Then $\lambda_r \in \Lambda$, where $\Lambda$ is the class of functions considered in the Skorokhod metric. Furthermore, $\gamma_r(t_{i+1}) = \tau_2$ and

$$\left|x \circ \gamma_r(t_{i+1}) - \hat{x}^{(r)}(t_{i+1})\right| = \left|x(\tau_2) - x(t_{i+1})\right| < 2\varepsilon.$$

For $t \in [t_{i-1}, t_{i+1})$, we have $t_{i-1} < \gamma_r(t) < \tau_2$ and

$$\left|x \circ \gamma_r(t) - \hat{x}^{(r)}\right| < 4\varepsilon,$$

since $\hat{x}^{(r)}(t)$ is either $x(t_{i-1})$ or $x(t_i)$ and the oscillation of $x$ on $[t_{i-1}\tau_2)$ is at most equal to $4\varepsilon$. This inequality also holds for $t \in [t_{i+1}, t_{i+2}]$, and it clearly holds for any interval $[t_i, t_{i+1}]$, $t_i, t_{i+1}$ adjacent points in $\Gamma(\varepsilon)$, if this interval does not contain any point in $\Gamma(\varepsilon)$. Clearly,

$$\left|\gamma_r(t) - t\right| \leq \delta_r.$$

Since $\varepsilon$ is arbitrary, we get $\rho(x, \hat{x}^{(r)}) \to 0$ $(r \to \infty)$. The corollary is easily verified. Of particular interest are the approximations $x^{(r)}$ and $\hat{x}^{(r)}$, when the distance between two adjacent points in $\Gamma_r$ is equal to $2^{-r}$. Then $\{x^{(r)}\}$ is the classical Schauder sequence. We shall use the notation Schauder sequence for any approximations $\{x^{(r)}\}$ belonging to Schauder nets $\Gamma_r$ on $[0,1]$. The classical Schauder sequence will be called special Schauder sequence. We call $\{\hat{x}^{(r)}\}$ the modified Schauder sequence (special if $\hat{x}^{(r)}$ belongs to a special $x^{(r)}$). Note that $\hat{x}^{(r)}(t) = x^{(r)}(t^{(r)})$, where $t^{(r)}$ is the point closest to the left of $t$ in $\Gamma_r$, $t^{(r)} = t$ if $t \in \Gamma_r$.

The special Schauder sequence and the special modified Schauder sequence can be represented by the following continuous functions, called Schauder functions, defined on $[0,1]$: $g_0(t) = t$, $g_{k2^{-r}}(t) = 0$ for $t \leq (k-1)2^{-r}$ and $t \geq (k+1)2^{-r}$, $n_r = 2^{-(r+1)/2}$ for $t = k2^{-r}$ and such that they are linear on $[(k-1)2^{-r}, k2^{-r}]$ and $[k2^{-r}, (k+1)2^{-r}]$, $k$ odd, $k < 2^r$. Note that $g_{k2^{-r}}(t)$ is different from 0 for exactly one value of $k$ if $t \in [0,1]$.

**Theorem 2.** *For any real-valued function $x$ on $[0,1]$, put*

$$c_0 = x(1) - x(0), \qquad c_1 = x(0)$$

(i) $$c_{k2^{-m}} = 2^{(m-1)/2}\{2x(k2^{-m}) - x[(k-1)2^{-m}] - x[(k+1)2^{-m}]\} \qquad \textit{for } k \textit{ odd} < 2^m,$$

(ii) $$x^{(r)}(t) = c_1 + c_0 g_0(t) + \sum_{m=1}^{r} \sum_{k \text{ odd} < 2^m} c_{k2^{-m}} g_{k2^{-m}}(t).$$

*Then $x^{(r)}$ is the polygon on $[0,1]$ with the corner points $(j2^{-r}, x(j2^{-r}))$, $j = 0, 1, \ldots, 2^r$, and $\hat{x}^{(r)}(t)$ is equal to $x(t^{(r)})$, where $t^{(r)}$ is the point closest to the left of $t$ in $\Gamma_r$, $t^{(r)} = t$, if $t \in \Gamma^{(r)}$. For $x \in C$, we have $\rho_0(x^{(r)}, x) \to 0$ $(r \to \infty)$, and for $x \in D$, $\rho(\hat{x}^{(r)}, x) \to 0$ $(r \to \infty)$.*

*Remark 1:* Let the points in $\Gamma_r$, $r = 1, 2, \ldots$, be ordered in some way. Then we can rewrite (ii) as

$$x^{(r)}(t) = \sum_{v \in \Gamma_r} c_v g_v(t), \qquad \text{where} \quad g_1(t) = 1 \quad \text{for all } t.$$

It follows by (i) that the $c_v$, $v \in \Gamma^{(r)}$ are uniquely determined by the $x(v)$, $v \in \Gamma_r$, and by (ii) that the $c_v$, $v \in \tilde{\Gamma}^{(r)}$, together with $x(0)$, determine the $x(v)$, $v \in \Gamma^{(r)}$. When the $x$ take any value in $C(D)$, then the $x^{(r)}(v)$, $v \in \Gamma_r$ are coordinates in a $2^r$-dimensional vector space, $\tilde{S}^{(r)}$ product space of spaces $R$. By the one-to-one correspondence of the $x(v)$ $(v \in \Gamma^{(r)})$ to the $c_v$ $(v \in \Gamma_r)$ and $x(0)$, a one-to-one linear (hence continuous) mapping of $\tilde{S}^{(r)}$ onto itself is given.

*Remark 2:* Note that $x^{(r)}$ is changed to $\hat{x}^{(r)}$ by changing $g_0$ to $\hat{g}$ and $g_{k2^{-m}}$ to $\hat{g}_{k2^{-m}}$ in (ii).

**Corollary:** $\min(s,t)$, $s, t \in [0,1]$ *has the infinite uniformly convergent Schauder series*

$$\min(s,t) = \sum_{v \in \tilde{\Gamma}} g_v(t) g_v(s), \qquad s, t \in [0,1].$$

*Proof:* We use induction with respect to $r$. For $r = 0$, we have

$$x^{(0)}(t) = x(0) + [x(1) - x(0)]t,$$

and thus $x^{(0)}(0) = x(0)$, $x^{(0)}(1) = x(1)$. Suppose that Theorem 2 holds for $r \le r_0 - 1$, and consider $r = r_0$. Any $t$ belongs to exactly one interval $[(k-1)2^{-r}, (k+1)2^{-r}]$. Since $(k-1)2^{-r}$ and $(k+1)2^{-r}$ can be written $k'2^{-r_1}$ and $k''2^{-r_2}$ with odd $k_1$ and $k_2$, and $r_1 \le r_0 - 1$, $r_2 \le r_0 - 1$, it follows by the assumption for the induction that

$$x^{(r)}[(k-1)2^{-r}] = x[(k-1)2^{-r}], \qquad x^{(r)}[(k+1)2^{-r}] = x[(k+1)2^{-r}].$$

Furthermore,

$$x^{(r)}(k2^{-r}) = x^{(r-1)}(k2^{-r}) + c_{k2^{-r}} g_{k2^{-r}}(k2^{-r}),$$

where $x^{(r-1)}(k2^{-r})$ is the mean value of $x[(k-1)2^{-r}]$ and $x[(k+1)2^{-r}]$, since $x^{(r-1)}$ is linear on $[(k-1)2^{-r}, (k+1)2^{-r}]$. Hence, by (i),

$$\begin{aligned} x^{(r)}(k2^{-r}) &= \tfrac{1}{2}\{x[(k-1)2^{-r}] + x[(k+1)2^{-r}] + 2^{-(r+1)/2} c_{k2^{-r}} \\ &= x(k2^{-r}). \end{aligned}$$

The statements about the convergence of $\{\hat{x}^{(r)}\}$ and $x^{(r)}$ follow by Theorem 1 and its corollary. The statements in the remark are obvious. In order to prove the corollary of Theorem 2, we let $s$ be a given number on $[0,1]$ and compute $c_0 = \min[s, 1] = 1$ and

$$c_k 2^{-m} = 2^{(m-1)/2}\{2\min(s, k2^{-m}) - \min[s, (k-1)2^{-m}] - \min s, (k+1)2^{-m}\}$$

and find that $c_{k2^{-m}} = g_{k2^{-m}}(s)$. □

**Theorem 3.** *$C$ is separable under the metric $\rho_0$ and $D$ separable under the metric $\rho$.*

*Proof:* The countable set of Schauder sequences

$$x^{(r)} = c_1 + c_0 g_0 + \sum_{m=1}^{r} \sum_{k \text{ odd} < 2^m} c_{k2^{-m}} g_{k2^{-m}}$$

with rational coefficients $c_i$ is dense in $C$. The corresponding set of modified Schauder sequences is dense in $D$. □

In (2.1), we have introduced the projection $\pi_r$ belonging to a finite set $\Gamma_r$. It maps $C$ onto a finite-dimensional vector space $\pi_r C$, and $D$ onto a finite-dimensional vector space $\pi_r D$. By Theorem 2.1, $\pi_r$ is a continuous mapping of $C$ onto $\pi_r C$ and a measurable mapping of $D$ onto $\pi_r D$. It follows by the definition of the Schauder sequence $\{x^{(r)}\}$ and the modified Schauder sequence $\{\hat{x}^{(r)}\}$ that the mappings of the vector space $\pi_r C$ into $C$ by

$$V_r\colon \{x(v)\}_{v \in \Gamma_r} \to x^{(r)} \tag{1}$$

and $\pi_r D$ into $D$ by

$$\hat{V}_r\colon \{x(v)\}_{v \in \Gamma_r} \to \hat{x}^{(r)} \tag{2}$$

are continuous. Hence, $T_r = V_r \pi_r$ is a continuous mapping of $C$ into $C$, and $\hat{T}_r = \hat{V}_r \pi_r$ a measurable mapping of $D$ into $D$. However, it follows by the corollary of Theorem 2.2 that for given $\sigma$-smooth measure $\mu$, there exist countable sets $\Gamma^{(\mu)}$ of points $t_i$ on $[0,1]$ such that any projection $\pi(t_1, t_2, \ldots, t_s)$ belonging to points in $\Gamma^{(\mu)}$ is a continuous mapping of $D$ almost everywhere $(\mu)$. Furthermore, the projections $\pi(0)$ and $\pi(1)$ are continuous. Then, of course, $\pi_r$ belonging to $\Gamma_r \in \Gamma^{(\mu)}$, and $\hat{V}_r$ determines a mapping $\hat{T}_r = \hat{V}_r \pi_r$ of $D$ into $D$, which is continuous almost everywhere $(\mu)$. Thus we have proved

**Theorem 4.** *Let $\{\Gamma_r\}$ be a sequence of Schauder nets, and let $\pi_r$ belonging to $\Gamma_r$ be the projection of the C-space onto $\pi_r C$ or of $D$ onto $\pi_r D$. Then $\pi_r$ is continuous on $C$ and measurable on $D$. If $\mu$ is a $\sigma$-smooth measure on $D$, the sequence $\{\Gamma_r\}$ may be chosen such that $\pi_r$ is continuous almost everywhere $(\mu)$.*

**Lemma 1.** *For any $x, y \in D$ and any dense countable set $\Gamma$ of points on $[0,1]$, $0 \in \Gamma$, $1 \in \Gamma$, we have*

$$\rho(x, y) \leq \sup_{t \in \Gamma} |x(t) - y(t)|.$$

[By definition, we have equality here for $x, y \in C$, since then $\rho(x, y) = \rho_0(x, y)$.]

*Proof:* We may form a sequence $\Gamma_r$ of Schauder nets such that $\Gamma_r \subset \Gamma$ for all $r$. Let $\hat{x}^{(r)}$ and $\hat{y}^{(r)}$ be the Schauder approximations of $x$ and $y$, respectively, and put

$$\sup_{t \in \Gamma} |x(t) - y(t)| = \delta.$$

We get

$$\rho(x, y) \leq \rho(x, \hat{x}^{(r)}) + \rho_0(\hat{x}^{(r)}, \hat{y}^{(r)}) + \rho(\hat{y}^{(r)}, y),$$

but $\hat{x}^{(r)}$ and $\hat{y}^{(r)}$ are uniquely determined by their values at the points $t \in \Gamma_r$ and, hence,

$$\rho_0(\hat{x}^{(r)}, \hat{y}^{(r)}) \leq \delta \qquad \text{for all} \quad r.$$

Furthermore, $\rho(x, \hat{x}^{(r)}) \to 0$, $\rho(y, \hat{y}^{(r)}) \to 0$ as $r \to \infty$. Hence, $\rho(x, y) \leq \delta$.

**Lemma 2.** *Let $\omega \to x(\omega, \cdot)$ be a measurable mapping of a measurable space $\Omega$ with $\sigma$-algebra $\mathscr{B}$ into $D$ under the metric $\rho$, and suppose that the mapping $\omega \to x(\omega, t)$ from $\omega$ into $R$ is measurable for $t \in \bigcup_{r=1}^{\infty} \Gamma_r$, where $\Gamma_r$ is the special sequence of Schauder nets. Then the mapping $\omega \to x(\omega, \cdot)$ is measurable.*

*Proof:* Let $\pi_r$ and $\hat{V}_r$ belong to $\Gamma_r$ as in Theorem 4. Since $\omega \to x(\omega, t)$ is measurable for any $t \in \bigcup_{r=1}^{\infty} \Gamma_r$, the mapping $\omega \to \pi_r x(\omega, \cdot)$ is measurable.

As in the proof of Theorem 4, we then conclude that $\omega \to \hat{V}_r \pi_r x(\omega, \cdot) = \hat{x}^{(r)}(\omega, \cdot)$ is measurable, but for any open set $G$ of $D$,

$$\{\omega : x(\omega, \cdot) \in G\} = \bigcap_{r=1}^{\infty} \{\omega : \hat{x}^{(r)}(\omega, \cdot) \in G\}$$

since $\rho(x, \hat{x}^{(r)}) \to 0$, $r \to \infty$. Hence, $\omega \to x(\omega, \cdot)$ is measurable. □

## 4. WEAK CONVERGENCE

Let $\{\Gamma_r\}_{r=1}^{+\infty}$ be a sequence of Schauder nets, $\Gamma_r \subset \Gamma_{r+1}$, and $\{x^{(r)}\}_{r=1}^{\infty}$ and $\{\hat{x}^{(r)}\}_{r=1}^{\infty}$ respectively, the Schauder sequence, the modified Schauder sequence, belonging to $\{\Gamma_r\}$ as in Section 3. In that section we introduced the projections $\pi_r$ of $C$ and $D$ onto finite-dimensional vector spaces and the continuous mappings $V_r$ and $\hat{V}_r$, which map $\pi_r C$ into $C$ and $\pi_r D$ into $D$, and consequently $T_r = V_r \pi_r$ maps $C$ into $C$ and $\hat{T}_r = \hat{V}_r \pi_r$ maps $D$ into $D$.

**Theorem 1.** *A sequence $\{\mu_n\}$ of σ-smooth measures on C converges weakly, and then to a σ-smooth measure, if and only if*

(i) $$\limsup_{n \to \infty} \mu_n(C) < \infty,$$

(ii) $\{\mu_n(\pi_r^{-1} \cdot)\}$ *converges weakly on* $\pi_r C$ *for* $r = 1, 2, \ldots,$

(iii) $$\lim_{r \to +\infty} \limsup_{n \to +\infty} \mu_n[\rho_0(x, x^{(r)}) \geq \varepsilon] = 0 \qquad \textit{for any} \quad \varepsilon > 0.$$

*If* (i)–(iii) *hold and $\mu$ is the weak limit of $\{\mu_n\}$, then $\mu(\pi_r^{-1} \cdot)$ is the weak limit of $\{\mu_n(\pi_r^{-1} \cdot)\}$.*

In $D$, under the metric $\rho$, we face the difficulty that $\pi_r$ of $D$ onto $\pi_r D$ may not be continuous. However, by Theorem 2.1, it is always measurable, and we have

**Theorem 2.** *A sequence $\{\mu_n\}$ of σ-smooth measures on D converges weakly to a σ-smooth measure if*

(i) $$\limsup_{n \to \infty} \mu_n(D) < \infty,$$

(ii) $\{\mu_n(\pi_r^{-1} \cdot)\}$ *converges weakly for* $r = 1, 2, \ldots,$

(iii) $$\lim_{r \to +\infty} \limsup_{n \to +\infty} \mu_n[\rho(x, \hat{x}^{(r)}) \geq \varepsilon] = 0 \qquad \textit{for} \quad r = 1, 2, \ldots,$$

*Conversely, if $\{\mu_n\}$ is a sequence of σ-smooth measures on D, converging weakly, then there exist infinitely many sequences $\{\pi_r\}$ of projections such that* (i)–(iii) *hold.*

*Remark 1:* (i)–(iii) in Theorem 2 imply the weak convergence of $\{\mu_n\}$ even if in (iii) $\hat{x}^{(r)}$ is changed to $x^{(r)}$.

*Remark 2:* If $\{\mu_n\}$ on $D$ converges weakly to $\mu$, there are at most countably many essential discontinuity points $\tau$ for $\mu$ on $(0, 1)$. If $\{\Gamma_r\}$ is a sequence of nets not containing essential discontinuity points for $\mu$, then (i)–(iii) necessarily hold if $\pi_r$ belongs to $\Gamma_r$.

*Proof:* We first prove that (i)–(iii) in Theorems 1 and 2 imply weak convergence. Then we carry out the proof for the $D$-space. It follows in the same way for the $C$-space. Let $\pi_r$ belong to a Schauder net $\Gamma_r$. Then $\pi_r$ is a measurable mapping (according to Theorem 2.1) which maps $D$ onto a finite-dimensional vector space $\tilde{S}^{(r)}$, which is obtained by a continuous mapping $\tilde{\pi}_r$ of $R^{(\infty)}$ such that $\pi_r x = \tilde{\pi}_r \tilde{x} = \tilde{x}^{(r)}$ for $x \in D$, $\tilde{x} \in R^{(\infty)}$, $\tilde{x}^{(r)} \in \tilde{S}^{(r)}$, whenever $\pi_r x = \tilde{x}^{(r)}$ or $\tilde{\pi}_r \tilde{x} = \tilde{x}^{(r)}$. By the definition of the Schauder sequence $\{\hat{x}^{(r)}\}$, there is a continuous mapping $\hat{V}_r$ such that $\hat{V}_r \tilde{x}^{(r)} = \hat{x}^{(r)}$. Thus the mapping $\hat{V}_r \pi_r$ is measurable, and then $x \to \rho(x, \hat{V}_r \pi_r x)$ is measurable according to Lemma I.8.4. Applying Lemma III.6.1, we get Theorem 2. Note that $V_r \pi_r$, mapping from $C$ into $C$, is continuous, and then $x \to \rho(x, V_r \pi_r x)$ is continuous according to Lemma I.8.4. Let a sequence $\{\mu_n\}$ converge weakly to $\mu$. Any projection $\pi_r$ of $C$ is continuous, and, hence, it follows by Lemma III.6.1 that (i)–(iii) in Theorem 1 are satisfied. For the $D$-space, we find by Theorem 2.2 and its corollary that we may choose the sequence $\Gamma_r$ such that the mappings $\pi_r$ are continuous a.s. $\mu$. Again applying Lemma III.6.1, we find that (i)–(iii) in Theorem 2 are satisfied. □

**Theorem 3.** *A $\sigma$-smooth measure $\mu$ on $D$ is the weak limit of any sequence $\{\mu(\pi_r^{-1} \hat{V}_r^{-1})\}$ belonging to a sequence $\Gamma_r$ of Schauder nets such that $\pi_r$ is continuous almost everywhere $\mu$ for all $r$ and $\rho(x, \hat{V}_r \pi_r x) \to 0$ $(r \to \infty)$. Furthermore, $\mu$ is uniquely determined by the projected measures $\mu(\pi^{-1} \cdot)$, where $\pi$ is any mapping of $S$ onto a finite-dimensional vector space. The theorem holds with $V_r$ instead of $\hat{V}_r$ and $\rho_0$ instead of $\rho$ on $C$. (Note that any projection from $C$ is continuous.)*

*Proof:* Consider the $D$-space. The first part of the theorem follows by Theorem 2 applied to the sequence $\{\mu_n\}$, with $\mu_n = \mu(\pi_n^{-1} \hat{V}_n^{-1} \cdot)$. Indeed, for $r < n$ we have

$$\pi_r \hat{V}_n \pi_n = \pi_r, \qquad \mu_n(\pi_r^{-1}) = \mu(\pi_r^{-1} \cdot),$$

and (i) and (ii) in Theorem 2 are self-evident. To show that (iii) in Theorem 2 is satisfied for the sequence $\mu_n$, we put

$$E_r = \{x : \rho(x, \hat{V}_r \pi_r x) \geq \varepsilon\}.$$

By the use of transformation of the integral and the dominated convergence theorem for integrals, we get

$$\mu_n\{x:\rho(x, \hat{V}_r\pi_r x) \geq \varepsilon\} = \int_D 1_{E_r}(x)\,\mu_n(dx) = \int_D 1_{E_r}(x)\,\mu(\pi_n^{-1}\hat{V}_n^{-1}\,dx)$$

$$= \int_D 1_{E_r}(\hat{V}_n\pi_n x)\,\mu(dx) \to \int_D 1_{E_r}(x)\,\mu(dx) \qquad (n \to \infty),$$

where the last integral tends to 0 as $r \to \infty$ since $1_{E_r}(x) \to 0$ $(r \to \infty)$ for any $x$. To prove that $\mu$ is uniquely determined by the measures $\mu(\pi_r^{-1}\cdot)$, we compare any two measures $\mu$ and $\nu$ such that $\mu(\pi_r^{-1}\cdot) = \nu(\pi_r^{-1}\cdot)$ for all $r$, where the $\pi_r$ belong to a sequence $\{\Gamma_r\}$ of Schauder nets. We choose the sequence $\{\Gamma_r\}$ such that $\pi_r$ is continuous a.s. both $\mu$ and $\nu$. Then, applying the first part of the theorem, we get $\mu = \nu$. The corresponding statements for the $C$-space follow in the same way. □

The verification of the condition (iii) in Theorem 1 and Theorem 2 may be difficult. We give two criteria for this condition.

**Lemma 1.** *Let $\{\Gamma_r\}_{r=1}^{\infty}$ be a sequence of Schauder nets on $[0,1]$, and let $\hat{x}^{(r)}$ belong to $x \in D$ as in Section 3. Then for a sequence $\{\mu_n\}$ of $\sigma$-smooth measures on $D$ and $\Gamma = \bigcup_{r=1}^{\infty}\Gamma_r$,*

(i) $$\lim_{r\to\infty}\limsup_{n\to\infty}\mu_n\{x:\rho(x, \hat{x}^{(r)}) \geq \varepsilon\}$$
$$\leq \lim_{r\to\infty}\limsup_{n\to\infty}\mu_n\left\{x:\sup_{t\in\Gamma}\left|x(t) - \hat{x}^{(r)}(t)\right|\right\} \geq \varepsilon.$$

On the $C$-space, this relation holds with $\rho_0$ instead of $\rho$ and $x^{(r)}$ instead of $\hat{x}^{(r)}$.

*Proof:* By Lemma 3.1, we have

$$\rho(x, \hat{x}^{(r)}) \leq \sup_{t\in\Gamma}\left|x(t) - \hat{x}^{(r)}(t)\right|$$

and equality with $\rho_0$ instead of $\rho$ on the $C$-space. □

**Lemma 2.** *Let $\{\Gamma_r\}$ be a sequence of Schauder nets consisting of the points $0 = t_0^{(r)} < t_1^{(r)} < \cdots < t_{n_r}^{(r)}$. Then for a sequence $\{\mu_n\}$ of $\sigma$-smooth measures on $D$ and $\Gamma = \bigcup_{r=1}^{\infty}\Gamma_r$,*

(i) $$\lim_{r\to\infty}\limsup_{n\to\infty}\mu_n[x:\rho(x, \hat{x}^{(r)}) \geq \varepsilon]$$
$$\leq \lim_{r\to\infty}\limsup_{n\to\infty}\mu_n\left\{\left[x:\max_{1\leq j\leq n_r}\left|x(t_j^{(r)}) - x(t_{j-1}^{(r)})\right| \geq \varepsilon\right]\right.$$
$$\left.+\left[\sup_{\substack{1\leq j\leq n_r,\, t_{j-1}<t<t_j\\ t\in\Gamma}}\min(\left|x(t) - x(t_{j-1}^{(r)})\right|, \left|x(t_j^{(r)}) - x(t)\right| \geq \varepsilon\right]\right\}$$

*for any $\varepsilon > 0$. On the C-space, this inequality holds with $\rho_0$ instead of $\rho$ and $x^{(r)}$ instead of $\hat{x}^{(r)}$.*

*Remark:* (i) is satisfied if

$$\text{(i')} \qquad \lim_{r\to\infty} \limsup_{n\to\infty} \sum_{j=1}^{n_r} \mu_n\{x: |x(t_j^{(r)}) - x(t_{j-1}^{(r)})| \geq \varepsilon\} = 0,$$

$$\text{(i'')} \qquad \lim_{r\to\infty} \limsup_{n\to\infty} \sum_{j=1}^{n_r} \mu_n \left\{x: \sup_{t_{j-1}^{(r)} < t < t_j^{(r)},\, t\in\Gamma} \min\{|x(t) - x(t_{j-1}^{(r)})|, |x(t_j^{(r)}) - x(t)|] \geq \varepsilon \right\} = 0$$

for any $\varepsilon > 0$. We often reduce (i'') to (i').

*Proof:* $\hat{x}^{(r)}$ agrees with $x$ at the points $t_j^{(r)}$ and is constant on $t_{j-1}^{(r)} \leq t < t_j^{(r)}$. Hence, we have, for $t_{j-1}^{(r)} \leq t \leq t_j^{(r)}$:

$$|x(t) - \hat{x}^{(r)}(t)| \leq |x(t) - x(t_{j-1}^{(r)})| + |x(t_j^{(r)}) - x(t_{j-1}^{(r)})|,$$
$$|x(t) - \hat{x}^{(r)}(t)| \leq |x(t) - x(t_j^{(r)})| + |x(t_j^{(r)}) - x(t_{j-1}^{(r)})|,$$

and consequently

$$|x(t) - x^{(r)}(t)| \leq |x(t_j^{(r)}) - x(t_{j-1}^{(r)})| + \min[|x(t) - x(t_{j-1}^{(r)})|, |x(t_j^{(r)}) - x(t)|].$$

Thus, applying Lemma 1, we get Lemma 2. Clearly (i') and (i'') imply (i) in the lemma. □

## 5. FLUCTUATIONS AND WEAK CONVERGENCE

According to the remark on Lemma 4.2, the crucial condition (iii) in Theorem 4.1 and Theorem 4.2 is satisfied if (i') and (ii'') are satisfied. In (i') we have to estimate the quantity

$$\mu_n\{x: |x(t_2) - x(t_1)| \geq \varepsilon\} \tag{1}$$

for a given interval $[t_1, t_2]$ on $[0, 1]$, and in (ii'') the quantity

$$\delta(t_1, t_2) = \mu_n \left\{x: \sup_{t_1 < t < t_2,\, t\in\Gamma} \min[|x(t) - x(t_1)|, |x(t_2) - x(t)|] \geq \varepsilon \right\} \tag{2}$$

for such an interval. Clearly

$$\delta(t_1, t_2) \leq \mu_n \left\{x: \sup_{t_1 < t \leq t_2,\, t\in\Gamma} |x(t) - x(t_1)| \geq \varepsilon \right\}. \tag{3}$$

We observe that $\delta_r(t_1, t_2)$ depends on the fluctuations $x$ in the interval $[t_1, t_2]$.

If there is a mapping $\omega \to X_n(\omega)$ from a probability space $(\Omega, \mathscr{B}, P)$ into $D$ and $\mu_n = P(X_n^{-1} \cdot)$, we can write (1)–(3) as

$$P\{\omega: |X_n(\omega, t_2) - X_n(\omega, t_2)| \geq \varepsilon\}, \tag{1'}$$

$$\delta_r(t_1, t_2) = P\left\{\omega: \sup_{t_1 < t < t_2,\, t \in \Gamma} \min[|X_n(\omega, t) - X_n(\omega, t)|, |X_n(\omega, t_2) - X_n(\omega, t)|] \geq \varepsilon\right\}, \tag{2'}$$

$$\delta_r(t_1, t_2) \leq P\left\{\omega: \sup_{t_1 < t \leq t_2,\, t \in \Gamma} |X_n(t) - X_n(t_1)| \geq \varepsilon\right\}. \tag{3'}$$

In our notation, we shall omit $x$; in (1)–(3) and in (1′)–(3′) we omit $\omega$, remembering that the sets under consideration are sets on $D$ and $\Omega$, respectively. We also write $X_n(t)$ instead of $X(\omega, t)$. In some applications, $X_n(t) - X_n(t_1)$ is the sum of real-valued random variables $\xi_i$, their number of terms increasing with $t$. Then we need the following lemma concerning fluctuations of sums.

**Lemma 1.** *Let $\{\xi_i\}_{i=1}^m$ be a sequence of real-valued independent random variables with mean values zero and $E[\xi_i^2] = \sigma_i^2$, $\sigma_i \geq 0$. Put*

$$S_j = \sum_{i=1}^{j} \xi_i, \qquad s_j^2 = \sum_{i=1}^{j} \sigma_i^2, \qquad s_j \geq 0, \qquad j \leq m.$$

*Then for any real number $\lambda$,*

$$P\left\{\max_{1 \leq j \leq m} |S_j| \geq \lambda s_m\right\} \leq 2P\{|S_m| \geq (\lambda - \sqrt{2})s_m.$$

*Proof:* We introduce the sets

$$E_1 = \{|S_1| \geq \lambda s_m\}, \qquad E_j = \left\{\max_{1 \leq i < j} |S_i| < \lambda s_m \leq |S_j|\right\}$$

for $j = 2, \ldots, m$, and get by this definition for $\lambda_1 < \lambda$

$$P\left\{\max_{1 \leq i \leq m} |S_i| \geq \lambda s_m\right\} \leq P\{|S_m| \geq (\lambda - \lambda_1)s_m\}$$
$$+ \sum_{j=1}^{m-1} P\{E_j \cap [||S_m|| < (\lambda - \lambda_1)s_m]\} \leq P\{|S_m| \geq (\lambda - \lambda_1)s_m\}$$
$$+ \sum_{j=1}^{m-1} P\{E_j \cap [|S_m - S_j| \geq \lambda_1 s_m]\}. \tag{4}$$

By the independence of the $\xi_i$, we have

$$P[E_j \cap \{|S_m - S_j| \geq \lambda_1 s_m\}] = P[E_j]P[\{|S_m - S_j| \geq \lambda_1 s_m\}],$$

and by Chebyshev's inequality

$$P[\{|S_m - S_j| \geq \lambda_1 s_m \leq \frac{1}{\lambda_1^2 s_m^2} E[S_m - S_j]^2 = \frac{1}{\lambda_1^2 s_m^2} \sum_{i=j+1}^{m} \sigma_i^2 \leq \frac{1}{\lambda_1^2}.$$

Clearly

$$\bigcup_{j=1}^{m-1} E_j \subset \left\{ \max_{1 \leq i \leq m} |S_i| \geq \lambda s_m \right\}.$$

Hence, we obtain by (4) and the last inequalities

$$P\left[\left\{ \max_{1 \leq i \leq m} |S_i| \geq \lambda s_m \right\}\right] \leq \left(1 - \frac{1}{\lambda_1^2}\right)^{-1} P[\{|S_m| \geq (\lambda - \lambda_1) s_m\}]. \tag{5}$$

For $\lambda_1 = \sqrt{2}$, it gives the inequality of the lemma.

In Section IV.16, we introduced $\varphi$-mixing random variables.

**Lemma 2.** *Let $\{\xi_i\}_{i=1}^{\infty}$ be a $\varphi$-mixing sequence of real-valued random variables with mean value zero and variances $E[\xi_i^2] = \sigma_i^2, \sigma_i > 0$, and*

$$1 + 2 \sum_{k=1}^{\infty} \varphi^{1/2}(k) = a < \infty.$$

*Then for $\lambda > \lambda_1 > \lambda_2 > 0$, $[a/(\lambda_1 - \lambda_2)^2] + \varphi(k) \leq \frac{1}{2}$, $m > k$ ($\lambda_1$ and $k$ sufficiently large), we have*

$$P\left[\left\{ \max_{1 \leq i \leq m} |S_i| \geq \lambda s_m \right\}\right] \leq 2P[\{|S_m| \geq (\lambda - \lambda_1) s_m\}] + 2k \sum_{i=1}^{m} P\left[\left\{ |\xi_i| \geq \frac{\lambda_2 s_m}{k} \right\}\right].$$

*Proof:* The inequality (4) also holds for dependent random variables. We now estimate the terms on the right-hand side of (4) as follows:

$$\begin{aligned} P[E_j \cap \{|S_m - S_j| \geq \lambda_1 s_m\}] \leq{} & P[E_j \cap \{|S_m - S_{j+k}| \geq (\lambda_1 - \lambda_2) s_m\}] \\ & + P[\{|S_{j+k} - S_j| \geq \lambda_2 s_m\}]. \end{aligned} \tag{6}$$

By the $\varphi$-mixing, we obtain

$$\begin{aligned} P[E_j \cap \{|S_m - S_{j+k}| \geq (\lambda_1 - \lambda_2) s_m\}] \leq{} & P(E_j) P[\{|S_m - S_{j+k}| \\ & \geq (\lambda_1 - \lambda_2) s_m\}] + \varphi(k) P(E_j), \end{aligned}$$

where, by Chebyshev's inequality for any $j$,

$$P[\{|S_m - S_j| \geq (\lambda_1 - \lambda_2) s_m\}] \leq \frac{1}{(\lambda_1 - \lambda_2)^2 s_m^2} E[S_m - S_j]^2,$$

and by the corollary of Lemma IV.16.1,

$$\boldsymbol{E}[S_m - S_j]^2 \leq \left[1 + 2\sum_{i=1}^{\infty} \varphi^{1/2}(k)\right] \sum_{i=j+1}^{m} \boldsymbol{E}(\xi_i^2) \leq a s_m^2.$$

Furthermore,

$$P[\{|S_{j+k} - S_j| \geq \lambda_2 s_m\}] \leq \sum_{i=j+1}^{j+k} P\left[\left\{|\xi_i| \geq \frac{\lambda_2 s_m}{k}\right\}\right].$$

Combining the last estimations with (6), we obtain

$$P[E_j \cap \{|S_m - S_j| \geq \lambda_1 s_m] \leq P(E_j)\left[\frac{a}{(\lambda_1 - \lambda_2)^2} + \varphi(k)\right] + \sum_{i=j+1}^{j+k} P\left[\left\{|\xi_i| \geq \frac{\lambda_2 s_m}{k}\right\}\right]. \tag{7}$$

As in the proof of Lemma 1, we have

$$\bigcup_{j=1}^{m-1} E_j \subset \left\{\max_{1 \leq j \leq m} |S_i| \geq \lambda s_m\right\}.$$

Thus (4) and (7) imply the inequality of the lemma.

We now give a lemma for the estimation of $\delta(t_1, t_2)$ in (2).

**Lemma 3.** *Let $\mu$ be a $\sigma$-smooth measure on $D(C)$ under the metric $\rho$ $(\rho_0)$, and suppose that there exists a nondecreasing bounded function $F$ from $[t_1, t_2]$ into $R$, $0 \leq t_1 < t_2 \leq 1$, such that*

(i) $\mu[\{|x(t) - x(t_1)| \geq \lambda], |x(t_2) - x(t)| \geq \lambda\} \leq \lambda^{-2\gamma}[F(t_2) - F(t_1)]^{\alpha}$

*for any $\lambda > 0$, given $\gamma \geq 0$, $\alpha > 1$. Then*

(ii) $$\mu\left[\left\{\sup_{t_1 < t < t_2,\, t \in \Gamma} \min(|x(t) - x(t_1)|, |x(t_2) - x(t)|) \geq \lambda\right\}\right] \leq K\lambda^{-2\gamma}[F(t_2) - F(t_1)]^{\alpha},$$

*with a constant $K$ depending only on $\gamma$ and $\alpha$.*

**Corollary:** *If $X$ is a real-valued random variable from a probability space $(\Omega, \mathscr{B}, P)$ into $D(C)$ and*

(i′) $P[\{|X(t) - X(t_1)| \geq \lambda, |X(t_2) - X(t)| \geq \lambda\}] \leq \lambda^{-2\gamma}[F(t_2) - F(t_1)]^{\alpha}$

*for any $\lambda > 0$, given $\gamma > 0$, $\alpha > 1$, then*

(ii′) $$P\left[\left\{\sup_{t_1 < t < t_2,\, t \in \Gamma} \min(|X(t) - X(t_1)|, |X(t_2) - X(t)|)\right\}\right] \leq K\lambda^{-2\gamma}[F(t_2) - F(t_1)]^{\alpha}.$$

*Furthermore,* (i′) *is satisfied if*

$$\text{(i}''\text{)} \qquad \boldsymbol{E}[|X(t) - X(t_1)|^\gamma |X(t_2) - X(t)|^\gamma] \le [F(t_2) - F(t_1)]^\alpha.$$

*Remark:* Note that in (i), we have fixed points $t_1$, $t$, and $t_2$.

*Proof:* Put

$$m_x(t_1, t, t_2) = \min\{|x(t) - x(t_1)|, |x(t_2) - x(t)|\}$$

for $t_1 < t < t_2$, and

$$M_x^{(k)}(t_1, t_2) = \max_{t_1 < \tau_i < t_2,\ i=1,\ldots,k} m_x(t_1, \tau_i, t_2)$$

for $k$ distinct points $\tau_i$ in $\Gamma$. Clearly $M_x^{k_1}(t_1, t_2)$ is nondecreasing as $k\uparrow$ if $\{\tau_i\}_{i=1}^\infty$ is a sequence of given points. Considering only points $\tau \in \Gamma$ and observing that $\Gamma_r$ contains finitely many points and $\Gamma_r \uparrow \Gamma$ as $r \uparrow \infty$, we conclude that $\mu[M_x^{(k)}(t_1, t_2) \ge \lambda]$ tends to the left-hand side of (ii). Hence, if we prove that

$$\mu[\{x : M_x^{(k)}(t_1, t_2)\} \ge \lambda] \le K\lambda^{-2\gamma}[F(t_2) - F(t_1)]^\alpha \tag{8}$$

for any $k = 1, 2, \ldots$, we obtain (ii) by the monotone convergence theorem for integrals when we let $k \uparrow \infty$ and let $k$ be the number of points in $\Gamma_r$. We prove (8) by the help of induction.

For $k = 0$, (8) reduces to (i). Suppose that (8) holds for $k \le k_0$ and consider $k = k_0 + 1$, and then let $\tau_1, \ldots, \tau_k$ be the $\tau$-points in $(t_1, t_2)$. Then there exists an index $n$ such that

$$F(\tau_{n-1}) - F(t_1) \le \tfrac{1}{2}[F(t_2) - F(t_1)], \qquad F(t_2) - F(\tau_n) \le \tfrac{1}{2}[F(t_2) - F(t_1)] \tag{9}$$

if we permit $\tau_{n-1}$ to be $t_1$ and $t_2$. There are $k_1 < k$ of the points $\tau_i$ on $(t_1, \tau_{n-1})$ and $k_2 < k$ of the points $\tau_i$ on $(\tau_{n-1}, t_2)$. To these intervals there belong the quantities $M_x^{k_1}(t_1, \tau_{n-1})$ and $M_x^{k_2}(\tau_n, t_2)$. First consider the case when $t_1 \le \tau_j \le \tau_{n-1}$. Then since

$$|x(t_2) - x(\tau_j)| \le |x(t_2) - x(\tau_{n-1})| + |x(\tau_{n-1}) - x(\tau_j)|$$

we get

$$m_x(t_1, \tau_j, t_2) \le M_x^{(k_1)}(t_1, \tau_{n-1}) + |x(t_2) - x(\tau_{n-1})|,$$

and since

$$|x(\tau_j) - x(t_1)| \le |x(\tau_{n-1}) - x(\tau_j)| + |x(\tau_{n-1}) - x(t_1)|,$$

we also get

$$m_x(t_1, \tau_j, t_2) \le M_x^{k_1}(t_1, \tau_{n-1}) + |x(\tau_{n-1}) - x(t_1)|.$$

Hence, for $t_1 \le \tau_j \le \tau_{n-1}$,

$$m_x(t_1, \tau_j, t_2) \le M_x^{k_1}(t_1, \tau_{n-1}) + m_x(t_1, \tau_{n-1}, t_2). \tag{10}$$

In the same way, we get for $\tau_n \le \tau_j \le t_2$,

$$m_x(t_1, \tau_j, \tau_2) \le M_x^{(k_2)}(\tau_n, t_2) + m_x(t_1, \tau_n, t_2). \tag{11}$$

Combining (10) and (11), we obtain

$$M_x^{(k)}(t_1, t_2) \le \max\{M_x^{(k_1)}(t_1, \tau_{n-1}) + m_x(t_1, \tau_{n-1}, t_2), M_x^{k_2}(\tau_n, t_2) + m_x(t_1, \tau_n, t_2)\}. \tag{12}$$

Putting $\lambda = \lambda_0 + \lambda_1$, $\lambda_0 > 0$, $\lambda_1 > 0$, we obtain by (12)

$$\begin{aligned}\mu[M_x^{(k)}(t_1, t_2) \ge \lambda] \le{}& \mu[M_x^{(k_1)}(t_1, \tau_{n-1}) \ge \lambda_0] \\ &+ \mu[m_x, (t_1, \tau_{n-1}, t_2) \ge \lambda_1] + \mu[M_x^{(k_2)}(\tau_n, t_2) \ge \lambda_0] \\ &+ \mu[m_x(t_1, \tau_n, t_2) \ge \lambda_1].\end{aligned} \tag{13}$$

By the assumption (i) and the assumption for the induction, we find that the right-hand side is at most equal to

$$\begin{aligned}&K\lambda_0^{-2\gamma}[F(\tau_{n-1}) - F(t_1)]^\alpha + \lambda_1^{-2\gamma}[F(t_2) - F(t_1)]^\alpha \\ &\quad + K\lambda_1^{-2\gamma}[F(t_2) - F(\tau_n)]^\alpha + \lambda_1^{-2\gamma}[F(t_2) - F(t_1)]^\alpha.\end{aligned}$$

By the choice of $\tau_n$ in (9), the inequality (13) then reduces to

$$\mu[M_x^{(k)}(t_1, t_2) \ge \lambda] \le K\lambda^{-2\gamma}[F(t_2) - F(t_1)]^\alpha 2^{-\alpha+1}(\lambda_0/\lambda)^{-2\gamma} + \frac{2}{K}(\lambda_1/\lambda)^{-2\gamma}.$$

For sufficiently large $K$, we have for suitable $\lambda_0$ and $\lambda_1$

$$2^{-\alpha+1}\left(\frac{\lambda_0}{\lambda}\right)^{-2\gamma} + \frac{2}{K}\left(\frac{\lambda_1}{\lambda}\right)^{-2\gamma} \le 1.$$

Thus the induction is fulfilled.

The corollary follows for $\mu = P(X^{-1}\cdot)$. Furthermore (i″) implies (i′) since

$$\begin{aligned}&P[|X(t) - X(t_1)| \ge \lambda, |X(t_2) - X(t)| \ge \lambda] \\ &\quad \le \lambda^{-2\gamma}E[|X(t) - X(t_1)|^\gamma |X(t_2) - X(t)|^\gamma]. \quad \square\end{aligned}$$

## 6. CONSTRUCTION OF PROBABILITY MEASURES ON THE C- AND D-SPACES

Let the mappings $\pi_r$, $\tilde{\pi}_r$, $V_r$, and $\hat{V}_r$ belong to a sequence of Schauder nets $\Gamma_r$ as in Section 4. By Theorem 4.3, any probability measure $\mu$ on $C$ is the weak limit of $\{\mu(\pi_r^{-1}V_r^{-1}\cdot)\}$ if $\rho(x, V_r\pi_r x) \to 0$ $(r \to \infty)$. Here $X^{(r)} = V_r\pi_r$ is a random function from $C$ with probability measure $\mu$. Thus $X^{(r)}$ is a random function from $C$ into $C$. However, it follows from the proof of Theorem 4.3 that the probability measures $\mu(\pi_r^{-1}\cdot)$ determine a probability measure $\tilde{\mu}$ on $R^{(\infty)}$ and that $V_r\tilde{\pi}_r$ is a random variable from $R^{(\infty)}$ into $C$. In the same way, Theorem 4.3 shows that $\hat{V}_r\pi_r$ is a random variable from $D$ into $D$ and $\hat{V}_r\tilde{\pi}_r$ is a random variable from $R^{(\infty)}$ into $D$. Furthermore, by Theorem 2.2, $\{\Gamma_r\}$ may be chosen such that $\pi_r$ and, hence, $\hat{V}_r\pi_r$ is continuous a.s. $\mu$. Hence, we have

**Theorem 1.** *Any probability measure on $C(D)$ is the weak limit of a sequence $\{P(X^{(r)-1}\cdot)\}$, where $X^{(r)}$ is a random variable from a probability space $(\Omega, \mathscr{B}, P)$ into $C(D)$.*

According to this theorem, we may always deal with sequences of random variables from a probability space $(\Omega, \mathscr{B}, P)$, when we consider weak limits of probability measures on $C$ and $D$. Now let $X$ be a random variable from a probability space $(\Omega, \mathscr{B}, P)$ into $C(D)$. Then $X(\cdot, t)$ is a real-valued random variable for any $t \in [0, 1]$, and $X(\omega, \cdot)$ a function into $C(D)$ for any $\omega \in \Omega$. When there can be no misunderstanding, we omit $\omega$ in our notations and put $X(t) = X(\cdot, t)$.

Consider a sequence $\{X^{(r)}\}$ from a probability space $(\Omega, \mathscr{B}, P)$ into $C(D)$. If this sequence converges a.s. $P$, then by Theorem I.13.2 there exists a random variable $\bar{X}$ from the completed probability space $(\Omega, \bar{\mathscr{B}}, \bar{P})$ such that $\{P(X^{(r)-1}\cdot)\}$ converges weakly to $\bar{P}(\bar{X}^{-1}\cdot)$.

Let $c_0$, $c_1$, and $c_{k2^{-m}}$, $k$ odd $< 2^m$, $m = 1, 2, \ldots$ be real-valued random variables from a probability space $(\Omega, \mathscr{B}, P)$. We form sequences

$$X^{(r)} = c_0 g_0 + \sum_{m=1}^{r} \sum_{k \text{ odd} < 2^m} c_{k2^{-m}} g_{k2^{-m}} \tag{1}$$

with the special Schauder functions, $g_v$ belonging to the special Schauder nets $\Gamma_r$. Then $X^{(r)}$ is a random variable (random function) from $(\Omega, \mathscr{B}, P)$ into $C$. If $\{X^{(r)}\}$ converges a.s., we conclude as above that $\{P[X^{(r)-1}\cdot]\}$ converges weakly to a probability measure on $C$. To show that $\{X^{(r)}\}$ converges a.s. may be difficult. We give the following criterion, which will be used in the next section.

**Lemma 1.** *Let* (1) *be given with random variables $c_0$ and $c_{k2^{-m}}$ from a probability space $(\Omega, \mathscr{B}, P)$ into $R$, and let the $a_m$ be positive numbers such that*

(i) $$\sum_{m=1}^{\infty} 2^{-m/2} a_m < \infty.$$

*If*

$$\text{(ii)} \qquad \sum_{m=1}^{\infty} 2^m \max_{1 \le k < 2^m} P\{|c_{k2^{-m}}| \ge a_m\} < \infty,$$

*then* $\{X^{(r)}\}$ *in* (1) *converges uniformly on C a.s. P.*

*Proof:* Put

$$E_m = \left\{\omega\colon \max_{1 \le k \le 2^m} |c_{k2^{-m}}| \ge a_m\right\}.$$

Then

$$P(E_m) \le 2^m \max_{1 \le k < 2^m} P\{|c_{k2^{-m}}| \ge a_m\}.$$

Hence, by (ii),

$$\sum_{m=1}^{\infty} P(E_m) < \infty.$$

According to the Borel–Cantelli lemma (Lemma II.9.1), we then have $P(\limsup_m E_m) = 0$, i.e., with probability 1, the events $E_m$ occur a.s. only finitely many times. This means that for almost all $\omega$, there exists $m_0(\omega)$ such that for $m > m_0(\omega)$,

$$\max_{1 \le k < 2^m} |c_{k2^{-m}}(\omega)| < a_m,$$

but if this relation holds, we have

$$\left| \sum_{k \text{ odd} < 2^m} c_{k2^{-m}}(\omega) g_{k2^{-m}}(t) \right| \le 2^{-(m+1)/2} a_m, \tag{2}$$

since $g_{k2^{-m}}(t)$ is different from 0 at most for one $k$, $k$ odd $< 2^m$, and $|g_{k2^{-m}}(t)| < 2^{-(m+1)/2}$ for all $k$ and $m$. Regarding (i), we conclude that the series with the terms equal to the left-hand side of (2) converge a.s. uniformly on $C$. □

*Remark:* Note that a probability measure on the $C$-space can be extended to a probability measure on the $D$-space. □

If $\hat{X}^{(r)}$ belongs to $X^{(r)}$ as in Theorem 2, where $X^{(r)}$ is defined by (1), then $\hat{X}^{(r)}$ is a random variable from $(\Omega, \mathscr{B}, P)$ into the $D$-space. It is easy to see that Lemma 1 holds for these $\hat{X}^{(r)}$ under the metric $\rho$ instead of $\rho_0$, since $\rho$ is not larger than $\rho_0$. Clearly the $\hat{X}^{(r)}$ may have discontinuity points at $t \in \Gamma_r$. General Schauder sequences can represent random functions as limits in the same way as the special Schauder sequences.

## 7. GAUSSIAN $\sigma$-SMOOTH MEASURES ON THE C- AND D-SPACES

A $\sigma$-smooth measure $\mu$ on $C(D)$ is called Gaussian if for any projection

$$\pi(\mathbf{t}), \mathbf{t} = (t_1, \ldots, t_k), \qquad t_i \in [0,1], \qquad 0 \le t_1 < t_2 < \cdots < t_k \le 1, \tag{1}$$

the measure

$$\mu^{\mathbf{t}} = \mu\{[\pi(\mathbf{t})]^{-1}\cdot\}$$

is a Gaussian measure on $\pi(t)C[\pi(\mathbf{t})D]$. Now a Gaussian measure on the finite-dimensional vector space $\pi(\mathbf{t})C[\pi(\mathbf{t})D]$ is uniquely determined by its mean-value vector

$$m(\mathbf{t}) = \int_{\pi(\mathbf{t})C} x^{\mathbf{t}}\mu^{\mathbf{t}}(dx^{\mathbf{t}}) \qquad [x^{\mathbf{t}} = \pi(\mathbf{t})x] \tag{2}$$

and its covariance function

$$r(t_i, t_j) = \int_{\pi(\mathbf{t})C} x(t_i)x(t_j)\,\mu^{\mathbf{t}}(dx^{\mathbf{t}}) = \int_C x(t_i)x(t_j)\,\mu(dx). \tag{3}$$

For any $t, s \in [0,1]$, the covariance function $r(t,s)$ on $C(D)$ is given by

$$r(t,s) = \int x(t)x(s)\,\mu(dx), \tag{4}$$

where the integration is taken over $C(D)$. In the following, we consider only Gaussian measures with mean-value vectors equal to the zero vector.

**Theorem 1.** *A Gaussian measure is uniquely determined by its covariance function when its mean values are zero.*

*Proof:* We carry out the proof for the $D$-space. Let $\mu$ be a Gaussian measure on $D$ and consider a sequence $\{\Gamma_r\}$ of Schauder nets such that the mapping $\pi_r$ belonging to $\Gamma_r$, as in Section 3, is continuous a.s. $\mu$. This choice is possible according to Theorem 2.2. Now $\pi_r D$ is mapped into $D$ by a continuous mapping $\hat{V}_r$ and, hence, $\pi_r\hat{V}_r$ is a mapping continuous a.s. $\mu$ of $D$ into $D$. By Theorem 4.3, $\mu$ is the weak limit of $\{\mu(\pi_r^{-1}\hat{V}_r^{-1}\cdot)\}$ as $r \to \infty$. Now $\mu(\pi_r^{-1}\cdot)$ is a probability measure on the finite-dimensional vector space $\pi_r D$, and as such, it is a Gaussian measure since $\mu$ is Gaussian. Hence, $\mu(\pi_r^{-1}\cdot)$ is uniquely determined by a covariance vector that is determined by the covariance function. Furthermore, the mapping $\hat{V}_r$ is uniquely determined by the vectors in $\pi_r D$. □

We construct Gaussian measures on the $C$- and $D$-spaces as in Section 6. Consider the $C$-space. Let $\{\Gamma_r\}$ be the sequence of special Schauder nets and $c_\nu$, $\nu \in \Gamma_r$, random variables from a probability space $(\Omega, \mathscr{B}, P)$ into $C$. We form the random function

$$X^{(r)} = \sum_{\nu \in \Gamma_r} c_\nu g_\nu \qquad (c_1 = 0). \tag{5}$$

It follows as in Section 6 that the uniform convergence a.s. $(P)$ of this series as $r \to \infty$ determines a probability measure $\mu$ on $D$, where $\mu$ is the weak limit of the sequence $\{\mu[X^{(r)-1}\cdot]\}$. The measure $\mu$ so constructed is Gaussian. Indeed, $\pi(\mathbf{t})X^{(r)}$ is a vector with coordinates

$$X^{(r)}(t_i) = \sum_{\nu \in \Gamma_r} c_\nu g(t_i) \qquad i = 1, 2, \ldots, s \qquad \text{for} \quad t = (t_1, t_2, \ldots, t_r). \tag{6}$$

This vector is Gaussian since the $c_\nu$ are Gaussian (Theorem IV.3.7). We give the following criterion for uniform convergence a.s. $(P)$ of $\{X^{(r)}\}$.

**Lemma 1.** *For $k$ odd $< 2^m$, let*

(i) $$\boldsymbol{E}(c_0) = 0, \qquad \boldsymbol{E}(c_1) = 0, \qquad \boldsymbol{E}(c_{k2^{-m}}) = 0,$$

(ii) $$\boldsymbol{E}[c_{k2^{-m}}^2] \le 2^{3m}\sigma_m^4 \exp -\frac{1}{\sigma_m^2}$$

*with positive numbers $\sigma_m$ such that*

(iii) $$\sum_{m=1}^{\infty} 2^m \sigma_m \exp -\frac{1}{2\sigma_m^2} < \infty.$$

*Then $\{X^{(r)}\}$ converges a.s. in the metric $\rho_0$, and thus $\{P(X^{(r)-1}\cdot)\}$ converges weakly to a Gaussian measure on $C$.*

*Remark 1:* According to our definition in Section III.1, $\{X^{(r)}\}$ converges a.s. to a random variable $X$ when we consider $X^{(r)}$ and $X$ as random variables from the completed probability space.

*Remark 2:* In order for the restriction by (ii) and (iii) to be as weak as possible, we must choose $\sigma_m$ as the largest value for which (ii) and (iii) hold.

Example: Choose $\sigma_m = (\alpha m)^{-1/2}(\ln 2)^{-1/2}$. Then

$$2^m \sigma_m \exp(-1/2\sigma_m^2) = 2^{-[(\alpha-2)/2]m}(\alpha m)^{-1/2}(\ln 2)^{-1/2}$$

and (iii) holds if $\alpha > 2$. Furthermore, (ii) is satisfied if

$$\boldsymbol{E}[c_{k2^{-m}}^2] \le 2^{(3-\alpha)m}(\alpha m)^{-2}(\ln 2)^{-2}.$$

*Proof of Lemma 1:* Let

$$\boldsymbol{E}[c_{k2^{-m}}^2] = b_m^2 \le 2^{3m}\sigma_m^4 \exp(-1/\sigma_m^2).$$

We shall apply Lemma 6.1, and show that for suitable $a_m$,

$$\sum_{m=1}^{\infty} 2^m \max_{1 \le k < 2^m} P\{|c_{k2^{-m}}| \ge a_m\} < \infty, \tag{7}$$

$$\sum_{m=1}^{\infty} 2^{-m/2} a_m < \infty. \tag{8}$$

Since $c_{k2^{-m}}$ is Gaussian with mean value 0 and variance $b_m^2$, we obtain

$$\begin{aligned} P\{|c_{k2^{-m}}| \ge a_m\} &\le \frac{1}{\sqrt{2\pi}\, b_m a_m} \int_{|x| \ge a_m} |x| \exp\left[-\frac{1}{2}\left(\frac{x}{b_m}\right)^2\right] dx \\ &= \frac{2}{\sqrt{2\pi}} \frac{b_m}{a_m} \exp\left[-\frac{1}{2}\left(\frac{a_m}{b_m}\right)^2\right]. \end{aligned}$$

For $a_m = b_m/\sigma_m$, we get

$$\begin{aligned} 2^m P\{|c_{k2^{-m}}| \ge a_m\} &\le (2/\sqrt{2\pi}) 2^m \sigma_m \exp(-1/2\sigma_m^2), \\ 2^{-m/2}(b_m/\sigma_m) &= 2^m \sigma_m \exp(-1/2\sigma_m^2). \end{aligned}$$

Hence, the series (7) and (8) converge when (i)–(iii) hold. □

We apply Lemma 1 to prove the existence of the famous Wiener measure and to give a representation of this measure. The Wiener measure $W$ on $C$ is characterized by the following properties:

(1°) $W$ is Gaussian,

(2°) $$\int_C x(t)\, W(dx) = 0,$$

(3°) $$\int_C x(t)x(s)\, v(dx) = \min[t, s] \qquad \text{for} \quad t, s \in [0, 1].$$

**Theorem 1.** *Let $c_0$ and $c_{k2^{-m}}$, $k$ odd $< 2^m$, $m = 1, 2, \ldots,$ be real-valued, independent Gaussian random variables from a probability space $(\Omega, \mathscr{B}, P)$ and suppose that $\mathbf{E}[c_0] = 0$, $\mathbf{E}[c_{k2^{-m}}] = 0$, $\mathbf{E}(c_0^2) = 1$, $\mathbf{E}(c_{k2^{-m}}^2) = 1$. Then*

(i) *The Schauder series*

$$c_0 g_0 + \sum_{m=1}^{\infty} \sum_{k \text{ odd} < 2^m} c_{k2^{-m}} g_{k2^{-m}}$$

*converges in the uniform metric $\rho_0$ a.s. $(P)$.*

(ii) *There exists a Gaussian random variable $X$ from the completed probability space $(\Omega, \mathscr{B}, \bar{P})$ into $C$ such that $X$ is equal to the series in* (i) *a.s. $\bar{P}$ and thus is continuous a.s. $\bar{P}$.*

(iii) *The measure $W = \bar{P}(X^{-1})$ is the Wiener measure on $C$.*

*Remark:* The random variable $X$ is called the Wiener process, and, if $t$ on $[0, 1]$ is considered as time point, the Brownian motion on $[0, 1]$.

*Proof:* Since $\boldsymbol{E}[c_{k2^{-m}}] = 0$ and $\boldsymbol{E}[c_{k2^{-m}}^2] = 1$, it follows by the lemma that the series in (i) converges a.s. $(P)$ to a random function and thus determines $X$ as in (ii). It remains to prove that $W$ so defined has the properties (2°) and (3°).

Since $\boldsymbol{E}[c_0] = 0$, $\boldsymbol{E}[c_{k2^{-m}}] = 0$, it follows that $\boldsymbol{E}[X^{(r)}(t)] = 0$ for any $t$. Furthermore, since the $c_0$ and $c_{k2^{-m}}$ are independent, we have

$$\boldsymbol{E}[c_i c_j] = 0 \qquad \text{for} \quad i \neq j,$$

and, hence,

$$\boldsymbol{E}[X^{(r)}(t)X^{(r)}(s)] = \boldsymbol{E}[c_0^2]g_0(s)g_0(t) + \sum_{m=1}^{r} \sum_{k \text{ odd} < 2^m} \boldsymbol{E}[c_{k2^{-m}}^2] g_{k2^{-m}}(t) g_{k2^{-m}}(s),$$

where $\boldsymbol{E}[c_0^2] = \boldsymbol{E}[c_{k2^{-m}}^2] = 1$. As $r \to \infty$, the right-hand side tends to $\min(t, s)$ according to the corollary of Theorem 3.2; but $\bar{P}(X^{(r)-1}\cdot)$ converges weakly to $\bar{P}(X^{-1}\cdot)$, and the mapping $x \to x(t)x(s)$ is continuous. Hence, as $r \to \infty$,

$$\boldsymbol{E}[X^{(r)}(t)X^{(r)}(s)] = \int_C x(t)x(s)\, \bar{P}(X^{(r)-1}dx) \to \int_C x(t)x(s)\, W(dx) = \boldsymbol{E}[X(t)X(s)],$$

which implies (3°). In the same way, (2°) follows. □

We may construct Gaussian measures on $D$ in the same way as on $C$, when the $c_\nu$ are real-valued random variables from a probability space $(\Omega, \mathscr{B}, P)$. Then we consider

$$\hat{X}^{(r)} = \sum_{\nu \in \Gamma_r} c_\nu \hat{g}_\nu \tag{9}$$

instead of $X^{(r)}$. If $\{\hat{X}^{(r)}\}$ converges uniformly a.s. $(P)$, then it also converges in the metric $\rho$ a.s. $(P)$. Hence, we may use Lemma 1 to show the convergence in the $\rho$-metric a.s. $(P)$.

## 8. EMBEDDING OF SUMS OF REAL-VALUED RANDOM VARIABLES IN RANDOM FUNCTIONS INTO THE *D*-SPACE

Let $\{\xi_i\}_{i=1}^{\infty}$ be a sequence of real-valued random variables from a probability space $(\Omega, \mathscr{B}, P)$ and let $\boldsymbol{E}[\xi_i^2] = \sigma_i^2$, $\sigma_i > 0$, $\boldsymbol{E}[\xi_i] = 0$, $i = 1, 2, \ldots$ .

For $t \in [0, 1]$, we define the integer-valued function $k_n(\cdot)$ by

$$k_n(1) = k_n, \qquad k_n(t) = \max j : \sum_{i=1}^{j} \sigma_i^2 \leq s_n^2 t, \qquad s_n^2 = \sum_{i=1}^{k_n} \sigma_i^2 \tag{1}$$

($s_n > 0$) and put

$$X_n(t) = \frac{1}{s_n} \sum_{j=1}^{k_n(t)} \xi_j. \tag{2}$$

According to Lemma 3.2, the function $X_n$ from $\Omega$ into $D$ is a random function. For the notations $X(t)$, $X(\omega, t)$, etc., compare Section 5. The random variables $X_n$ determine the measures $P(X_n^{-1}\cdot)$ on $D$. Our main problem will be to determine weak limits of $\{P(X_n^{-1}\cdot)\}$ as $n \to \infty$. Then we apply the theorems and lemmas in Sections 4 and 5. We approximate $X_n$ by the modified Schauder sequence

$$\hat{X}_n^{(r)} = X_n(0) + \sum_{m=1,\, k \text{ odd} < 2^m}^{r} c_{k2^{-m}}^{(m)} \hat{g}_{k2^{-m}}, \tag{3}$$

where $X_n(0)$ and the $c_{k2^{-m}}^{(n)}$ are then real-valued random variables. Remember that a Schauder approximation belongs to a Schauder net $\Gamma_r$, in this case the special Schauder net. To $\Gamma_r$ belongs the projection $\pi_r$, which is measurable, and the continuous mapping $\hat{V}_r$ of $\pi_r D$ into $D$. Hence, $\pi_r X_n$ is a random vector into the finite-dimensional vector space $\pi_r D$, and $\hat{V}_r \pi_r X_n$ is a random variable (random function) into $D$.

Considering weak convergence of the sequence $\{\mu_n\}$, $\mu_n = P(X_n^{-1}\cdot)$, we apply Theorem 4.2. Since $\mu_n$ is a probability measure, (i) in this theorem is satisfied. To show (ii), we have to prove that the distribution of the real-valued random vectors with coordinates $X_n[t_i^{(r)}]$, $t_i^{(r)}$, $i = 1, \ldots, n_r$, $t_i^{(r)} \in \Gamma_r$, converge weakly. By Theorem IV.5.2, this is the case if and only if the distribution of the real-valued random variables

$$\eta_n^{(r)} = \sum_{i=1}^{n_r} \alpha_i X_n[t_i^{(r)}] \tag{4}$$

converges weakly as $n \to \infty$ for any $r = 1, 2, \ldots$ and any real numbers $\alpha_i$. We have criteria for such convergence in Section IV. For the fulfillment of the condition (iii) in Theorem 4.2, we may use the lemmas in Sections 4 and 5. As a first application of Theorem 4.2, we prove a generalization of Donsker's famous theorem.

**Theorem 1.** *Let $X_n(t)$ belong to the sequence $\{\xi_i\}$ as above. Under the Lindeberg condition for the independent sequence $\{\xi_i\}$, the sequence $\{P(X_n^{-1}\cdot)\}$ converges weakly to the Wiener measure on $D$.*

*Remark:* In his original theorem, Donsker considers the $C$-space and identically distributed random variables.

For the proof, we shall use

**Lemma 1.** *If $\{\xi_i\}$ is an independent sequence satisfying the L condition, then for $0 \le s \le t \le 1$, $E(\xi_i) = 0$, we have*

$$\text{(i)} \qquad \lim_{n\to\infty} \frac{1}{s_n^2} \sum_{i=1}^{k_n(t)} \boldsymbol{E}(\xi_i^2) = t \qquad \textit{uniformly for} \quad 0 \le t \le 1,$$

$$\text{(ii)} \qquad \lim_{n\to\infty} \boldsymbol{E}X_n(t)X_n(s) = s,$$

$$\text{(iii)} \qquad \lim_{n\to\infty} \boldsymbol{E}[[X_n(t) - X_n(s)]^2] = t - s.$$

*Proof:* By the definition of $k_n(t)$, we get

$$t - \frac{1}{s_n^2} \boldsymbol{E}[\xi_{k_n(t)}^2] \le \frac{1}{s_n^2} \boldsymbol{E} \sum_{i=1}^{k_n(t)} \xi_i^2 \le t.$$

Hence, (i) holds according to the Lindeberg condition. Furthermore,

$$\boldsymbol{E}X_n(t)X_n(s) = \frac{1}{s_n^2} \boldsymbol{E}\left[\sum_{i=1}^{k_n(s)} \xi_i^2\right] \to s \qquad (n \to \infty)$$

since $\boldsymbol{E}(\xi_i) = 0$, and then $\boldsymbol{E}[\xi_i \xi_j] = 0$ for $i \ne j$, according to independence. Clearly (iii) follows from (ii). □

*Proof of Theorem 1:* For given $r$ and $\alpha_i$, we may write (4) as

$$\eta_n^{(r)} = \frac{1}{s_n} \sum_{i=1}^{k_n(1)} b_j \xi_j, \tag{5}$$

where some $b_j$ may be 0. Clearly, the sequence $\{1/s_n)b_j\xi_j\}_{i=1}^{k_n(1)}$ satisfies the Lindeberg condition, and by (ii) in the lemma we get

$$\lim_{n\to\infty} \boldsymbol{E}[\eta_n^{(r)}]^2 = \lim_{n\to\infty} \sum_{i=1}^{n_r} \{\alpha_i^2 \boldsymbol{E}X_n^2(t_i^{(r)})\} + \sum_{i<j\le n_r} \alpha_i\alpha_j \boldsymbol{E}X_n(t_i^{(r)})X_n(t_j^{(r)})$$

$$= \sum_{i=1}^{n_r} \alpha_i^2 (t_i^{(r)})^2 + 2 \sum_{i<j\le n_r} \alpha_i\alpha_j (t_i^{(r)})^2. \tag{6}$$

Applying Theorem IV.16.1 for independent sequences, we conclude that the distribution function of $\eta_n^{(r)}$ converges weakly to the Gaussian measure with mean value 0 and variance given by the right-hand side of (6). It remains to prove (iii) in Theorem 4.2. According to the remark on Lemma 4.2, condition (iii) holds if for any $\varepsilon > 0$

$$\lim_{r\to\infty} \limsup_{n\to\infty} \sum_{j=1}^{n_r} P\{|X_n(t_j^{(r)}) - X_n(t_{j-1}^{(r)})| \geq \varepsilon\} = 0, \tag{7}$$

$$\lim_{r\to\infty} \limsup_{n\to\infty} \sum_{j=1}^{n_r} P\left\{\sup_{t_{j-1}^{(r)} < t \leq t_j^{(r)},\, t\in\Gamma} |X_n(t) - X_n(t_{j-1}^{(r)})| \geq \varepsilon\right\} = 0, \tag{8}$$

$$X_n(t_j^{(r)}) - X_n(t_{j-1}^{(r)}) = \sum_{i=k_n(t_{j-1}^{(r)})+1}^{k_n(t_j^{(r)})} \frac{\xi_i}{s_n}, \tag{9}$$

and there are only finitely many terms in this sum (at most $k_n$ in number). Thus in (8) we may write "max" instead of "sup." Consider first (7). As we have just found [compare (4)], the distribution of $X_n(t_j^{(r)}) - X_n(t_{j-1}^{(r)})$ converges weakly to the Gaussian measure with mean value 0 and variance $t_j^{(r)} - t_{j-1}^{(r)} = 2^{-r}$. Now if a Gaussian measure with mean value 0 has the variance $\sigma^2$, we get the estimation

$$\begin{aligned}\frac{1}{\sqrt{2\pi}\sigma} \int_{|t|\geq\varepsilon} \exp\left(-\frac{1}{2}\frac{t^2}{\sigma^2}\right) dt &\leq \frac{1}{\sqrt{2\pi}\sigma\varepsilon} \int_{|t|\geq\varepsilon} |t| \exp\left(-\frac{t^2}{2\sigma^2}\right) dt \\ &\leq \frac{2\sigma}{\sqrt{2\pi}\varepsilon} \exp\left(-\frac{\varepsilon^2}{2\sigma^2}\right) \leq c_k\left(\frac{\sigma}{\varepsilon}\right)^k \end{aligned}\tag{10}$$

with a constant $c_k$ dependent only on $k$ for $k = 1, 2, \ldots, \sigma/\varepsilon \leq 1$. Using these inequalities, we get

$$\limsup_{n\to\infty} P[|X_n(t_j) - X_n(t_{j-1})| \geq \varepsilon] \leq c_k[2^{-r/2}/\varepsilon]^k. \tag{11}$$

Since $n_r = 2^r$ is the number of terms in (7) and

$$2^r \cdot c_k[2^{-r/2}/\varepsilon]^k \to 0 \qquad (r\to\infty)$$

for $k = 3$, relation (7) holds. To prove (8), we apply Lemma 5.1 and observe that we deal with the sum

$$\sum_{k_n(t_{i-1}^{(r)}) < i \leq k_n(t_i^{(r)})} \frac{\xi_i}{s_n}$$

and its partial sums. Thus, applying the inequality in this lemma, with $\lambda > \sqrt{2}$ and

$$\tilde{s}_n^2 = \sum_{k_n(t_{i-1}^{(r)}) < i \leq k_n(t_i^{(r)})} \boldsymbol{E}\left[\frac{\xi_i^2}{s_n^2}\right],$$

we get

$$\limsup_{n\to\infty} P\left[\left\{\max_{k_n(t_{i-1}^{(r)})<k_n(t)\le k_n(t_j^{(r)})} |X_n(t) - X_n(t_{j-1}^{(r)})| \ge \lambda\tilde{s}_n\right\}\right]$$
$$\le 2\limsup_{n\to\infty} P[\{|X_n(t_j^{(r)}) - X_n(t_{j-1}^{(r)})| \ge (\lambda - \sqrt{2})\tilde{s}_n\}]. \qquad (12)$$

As $n \to \infty$, $\tilde{s}_n^2 \to t_j^{(r)} - t_{j-1}^{(r)} = 2^{-r}$. Hence, for any $\varepsilon > 0$, we may choose $\lambda = \lambda_r$ such that $\lambda_r 2^{-r} > 2\varepsilon$, $\sqrt{2}2^{-r} < \varepsilon$. We then find, as in the proof of (7), that (8) also holds. □

Next we deal with $\varphi$-mixing sequences $\{\xi_i\}$ of real-valued random variables under the Lindeberg condition ($L$) and Ibragimov condition ($I$). (Compare Section IV.16.)

**Theorem 2.** *Let $X_n(t)$ belong to the sequence $\{\xi_i\}$ as above. If $\{\xi_i\}$ satisfies the I and L conditions, and if*

(i) $$\lim_{n\to\infty} \boldsymbol{E}[X_n(t)]^2 = \sigma^2(t)$$

*exists and is finite for $0 \le t \le 1$, then the sequence $P(X_n^{-1}\cdot)$ converges weakly to the Gaussian measure on D with mean-values zero and covariance function $\sigma^2[\min(s,t)]$.*

For the proof, we shall use

**Lemma 2.** *If $\{\xi_i\}$, $\boldsymbol{E}(\xi_i) = 0$, satisfies the I and L conditions and condition* (i) *in Theorem 2, then for $0 \le s \le t \le 1$,*

(i) $$\lim_{n\to\infty} \frac{1}{s_n^2}\sum_{i=1}^{k_n(t)} \boldsymbol{E}[\xi_i^2] = t,$$

(ii) $$\lim_{n\to\infty} \boldsymbol{E}[X_n(t)X_n(s)] = \sigma^2(s),$$

(iii) $$\lim_{n\to\infty} \boldsymbol{E}[X_n(t) - X_n(s)]^2 = \sigma^2(t) - \sigma^2(s) \le a(t-s)$$

*with*

$$a = 1 + 2\sum_{v=1}^{\infty} \varphi^{1/2}(v).$$

*Proof:* (i) follows as (i) in Lemma 1. Observing that for $0 \le s < t < u \le 1$ the random variables $X_n(s)$ and $X_n(u) - X_n(t)$ are measurable with respect to the $\sigma$-algebras generated by $\xi_i$, $i \le k_n(s)$, and $i \ge k_n(t)$, respectively, we get by Lemma IV.16.1

$$|\boldsymbol{E}X_n(s)[X_n(u) - X_n(t)]| \le \boldsymbol{E}^{1/2}[X_n^2(s)]\boldsymbol{E}^{1/2}[X_n(u) - X_n(t)]^2\varphi^{1/2}[k_n(t) - k_n(s)].$$

Here $k_n(t) - k_n(s) \to \infty$ $(n \to \infty)$ according to the $L$ condition and (i) since $t > s$. Hence, for $t > s$,

$$\lim_{n \to \infty} \boldsymbol{E}\{X_n(s)[X_n(u) - X_n(t)]\} = 0. \tag{13}$$

By the corollary of Lemma IV.16.1 and (i), we get

$$\boldsymbol{E}[X_n(t) - X_n(s)]^2 \leq a \frac{1}{s_n^2} \sum_{i=k_n(t)+1}^{k_n(s)} \boldsymbol{E}[\xi_i^2] \to a(t-s) \qquad (n \to \infty). \tag{14}$$

Hence,

$$\begin{aligned} &|\boldsymbol{E}\{X_n(s)[X_n(t) - X_n(s)]\}| \leq \boldsymbol{E}^{1/2}[X_n^2(s)]\boldsymbol{E}^{1/2}[X_n(t) - X_n(s)]^2 \\ &\to 0 \qquad (n \to \infty,\ t \to s) \end{aligned}$$

and thus (13) holds also for $t = s$. Then (ii) follows since $\boldsymbol{E}[X_n^2(s)] \to \sigma^2(s)$ by definition. Furthermore, (ii) implies (iii). The proof of Theorem 2 now runs as the proof of Theorem 1. Applying Theorem IV.16.1, we conclude that the distribution of $\eta_n^{(r)}$ converges weakly to a Gaussian measure for any $\alpha_i$. Thus the condition (ii) in Theorem 4.2 is satisfied, and (i) in this theorem is obvious. Again, to show (iii) in Theorem 4.2 we have to prove that (7) and (8) are satisfied. Now the distribution of $X_n(t) - X_n(s)$ converges weakly according to Theorem IV.16.1, and to the Gaussian measure with mean value 0 and variance

$$\lim_{n \to \infty} \boldsymbol{E}[X_n(t) - X_n(s)]^2 = \sigma^2(t) - \sigma^2(s) \leq a(t-s).$$

Proceeding then as in the proof of Theorem 1, but with a variance $\leq a(t-s)$ instead of $t - s$, we find that (7) holds. To prove (8), we apply Lemma 5.2 and get, with the same notations as above,

$$\begin{aligned} &P\left[\max_{k_n(t_{j-1}^{(r)}) \leq k_n(t) \leq k_n(t_{j-1}^{(r)})} |X_n(t) - X_n(t_{j-1}^{(r)})| \geq \lambda \tilde{s}_n\right] \\ &\quad \leq 2P[|X_n(t_j^{(r)}) - X_n(t_{j-1}^{(r)})| \geq (\lambda - \lambda')\tilde{s}_n] + 2k \sum_{i=k_n(t_{j-1}^{(r)})}^{k_n(t_j^{(r)})} P\left[\left|\frac{\xi_i}{s_n}\right| \geq \frac{1}{k}\lambda''\tilde{s}_n\right], \end{aligned} \tag{15}$$

provided that $\lambda > \lambda' > \lambda''$ and

$$a/(\lambda' - \lambda'')^2 + \varphi(k) \leq \tfrac{1}{2}. \tag{16}$$

Now $(\tilde{s}_n)^2 \to 2^{-r}$ as $n \to \infty$. We may choose $\lambda$, $\lambda'$, and $\lambda''$ to given $\varepsilon > 0$ such that $\lambda 2^{-r} = 3\varepsilon, \lambda' 2^{-r} = 2\varepsilon, \lambda'' 2^{-r} = \varepsilon$, and $a/(\lambda' - \lambda'') < \frac{1}{4}$. Furthermore, we choose $k$ so large that $\varphi(k) < \frac{1}{4}$, so that (16) is satisfied. As $n \to \infty$, the sum on the right-hand side of (15) tends to 0 according to the Lindeberg condition. Then the right-hand side of (15) tends to

$$2 \limsup_{n \to \infty} P[\{|X_n(t_j^{(r)}) - X_n(t_{j-1}^{(r)})| \geq \varepsilon\}]$$

as $n \to \infty$. Using this inequality, we prove as above that (8) holds also for the $X_n(t)$ considered here.

## 9. EMPIRICAL DISTRIBUTION FUNCTIONS

Let $\xi_i$, $i = 1, 2, \ldots$ be independent random variables from a probability space $(\Omega, \mathscr{B}, P)$ into $[0, 1]$, and define the function $u$ from $R$ into $R$ by $u(t) = 0$ for $t \leq 0$, $u(t) = 1$ for $t > 0$. The function

$$\omega \to \frac{1}{n} \sum_{i=1}^{n} u[\cdot - \xi_i(\omega)] \qquad (n = 1, 2, \ldots) \tag{1}$$

is called the empirical distribution function. Clearly it is a mapping from $\Omega$ into the $D$-space.

**Lemma 1.** *The empirical distribution function is a random variable from $(\Omega, \mathscr{B}, P)$ into $D$ under the Skorokhod metric $\rho$.*

*Proof:* It is sufficient to show that

$$\omega \to u(\cdot - \xi(\omega)] \tag{2}$$

is a random variable from $(\Omega, \mathscr{B}, P)$, i.e., that this mapping is measurable. Since

$$\omega \to u[t - \xi(\omega)]$$

is a real-valued random variable for any $t \in [0, 1]$, it follows by Lemma 3.2 that the function in (2) is measurable. Consider the random variable

$$\omega \to \frac{1}{n} \sum_{i=1}^{n} u[t - \xi_i(\omega)]. \tag{3}$$

It takes on the value $k/n$ if exactly $k$ values $\xi_i(\omega)$ are less than $t$. We assume that the distribution function $F$ of $\xi_i$

$$P\{\omega : \xi(\omega) \leq t\} = F(t) \tag{4}$$

is continuous so that $P[\omega : \xi(\omega) = t]$ is equal to 0. $\square$

We now deal with the random variable

$$\omega \to X_n(\omega, \cdot) = \frac{1}{\sqrt{n}} \sum_{i=1}^{n} \{u[\cdot - \xi_i(\omega)] - F\}$$

from $(\Omega, \mathscr{B}, P)$ into $D$. As in Section 5, we write $X_n$ instead of $X_n(\cdot, \cdot)$ and $X_n(t)$ instead of $X_n(\omega, t)$, remembering how $X_n$ and $X_n(t)$ depend on $\omega \in \Omega$. Clearly

$$\boldsymbol{E}[X_n(t)] = 0 \qquad \text{for} \quad t \in [0, 1].$$

Since $X_n$ is a measurable mapping of $(\Omega, \mathscr{B}, P)$ into $D$, it determines a probability measure $P(X_n^{-1} \cdot)$ on $D$.

**Theorem 1.** *The sequence $\{P(X_n^{-1} \cdot)\}$ converges weakly to the Gaussian measure $\mu$ with mean values zero and covariance function*

$$\lim_{n \to \infty} \boldsymbol{E}[X_n(t) X_n(s)] = F(s) - F(s)F(t) \qquad \textit{for} \quad s < t.$$

*Proof:* By definition of $u$, we have

$$u(t - \xi)u(s - \xi) = u(s - \xi) \qquad \text{for} \quad s \leq t. \tag{5}$$

Hence,

$$\boldsymbol{E}[u(t - \xi)u(s - \xi)] = \boldsymbol{E}[u(s - \xi)] = F(s)$$

for $s \leq t$, and thus

$$\boldsymbol{E}\left\{\frac{1}{n}[u(t - \xi) - F(t)]^2\right\} = \frac{1}{n}[F(t) - F^2(t)], \tag{6}$$

and since the $\xi_i$ are independent,

$$\boldsymbol{E}[u(t - \xi_i) - F(t)][u(s - \xi_j) - F(s)] = \begin{cases} 0 & (i \neq j), \\ F(s) - F(t)F(s) & (i = j) \end{cases}. \tag{7}$$

Using these relations, we obtain

$$\begin{aligned} \boldsymbol{E}[X_n(t)X_n(s)] &= \frac{1}{n} \sum_{i=1}^{n} \sum_{j=1}^{n} \boldsymbol{E}[u(t - \xi_i) - F(t)][u(s - \xi_j) - F(s)] \\ &= F(s) - F(t)F(s) \qquad \text{for} \quad s \leq t. \end{aligned} \tag{8}$$

To prove that $\{P(X_n^{-1} \cdot)\}$ converges weakly, we shall use Theorem 4.2 and then show that the conditions (i)–(iii) are satisfied. Now $P(X_n^{-1} D) = P(\Omega) = 1$, so that (i) holds. In order to prove (ii), we have to show that the distribution of any random vector

$$[X_n(t_1^{(r)}), \ldots, X_n(t_{n_r}^{(r)})], \qquad t_i^{(r)} \in \Gamma_r$$

converges weakly as $n \to \infty$. By Theorem IV.5.2 (compare Section 8), this is the case if the distribution of

$$\eta_n^{(r)} = \sum_{i=1}^{n_r} \alpha_i X_n(t_i^{(r)})$$

converges weakly as $n \to \infty$ for any real numbers $\alpha_i$, $i = 1, \ldots, n_r$. Putting

$$Y_j = \frac{1}{\sqrt{n}} \sum_{i=1}^{n_r} \alpha_i [u(t_i^{(r)} - \xi_j) - F(t_i^{(r)})],$$

we have

$$\eta_n^{(r)} = \sum_{j=1}^{n} Y_i,$$

where the random variables are independent and identically distributed. Furthermore, $\boldsymbol{E}(Y_j) = 0$, and using (8), we get

$$\begin{aligned}
\boldsymbol{E}[\eta_n^{(r)}]^2 &= \sum_{i=1}^{n_r} \sum_{j=1}^{n_r} \alpha_i \alpha_j \boldsymbol{E} X_n(t_i^{(r)}) X_n(t_j^{(r)}) \\
&= \sum_{i=1}^{n_r} \alpha_i^2 \{F(t_i^{(r)}) - [F(t_i^{(n)})]^2\} \\
&\quad + 2 \sum_{i<j\le n_r} \alpha_i \alpha_j [F(t_i^{(r)}) - F(t_i^{(r)}) F(t_j^{(r)})]. \qquad (9)
\end{aligned}$$

Hence by Theorem IV.16.1, the distribution of $\eta_n^{(r)}$ converges weakly, as $n \to \infty$, to the Gaussian measure with mean value 0 and the variance given by (9). This implies (ii) in Theorem 4.2.

Consider the condition (iii) in this theorem. By the remark on Lemma 4.2, this condition holds if for any $\varepsilon > 0$

$$\lim_{r\to\infty} \limsup_{n\to\infty} \sum_{j=1}^{n_r} P\{|X_n(t_j^{(r)}) - X_n(t_{j-1}^{(r)})| \ge \varepsilon\} = 0, \qquad (10)$$

$$\lim_{r\to\infty} \limsup_{n\to\infty} \sum_{j=1}^{n_r} P\left\{ \sup_{t_{j-1}^{(r)} < t < t_j^{(r)},\, t\in\Gamma} \min[|X_n(t) - X_n(t_{j-1}^{(r)})|, |X_n(t_j^{(r)}) - X_n(t)|] \ge \varepsilon \right\} = 0. \qquad (11)$$

As for $\eta_n^{(r)}$, we find that the distribution of $X_n(t_j^{(r)}) - X_n(t_{j-1}^{(r)})$ converges weakly to the Gaussian measure with mean value 0 and variance

$$\begin{aligned}
\boldsymbol{E}[X_n(t_j^{(r)}) - X_n(t_{j-1}^{(r)})]^2 &= F(t_j^{(r)}) - [F(t_j^{(r)})]^2 + F(t_{j-1}^{(r)}) \\
&\quad - [F(t_{j-1}^{(r)})]^2 - 2[F(t_{j-1}^{(r)}) - F(t_j^{(r)}) F(t_{j-1}^{(r)})] = F(t_j^{(r)}) - F(t_{j-1}^{(r)}) \\
&\quad - [F(t_j^{(r)}) - F(t_{j-1}^{(r)})]^2 \le F(t_j^{(r)}) - F(t_{j-1}^{(r)}).
\end{aligned}$$

Hence, using estimation V.8.10, we obtain

$$\lim_{n\to\infty} P[|X_n(t_j^{(r)}) - X_n(t_{j-1}^{(r)})| \geq \varepsilon] \leq \frac{c_4}{\varepsilon^4}[F(t_j^{(r)}) - F(t_{j-1}^{(r)})]^2. \tag{12}$$

Then

$$\limsup_{r\to\infty}\limsup_{n\to\infty}\sum_{j=1}^{n_r} P[|X_n(t_j^{(r)}) - X_n(t_{j-1}^{(r)})| \geq \varepsilon]$$

$$\leq \frac{c_4}{\varepsilon^4}\limsup_{r\to\infty}\sup_{1\leq j\leq n_r}|F(t_j^{(r)}) - F(t_{j-1}^{(r)})|F(1) = 0, \tag{13}$$

since $F$ is continuous (hence, uniformly continuous) and $t_j^{(r)} - t_{j-1}^{(r)} = 2^{-r}$. Thus, (10) holds. To prove (11), we use the corollary of Lemma 5.3 and show that

$$\boldsymbol{E}[X_n(t) - X_n(t_1)]^2[X_n(t_2) - X_n(t)]^2 \leq c[F(t_2) - F(t_1)]^2 \tag{14}$$

with a constant $c$ for $0 \leq t_1 < t < t_2 \leq 1$. The corollary then states that

$$P\left[\sup_{t_1<t<t_2,\, t\in\Gamma} \min\{|X_n(t) - X_n(t_1)|, |X_n(t_2) - X_n(t)|\} \geq \lambda\right]$$
$$\leq K\lambda^{-4}[F(t_2) - F(t_1)]^2$$

with a constant $K$, independent of $t_1$ and $t_2$ for $0 \leq t_1 < t_2 \leq 1$. Applying this inequality to the terms in the sum of (11), we find, as in (13), that (11) holds.

It remains to prove (14). Putting

$$Y_i = u(t - \xi_i) - u(t_1 - \xi_i) - F(t) + F(t_1),$$
$$\tilde{Y}_i = u(t_2 - \xi_i) - u(t - \xi_i) - F(t_2) + F(t),$$

we get

$$X_n(t) - X_n(t_1) = \frac{1}{\sqrt{n}}\sum_{i=1}^n Y_i, \qquad X_n(t_2) - X_n(t) = \frac{1}{\sqrt{n}}\sum_{i=1}^n \tilde{Y}_i.$$

Here $Y_i$ is independent of $Y_j$ and $\tilde{Y}_j$, $\tilde{Y}_i$ is independent of $\tilde{Y}_j$ for $i \neq j$, and $\boldsymbol{E}(Y_i) = \boldsymbol{E}(\tilde{Y}_i) = 0$. Using these facts, we obtain

$$\boldsymbol{E}[X_n(t) - X_n(t_1)]^2[X_n(t_2) - X_n(t)]^2 = \frac{1}{n^2}\sum_{i=1}^n\sum_{j=1}^n E[Y_i^2\tilde{Y}_j^2]$$
$$+ \frac{4}{n^2}\sum\sum_{1\leq i<j\leq n} E[Y_iY_j\tilde{Y}_i\tilde{Y}_j]. \tag{15}$$

Since $0 \le t_1 < t < t_2 \le 1$, we have

$$u(t - \xi_i)u(t_1 - \xi_i) = u(t_1 - \xi_i),$$
$$u(t_2 - \xi_i)u(t - \xi_i) = u(t - \xi_i),$$
$$[u(t - \xi_i) - u(t_1 - \xi_i)][u(t_2 - \xi_i) - u(t - \xi_i)] = 0.$$

Hence we get

$$Y_i^2 = [u(t - \xi_i) - u(t_1 - \xi_i)]\{1 - 2[F(t) - F(t_1)]\} + [F(t) - F(t_1)]^2,$$
$$\tilde{Y}_i^2 = [u(t_2 - \xi_i) - u(t - \xi_i)]\{1 - 2[F(t_2) - F(t)]\} + [F(t_2) - F(t)]^2,$$
$$Y_i\tilde{Y}_i = -[F(t_2) - F(t)][u(t - \xi_i) - u(t_1 - \xi_i)] - [F(t) - F(t_1)][u(t_2 - \xi_i) - u(t - \xi_i)],$$
$$[Y_i\tilde{Y}_i]^2 = [F(t_2) - F(t)]^2[u(t - \xi_i) - u(t_1 - \xi_i)] + [F(t) - F(t_1)]^2[u(t_2 - \xi_i) - u(t - \xi_i)],$$
$$\boldsymbol{E}[Y_i\tilde{Y}_i]^2 = [F(t_2) - F(t)]^2[F(t) - F(t_1)] + [F(t) - F(t_1)]^2 \times [F(t_2) - F(t)] \le [F(t_2) - F(t_1)]^3.$$

Since $Y_i^2$ and $\tilde{Y}_j^2$ are independent for $i \ne j$, we further obtain for $i \ne j$,

$$\begin{aligned}\boldsymbol{E}[Y_i^2\tilde{Y}_j^2] = \boldsymbol{E}[Y_i^2]\boldsymbol{E}[\tilde{Y}_j^2] &= ([F(t) - F(t_1)]\{1 - 2[F(t) - F(t_1)]\} \\ &\quad + [F(t) - F(t_1)]^2)([F(t_2) - F(t)]\{1 - 2[F(t_2) - F(t)]\} \\ &\quad + [F(t_2) - F(t)]^2) \\ &\le [F(t_2) - F(t_1)]^2.\end{aligned}$$

Since $Y_i\tilde{Y}_i$ and $Y_j\tilde{Y}_j$ are independent for $i \ne j$, we obtain

$$\begin{aligned}\boldsymbol{E}[Y_i\tilde{Y}_iY_j\tilde{Y}_j] &= \boldsymbol{E}[Y_i\tilde{Y}_i]\boldsymbol{E}[Y_j\tilde{Y}_j] \\ &= 4[F(t_2) - F(t)]^2[F(t) - F(t_1)]^2 \le [F(t_2) - F(t_1)]^4.\end{aligned}$$

Using these estimations of the expectations in (15), we find that the right-hand side of (15) is at most equal to

$$\frac{1}{n}[F(t_2) - F(t_1)]^3 + [F(t_2) - F(t_1)]^2 + 4[F(t_2) - F(t_1)]^4.$$

Thus (14) holds. □

## 10. EMBEDDING OF SEQUENCES OF MARTINGALE DIFFERENCES IN RANDOM FUNCTIONS

Sequences of martingale differences were discussed in Section IV.15. We refer the reader to that section.

**Theorem 1.** *Let $\{\xi_i\}_{i=1}^{\infty}$ be a stationary sequence of martingale differences with $\boldsymbol{E}(\xi_i) = 0$, $\boldsymbol{E}(\xi_i^2) = \sigma^2$, and let $\{\xi_i^2\}_{i=1}^{\infty}$ be mean ergodic. Then the distribution of $\{X_n\}$,*

$$X_n(t) = \frac{1}{\sigma\sqrt{n}} \sum_{i \le nt} \xi_i,$$

*converges weakly to the Wiener measure on the D-space.*

*Proof:* We apply Theorem 4.2. Proceeding as in Section 8, we have to show that the conditions (ii) and (iii) are satisfied. Clearly (i) is obvious. As we have shown in Section 8, the condition (ii) holds if

$$\sum_{i=1}^{n_r} \alpha_i X_n(t_i^{(r)})$$

converges weakly in distribution to a Gaussian measure for any $r$ and any real numbers $\alpha_i$, where the $t_i^{(r)}$ belong to $\Gamma$, and $r$ is any positive integer. This, however, is true according to the corollary of Theorem IV.15.1. Proceeding as in Section 8, we shall show that (iii) is also satisfied.

By the corollary just mentioned, the distribution of $X_n(t_j^{(r)}) - X_n(t_{j-1}^{(r)})$ converges weakly to the Gaussian measure with mean value 0 and variance

$$\lim_{n\to\infty} \boldsymbol{E}[X_n(t_j^{(r)}) - X_n(t_{j-i}^{(r)})]^2 = 2^{-r}. \tag{1}$$

Using this relation, we find as in the proof of (7) in Section 8 that this relation also holds here. We show that (8) in Section 8 is also satisfied. Then we consider

$$P\left[\max_{t_{j-1}^{(r)} < t \le t_j^{(r)}} |X_n(t) - X_n(t_{j-1}^{(r)})| \ge \varepsilon\right]. \tag{2}$$

We write "max" instead of "sup," since $X_n(t) - X_n(t_{j-1}^{(r)})$ is a finite sum of random variables $\xi_i/\sigma\sqrt{n}$. Our proof is a consequence of

**Lemma 1.** *Let $\{\xi_i\}$ be a stationary sequence of martingale differences with $\boldsymbol{E}(\xi_i) = 0$, $\boldsymbol{E}(\xi_i^2) = \sigma^2 > 0$, and put $S_j = \sum_{i=1}^{j} \xi_i$, $s_m^2 = \sum_{i=1}^{m} \boldsymbol{E}(\xi_i^2)$. Then for $\lambda > \lambda_1$, we have*

$$\text{(i)} \qquad P\left[\max_{1\le j\le m} |S_j| \ge \lambda s_m\right] \le (1 - \lambda_1^{-2})^{-1} P[|S_m| \geqq (\lambda - \lambda_1)s_m]$$
$$+ \frac{1}{\sigma^2(\lambda_1^2 - 1)} \boldsymbol{E}\left\{\left|\frac{1}{m}\sum_{i=1}^{m} \xi_i^2 - \sigma^2\right|\right\}.$$

*Proof:* By V.5.4, which holds also for dependent random variables, we have

$$P\left\{\max_{1\le i\le m}|S_i|\ge\lambda s_m\right\}\le P\{|S_m|\ge(\lambda-\lambda_1)s_m\}$$
$$+\sum_{i=1}^{m}P[E_i\cap\{|S_m-S_i|\ge\lambda_1 s_m\}],\tag{3}$$

where

$$E_i=\left\{\max_{j<i}|S_j|<\lambda s_m\le|S_i|\right\},$$
$$\bar{E}=\bigcup_{i=1}^{m-1}E_i\subset\left\{\max_{1\le i\le m}S_i\right\},\qquad E_i\cap E_j=\varnothing\qquad\text{for}\quad i\ne j.\tag{4}$$

By Chebyshev's inequality, we obtain

$$P[E_i\cap\{|S_m-S_i|\ge\lambda_1 s_m\}]\le\frac{1}{\lambda_1^2 s_m^2}\boldsymbol{E}[1_{E_i}(S_m-S_i)^2]$$
$$=\frac{1}{\lambda_1^2 s_m^2}\left\{\boldsymbol{E}\left[1_{E_i}\sum_{j=i+1}^{m}\xi_j^2\right]\right.$$
$$\left.+2\sum_{i<j_1<j_2\le m}\boldsymbol{E}[1_{E_i}\xi_{j_1}\xi_{j_2}]\right\}.\tag{5}$$

Let $\mathscr{B}_i$ be the $\sigma$-algebra generated by the $\xi_j$, $j\le i$. Since $\{\xi_i\}$ is a sequence of martingale differences, we get for $i<j_1<j_2$,

$$\boldsymbol{E}[1_{E_i}\xi_{j_1}\xi_{j_2}]=\boldsymbol{E}[1_{E_i}\xi_{j_1}\boldsymbol{E}(\xi_{j_2}|\mathscr{B}_{j_1})]=0.$$

Hence (3) reduces to

$$P[E_i\cap\{|S_m-S_i|\ge\lambda_1 s_m\}]\le\frac{1}{\lambda_1^2 s_m^2}\boldsymbol{E}\left[1_{E_t}\sum_{j=1}^{m}\xi_j^2\right],\tag{6}$$

and by (5) we get

$$P\left[\left\{\max_{1\le i\le m}|S_i|\ge\lambda s_m\right\}\right]\le P\{|S_m|\ge(\lambda-\lambda_1)s_m\}+\frac{1}{\lambda_1^2 s_m^2}\boldsymbol{E}\left[1_E\sum_{i=1}^{m}\xi_i^2\right].\tag{7}$$

Here

$$E\left[1_{\bar{E}} \sum_{j=1}^{m} \xi_j^2\right] \le E\left[1_{\bar{E}} \left|\sum_{j=1}^{m} \xi_j^2 - m\sigma^2\right|\right] + m\sigma^2 P(\bar{E})$$
$$\le mE\left|\frac{1}{m} \sum_{j=1}^{m} \xi_j^2 - \sigma^2\right| + m\sigma^2 P(\bar{E}).$$

Choosing $\lambda_1 < \lambda_2$ and observing (4) and the fact that $s_m^2 = m\sigma^2$, we find by (7) and the last inequalities that the inequality in the lemma holds. □

*Completion of the Proof of the Theorem:* Consider (2). Here

$$X_n(t_j^{(r)}) - X_n(t_{j-1}^{(r)}) = \sum_{nt_{j-1}^{(r)} < i < nt_j^{(r)}} \xi_i / \sigma\sqrt{n}. \tag{8}$$

Denote the number of terms in the sum by $m_n^{(r)}$. Then

$$n2^{-r} - 2^{-r} \le m_n^{(r)} \le n2^{-r}.$$

Furthermore, $X_n(t) - X_n(t_{j-1}^{(r)})$ is a partial sum of (8) and $E[\xi_i^2/\sigma^2 n] = 1/n$. Applying the lemma, we obtain for $\lambda_1 = \sqrt{2} < \lambda$

$$P\left[\left\{\max_{t_{j-1}^{(r)} < t \le t_j^{(r)}} |X_n(t) - X_n(t_{j-1}^{(r)})| \ge \left(m_n^{(r)} \frac{1}{n}\right)^{1/2} \lambda\right\}\right]$$
$$\le 2P\left[\left\{|X_n(t_j^{(r)}) - X_n(t_{j-1}^{(r)})| \ge \left(m_n^{(r)} \frac{1}{n}\right)^{1/2} (\lambda - \sqrt{2})\right\}\right]$$
$$+ E\left|\frac{1}{m_n^{(r)}} \sum_{nt_{j-1}^{(r)} < i \le nt_j^{(r)}} \xi_i^2 - \sigma^2\right|. \tag{9}$$

As $n \to \infty$, the second term on the right-hand side of (9) tends to 0, since the sequence $\{\xi_i\}$ is stationary and the sequence $\{\xi_i^2\}$ is mean ergodic by assumption. Furthermore,

$$(m_n^{(r)}/n)^{1/2} \to 2^{-r/2},$$

so that

$$\limsup_{n\to\infty} P\left\{\max_{t_{j-1}^{(r)} < t \le t_j^{(r)}} |X_n(t) - X_n(t_{j-1}^{(r)})| \ge 2^{-r/2}\lambda\right\}$$
$$\le 2 \limsup_{n\to\infty} P\{|X_n(t_j^{(r)}) - X_n(t_{j-1}^{(r)})| \ge 2^{-r/2}(\lambda - \sqrt{2})\}. \tag{10}$$

We have found above that $X_n(t_j^{(r)}) - X_n(t_{j-1}^{(r)})$ as $n \to \infty$ converges weakly to the Gaussian measure with mean value 0 and variance $2^{-r}$. Thus we find by (9), as in the proof of Theorem 8.1, that (iii) in Lemma 4.2 holds for the $X_n(t)$ considered here. Indeed, we may choose $\lambda = 2^{r/4}$ and use the estimation in V.8.10.

CHAPTER VI

# WEAK CONVERGENCE IN SEPARABLE HILBERT SPACES

## 1. $\sigma$-SMOOTH MEASURES ON $l^2$-SPACE

The Hilbert space $l^2$ was introduced in Section I.11, and there we also discussed general Hilbert spaces. Any separable Hilbert space is isometric to $l^2$ and, hence, the $l^2$-space is a representative of a rather general class of Hilbert spaces. A weak limit problem for sequences of measures on a separable Hilbert space can then be transformed to a corresponding limit problem on $l^2$. Therefore, we shall deal only with the Hilbert space $l^2$.

A projection $\pi$ of $l^2$ is a mapping of $l^2$ onto $R^{(k)}$ for some $k$. By Lemma I.8.1, it is continuous. Indeed, if $x \in l^2$ and $\{x_n\}$ is a sequence in $l^2$ such that $\|x_n - x\| \to 0$ $n \to \infty$, then clearly $x_n^{(i)} \to x^{(i)}$ $(n \to \infty)$ for any coordinate $x_n^{(i)}$ of $x_n$ and $x^{(i)}$ of $x$.

The weak limit problem for a sequence $\{\mu_n\}$ of $\sigma$-smooth measures on $l^2$ can be reduced by Lemma III.6.1. Indeed, if $\pi_r$ maps $l^2$ onto the finite-dimensional vector space $R^{(r)}$ of vectors $(x^{(1)}, \ldots, x^{(r)})$,

$$x \to (x^{(1)}, \ldots, x^{(r)}), \tag{1}$$

then there is a continuous mapping $V_r$ of $R^{(r)}$ into $l^2$, where

$$V_r(x^{(1)}, \ldots, x^{(r)}) = (x^{(1)}, \ldots, x^{(r)}, 0, 0 \ldots), \tag{2}$$

and clearly

$$\|x - V_r \pi_r x\| \to 0 \qquad (r \to \infty). \tag{3}$$

Applying Lemma III.6.1, we get

**Theorem 1.** *A sequence* $\{\mu_n\}$ *of probability measures on* $l^2$ *converges weakly to a probability measure* $\mu$ *on* $l^2$ *if and only if*

(i) $\{\mu_n(\pi_r^{-1}\cdot)\}$ *converges weakly on* $R^{(r)}$ *for any projection* (1),

(ii) $$\lim_{r\to\infty} \limsup_{n\to\infty} \mu_n\{\|x - V_r\pi_r x\| \geq \varepsilon\} = 0 \qquad \textit{for any} \quad \varepsilon > 0.$$

As a corollary of this theorem, we get

**Theorem 2.** *Any probability measure* $\mu$ *on* $l^2$ *is uniquely determined by its projections* $\{\mu(\pi_r^{-1}\cdot)\}$ *and is the weak limit of a sequence of such projections. Indeed, a probability measure* $\mu$ *determines the probability measure* $\mu(\pi_r^{-1}\cdot)$ *on* $\pi_r l^2$, *and by Theorem* 1, $\{\mu(\pi_r^{-1}V_r^{-1}\cdot)\}$ *converges weakly to* $\mu$.

*Remark:* In the theorems above, we considered only probability measures, but the weak limit problem for any sequence of $\sigma$-smooth measures can be reduced to a corresponding limit problem for probability measures, for if a sequence $\{\mu_n\}$ of $\sigma$-smooth measures on $l^2$ should converge weakly, then $\{\mu_n(l^2)\}$ must converge to a finite number. If this number is zero, then $\{\mu_n\}$ converges to the zero measure. Otherwise $\mu_n(l^2) > 0$ for sufficiently large $n$ and $\{(1/\mu_n(l^2)]\mu_n\}$ is a sequence of probability measures that must converge weakly.

Let $\mathscr{M}$ be the class of $\sigma$-smooth measures $\mu$ on $l^2$, where $\mu$ is different from the zero measure. The convolution $\mu * \nu$ of any two $\sigma$-smooth measures $\mu$ and $\nu$ on $l^2$ is defined, and clearly $\mathscr{M}$ is a commutative semigroup, that contains the unit measure $e$, having all its mass at the zero point and this mass being equal to 1. Furthermore, $\mathscr{M}$ contains the class $\mathscr{M}_0$ of probability measures, and $e \in \mathscr{M}_0$.

When we deal with convolutions of probability measures on $l^2$, the following lemma is of importance.

**Lemma 1.** *Let* $\pi_r$ *and* $V_r$ *be defined as above, and put* $\Delta_r x = x - V_r\pi_r x$. *If* $\mu$ *and* $\nu$ *are probability measures on* $l^2$ *and* $\lambda = \mu * \nu$ *and*

(i) $$\lambda\{\|\Delta_r x\| \geq \varepsilon\} < \varepsilon_r$$

*for some* $\varepsilon_r > 0$, *then there exists* $x_r$ *in* $l^2$ *such that*

(ii) $$\mu\{\|\Delta_r(x + x_r)\| \geq \varepsilon\} < \varepsilon_r.$$

*If* $\mu$ *is symmetrical, moreover,*

(iii) $$\mu\{\|\Delta_r x\| \geq \varepsilon\} < 2\varepsilon_r.$$

*Proof:* Put $E_0 = \{x : \|\Delta_r x\| < \varepsilon\}$. The function $x \to \mu(E_0 - x)$ is measurable, even continuous, since $E_0$ is an open set. Indeed, for any sequence

$\{x_n\}$ with $\|x_n - x\| \to 0$, $1_{E_0 - x_n} \to 1_{E_0 - x}$ $(n \to \infty)$, and hence, by the dominated convergence theorem,

$$\mu(E_0 - x_n) = \int 1_{E_0 - x_n} \mu(dx) \to \int 1_{E_0 - x} \mu(dx) = \mu(E_0 - x).$$

Thus the continuity of $x \to \mu(E_0 - x)$ follows by Lemma I.8.1. For $\lambda = \mu * v$ and

$$E = \{x : \mu(E_0 - x) \geq 1 - \varepsilon_r\},$$

we get

$$0 < 1 - \varepsilon_r \leq \lambda(E_0) = \int_{l^2} \mu(E_0 - x)\, v(dx) < v(E) + (1 - \varepsilon_r)v(\complement E) = 1 - \varepsilon_r v(\complement E).$$

Hence $v(\complement E) < 1$, $v(E) > 0$, i.e., $E$ is not empty. Thus there exists $x_r$ such that $\mu(E_0 - x_r) \geq 1 - \varepsilon_r$. This implies (ii). Let $\mu$ be symmetrical. By (ii), we have

$$\mu\{\|\Delta_r(x + x_r)\| \geq \varepsilon\} < \varepsilon_r,$$

but

$$\mu\{\|\Delta_r(x + x_r)\| \geq \varepsilon\} = \mu[\|\Delta_r(x - x_r)\| \geq \varepsilon]$$

according to the symmetry of $\mu$. Now

$$2\|\Delta_r x\| \leq \|\Delta_r(x + x_r)\| + \|\Delta_r(x - x_r)\|.$$

Thus

$$\|\Delta_r x\| < \varepsilon \qquad \text{if} \quad \|\Delta_r(x + x_r)\| < \varepsilon \quad \text{and} \quad \|\Delta_r(x - x_r)\| < \varepsilon,$$

i.e., $\{x : \|\Delta_r x\| < \varepsilon\}$ contains

$$\{x : \|\Delta_r(x + x_r)\| < \varepsilon\} \cap \{x : \|\Delta_r(x - x)\| < \varepsilon\}$$

and (iii) holds. □

The notation

$$\Delta_r x = x - V_r \pi_r x$$

will be used, and we put $\Delta_0 x = x$. Also, we use the notation

$$\bar{v}_r = \int_E g(x)\, \Delta_r x\, \mu(dx)$$

for the element $\bar{v}_r$ in $l^2$ with the coordinates

$$v_r^{(i)} = \begin{cases} \int_E g(x) x^{(i)}\, \mu(dx) & \text{for} \quad i > r, \\ 0 & \text{for} \quad i \leq r, \end{cases}$$

provided that the integrals are defined and finite, $g$ being a real-valued measurable function, $E$ a measurable set, and

$$\sum_{i=1}^{\infty} [v_r^{(i)}]^2 < \infty,$$

so that $\bar{v}_r \in l^2$.

**Lemma 2.** *Let $g$ and $h$ be measurable functions from $l^2$ and $E$ a measurable set. Then for any $\sigma$-smooth, $\sigma$-finite measure on $l^2$,*

$$\text{(i)} \qquad \left\| \int_E g(x)h(x)\, \Delta_r x\, \mu(dx) \right\|^2 \le \int_E g^2(x) \|\Delta_r x\|^2\, \mu(dx) \times \int_E h^2(x)\, \mu(dx),$$

*and*

$$\bar{v}_r = \int_E g(x)h(x)\, \Delta_r x\, \mu(dx) \in l^2$$

*when the right-hand side of* (i) *is finite. The statement also holds if we change $\Delta_r x$ to $V_r \pi_r x$.*

*Proof:* By Lagrange's inequality, we get

$$[v^{(i)}]^2 = \left[ \int_E g(x)h(x)x^{(i)}\, \mu(dx) \right]^2 \le \int_E g^2(x)[x^{(i)}]^2\, \mu(dx) \int_E h^2(x)\, \mu(dx).$$

Hence,

$$\sum_{i=1}^{\infty} [v^{(i)}]^2 \le \int_E g^2(x) \|\Delta_r x\|^2\, \mu(dx) \int_E h^2(x)\, \mu(dx).$$

Clearly these relations hold also if we change $\Delta_r x$ to $V_r \pi_r x$. □

## 2. WEAK CONVERGENCE OF CONVOLUTION PRODUCTS OF PROBABILITY MEASURES ON $l^2$

As in Chapter IV, we shall deal with convolution products in the standard form

$$\nu_n = \prod_{j=1}^{k_n *} \mu_{nj}, \qquad k_n \uparrow \infty \qquad \text{as} \quad n \uparrow \infty, \tag{1}$$

where the $\mu_{nj}$ are now probability measures on $l^2$. If $\pi_r$ is the projection considered in Section 1 and the mapping $V_r$ is defined as there, we find by

Theorem 1.1 that $\nu_n$ converges weakly if and only if

$$\nu_n(\pi_r^{-1}\cdot) = \prod_{j=1}^{k_n *} \mu_{nj}(\pi_r^{-1}\cdot) \tag{2}$$

converges weakly for $r = 1, 2, \ldots,$

$$\lim_{r\to\infty} \limsup_{n\to\infty} \nu_n\{\|\Delta_r x\| \geq \varepsilon\} = 0 \tag{3}$$

for any $\varepsilon > 0$, and $\Delta_r x = x - V_r \pi_r x$. However, condition (3) is difficult to handle directly. We now state

**Theorem 1.** *In order for the convolution product* (1) *to converge weakly, it is sufficient that the following conditions be satisfied:*

(i) $\{\mu_n(\pi_r^{-1}\cdot)\}$ *converges weakly as* $n \to \infty$ *for* $r = 1, 2, \ldots,$

(ii) $$\lim_{r\to\infty} \limsup_{n\to\infty} \sum_{j=1}^{k_n} \int_{\|\Delta_r x\| \geq \varepsilon} \mu_{nj}(dx) = 0 \qquad \text{for} \quad \varepsilon > 0,$$

(iii) $$\lim_{r\to\infty} \limsup_{n\to\infty} \sum_{j=1}^{k_n} \left\| \int_{\|\Delta_r x\| < \varepsilon} \Delta_r x \, \mu_{nj}(dx) \right\| = 0 \qquad \text{for} \quad \varepsilon > 0,$$

(iv) $$\lim_{r\to\infty} \limsup_{n\to\infty} \sum_{j=1}^{k_n} \int_{\|\Delta_r x\| < \varepsilon} \|\Delta_r x\|^2 \, \mu_{nj}(dx) = 0 \qquad \text{for} \quad \varepsilon > 0.$$

For the notation in (iii), compare the remarks at the end of Section 1. For the proof of the theorem, we shall use the function $x \to \exp(-\|x\|^2)$ from $l^2$ into $R$ and the following lemma, which also gives estimations for later use.

**Lemma 1.** *Put*

$$g(x, y) = 2\exp(-\|x\|^2) - \exp(-\|x - y\|^2 - \exp(-\|x + y\|^2),$$
$$h(x, y) = \exp(-\|x\|^2) - \exp(-\|x + y\|^2) - 2x \cdot y \exp(-\|x\|^2).$$

*The following inequalities hold for* $x, y \in l^2$ *with finite numbers* $a(\eta) > 0$ *and* $c_1(\eta) > 0$, *depending on* $\eta$, *and positive numbers* $c_2$ *and* $c_3$,

(i) $|h(x, y)| \leq a(\eta)\|y\|^2$ *for all* $x$ *and* $\|y\| \leq \eta$,

*where* $a(\eta)$ *is finite for finite* $\eta$ *and tends to* 0 *as* $\eta \to 0$,

(ii) $|g(x, y)| \leq 2$ *for all* $x$ *and* $y$,

(iii) $g(x, y) \geq c_1(\eta)$ *for* $\|x\| \leq \frac{1}{4}\eta, \quad \|y\| \geq \eta$,

(iv) $|g(x, y)| \leq c_2\|y\|^2$ *for* $\|y\| \leq \frac{1}{2}$,

(v) $g(x, y) \geq c_3\|y\|^2$ *for* $\|x\| \leq \frac{1}{2}, \quad \|y\| \leq \frac{1}{2}$.

*Remark:* This lemma holds for any Hilbert space.

*Proof:*

(i) $$\|x\|^2 - \|x + y\|^2 = -\|y\|^2 - 2x \cdot y$$

and, hence, for $\|y\| \le \eta$ we get by the Taylor expansion and by observing that $|-\|y\|^2 - 2x \cdot y| \le \eta^2 + 2\eta\|x\|$,

$$\exp(-\|x\|^2)|1 - \exp[\|x\|^2 - \|x + y\|^2] - 2x \cdot y| \le a(\eta)\|y\|^2.$$

(ii) This is obvious.
(iii) For $\|x\| \le \frac{1}{4}\eta$, $\|y\| \ge \eta$, we have

$$g(x, y) = [\exp(-\|x\|^2)][2 - \exp(-\|y\|^2 + 2|x \cdot y|) - \exp(-\|y\|^2 - 2|x \cdot y|)], \tag{4}$$

where

$$\|y\|^2 - 2|x \cdot y| \ge \tfrac{1}{2}\eta^2, \qquad \|y\|^2 + 2|x \cdot y| \ge \eta^2.$$

Hence,

$$g(x, y) \ge [\exp(-\tfrac{1}{16}\eta^2)]\{2 - 2\exp(-\tfrac{1}{2}\eta^2)\} = c_1(\eta).$$

(iv) This follows by (i), considered for $y$ and $-y$.
(v) We write (4):

$$g(x, y) = 2[\exp -(\|x\|^2 + \|y\|^2)]\{\exp\|y\|^2 - \cosh 2|x \cdot y|\}.$$

By Taylor's expansion, we get for $\|x\| \le \frac{1}{2}$, $\|y\| \le \frac{1}{2}$,

$$\cosh(2|x \cdot y|) - 1 \le \tfrac{1}{2}\|y\|^2 \cosh(\tfrac{1}{2}),$$
$$\exp\|y\|^2 \ge 1 + \|y\|^2,$$

and thus

$$g(x, y) \ge 2\|y\|^2[1 - \tfrac{1}{2}\cosh(\tfrac{1}{2})]\exp(-\tfrac{1}{2}) = c_3\|y\|^2. \quad \square$$

The proof of the theorem will follow from

**Lemma 2.** *Let $\{\mu_{nj}\}_{j=1}^{k_n}$, $n = 1, 2, \ldots$ be sequences of probability measures. Then*

(i) $$\sup_n \int_{l^2} [1 - \exp(-\|\Delta_r y\|^2)] \prod_{j=1}^{k_{n*}} \mu_{nj}(dy) \le \sup_n \sum_{j=1}^{k_n} \left\{ \int_{\|\Delta_r y\| \ge \eta} \mu_{nj}(dy) + \left\| \int_{\|\Delta_r y\| < \eta} \Delta_r y \, \mu_{nj}(dy) \right\| + a(\eta) \int_{\|\Delta_r y\| < \eta} \|\Delta_r y\|^2 \mu_{nj}(dy) \right\}.$$

*Proof:* By using the identity

$$e - \prod_{j=1}^{k_{n*}} \mu_{nj} = \sum_{j=1}^{k_n} (e - \mu_{nj}) * \prod_{i=0}^{j-1} \mu_{ni} \qquad (\mu_{n0} = e), \tag{5}$$

where $e$ is the unit measure, we form the convolution of $\exp(-\|\Delta_r x\|^2)$ with both sides of (5) and get, also applying Fubini's relation,

$$\begin{aligned}
&\int_{l^2} [\exp(-\|\Delta_r x\|^2) - \exp(-\|\Delta_r(x - y)\|^2)] \prod_{j=1}^{k_{n*}} \mu_{nj}(dy) \\
&= \sum_{j=1}^{k_n} \int_{l^2} \left\{ \int_{l^2} [\exp(-\|\Delta_r(x - z)\|^2) - \exp(-\|\Delta_r x - y - z\|^2)] \right. \\
&\quad \left. \times \mu_{nj}(dy) \right\} \prod_{i=0}^{j-1 *} \mu_{nj}(dz) \Big\}.
\end{aligned} \tag{6}$$

Choosing $x$ as the zero element, we get from this identity

$$\begin{aligned}
&\int_{l^2} [1 - \exp(-\|\Delta_r y\|^2)] \prod_{j=1}^{k_{n*}} \mu_{nj}(dy) \\
&\leq \sum_{j=1}^{k_n} \sup_z \left| \int_{l^2} [\exp(-\|\Delta_r z\|^2) - \exp(-\|\Delta_r(z + y)\|^2)] \mu_{nj}(dy) \right|.
\end{aligned} \tag{7}$$

By Lemma 1, we get

$$\begin{aligned}
&\sup_z \left| \int_{l^2} [\exp(-\|\Delta_r z\|^2) - \exp(-\|\Delta_r(z + y)\|^2)] \, \mu_{nj}(dy) \right| \\
&\quad \leq \int_{\|\Delta_r y\| \geq \eta} \mu_{nj}(dy) + 2 \sup_z \|\Delta_r z\| \exp(-\|\Delta_r z\|) \\
&\qquad \times \left\| \int_{\|\Delta_r y\| < \eta} \Delta_r y \, \mu_{nj}(dy) \right\| + a(\eta) \int_{\|\Delta_r y\| < \eta} \|\Delta_r y\|^2 \, \mu_{nj}(dy).
\end{aligned}$$

Since

$$2\tau \exp(-\tau^2) < 1$$

for any real number $\tau$, we obtain the inequality of the lemma from (7) and the last inequality. □

*Proof of the theorem:* If (ii)–(iv) hold, we get by the lemma for $v_n$ defined by (1)

$$[1 - \exp(-\beta^2)] \int_{\|\Delta_r y\| \geq \beta} v_n(dy) \to 0 \qquad (n \to \infty, r \to \infty)$$

(repeated limit). □

## 3. NECESSARY AND SUFFICIENT CONDITIONS FOR THE WEAK CONVERGENCE OF CONVOLUTION PRODUCTS OF SYMMETRICAL PROBABILITY MEASURES

**Theorem 1.** *If the probability measures* $\mu_{ni}$ *in Theorem* 2.1 *are symmetrical, the conditions* (i)–(iv) *in this theorem are both necessary and sufficient.*

A measure $\mu$ on $l^2$ is called symmetrical if $\mu(E) = \mu(-E)$ for any measurable set $E$ on $l^2$. To prove the theorem, we shall again use the identity (2.6) for symmetrical $\mu_{ni}$. Changing $y$ to $-y$ and regarding the symmetry of $\mu_{ni}$, we get a corresponding identity with $\Delta_r(x - y)$ changed to $\Delta_r(x + y)$ and $\Delta_r(x - y - z)$ changed to $\Delta_r(x + y + z)$. Combining these identities, we get, with the function $g$ introduced in Lemma 2.1,

$$\int_{l^2} g(\Delta_r x, \Delta_r y) \prod_{j=1}^{k_n *} \mu_{nj}(dy)$$
$$= \sum_{j=1}^{k_n} \int_{l^2} \left\{ \int_{l^2} g(\Delta_r x - \Delta_r z, \Delta_r y)\, \mu_{nj}(dy) \right\} \int_{l^2} \prod_{i=1}^{j-1 *} \mu_{ni}(dz). \tag{1}$$

Choosing $x$ as the zero element and changing the order of integration, we can write

$$\int_{l^2} g(0, \Delta_r y) \prod_{j=1}^{k_n *} \mu_{nj}(dy) = \sum_{j=1}^{k_n} \int_{l^2} f_{nj}^{(r)}(\Delta_r y)\, \mu_{nj}(dy) \tag{2}$$

with

$$f_{nj}^{(r)}(\Delta_r y) = \int_{l^2} g(\Delta_r z, \Delta_r y) \prod_{i=1}^{j-1 *} \mu_{ni}(dz).$$

Now assume that the convolution product converges weakly and necessarily to a probability measure $v$ since the $\mu_{ni}$ are probability measures. Since $\|\Delta_r z\| \downarrow 0$ as $r \uparrow \infty$, we then have

$$\int_{\|\Delta_r z\| > \varepsilon} \prod_{j=1}^{k_n *} \mu_{ni}(dz) < \varepsilon$$

for any given $\varepsilon$ and $r \geq r(\varepsilon)$, $n \geq n(r)$. Remembering that the $\mu_{ni}$ are symmetrical, we obtain by Lemma 1.1

$$\int_{\|\Delta_r z\| \geq \varepsilon} \prod_{i=1}^{j-1 *} \mu_{ni}(dz) < 2\varepsilon \tag{3}$$

for $j = 1, \ldots, k_n$, $r > r(\varepsilon)$, $n > n(r)$. Using this inequality, we estimate $f_{nj}^{(r)}$ by Lemma 2.1 for $r > r_0(\varepsilon)$, $n \geq n_0(r)$.

For $\varepsilon < \frac{1}{4}\eta$, $\|\Delta_r y\| \geq \eta$, $c_1(\eta)(1 - 2\varepsilon) - 4\varepsilon \geq \frac{1}{2}c_1(\eta)$, we get

$$\begin{aligned} f_{nj}^{(r)}(\Delta_r y) &\geq c_1(\eta) \int_{\|\Delta_r z\| < \varepsilon} \prod_{i=1}^{j-1}{}^{*} \mu_{ni}(dz) \\ &\quad - 2 \int_{\|\Delta_r z\| \geq \varepsilon} \prod_{i=1}^{j-1}{}^{*} \mu_{ni}(dz) \\ &\geq c_1(\eta)(1 - 2\varepsilon) - 4\varepsilon \geq \tfrac{1}{2}c_1(\eta). \end{aligned}$$

For $\|\Delta_r y\| < \eta$, $\eta < \frac{1}{2}$, $c_3(1 - 2\varepsilon) - 2\varepsilon c_2 > \frac{1}{2}c_3$, we get

$$f_{nj}^{(r)}(\Delta_r y) \geq [c_3(1 - 2\varepsilon) - 2\varepsilon c_2]\|\Delta_r y\|^2 \geq \tfrac{1}{2}c_3\|\Delta_r y\|^2.$$

Regarding these inequalities, we obtain

$$\begin{aligned} \int_{l^2} f_{nj}^{(r)}(\Delta_r y)\, \mu_{nj}(dy) &\geq \tfrac{1}{2}c_1(\eta) \int_{\|\Delta_r y\| \geq \eta} \mu_{nj}(dy) \\ &\quad + \tfrac{1}{2}c_3 \int_{\|\Delta_r y\| < \eta} \|\Delta_r y\|^2\, \mu_{nj}(dy). \end{aligned} \tag{4}$$

Combining (4) for $j = 1, 2, \ldots, k_n$ with (2) and observing that

$$g(0, \Delta_r y) = 2[1 - \exp(-\|\Delta_r y\|^2)],$$

we get

$$\begin{aligned} 2 \int_{l^2} [1 - \exp(-\|\Delta_r y\|^2)] \prod_{j=1}^{k_n *} \mu_{nj}(dy) &\geq \tfrac{1}{2}c_1(\eta) \sum_{j=1}^{k_n} \int_{\|\Delta_r y\| \geq \eta} \mu_{nj}(dy) \\ &\quad + \tfrac{1}{2}c_3 \sum_{j=1}^{k_n} \int_{\|\Delta_r y\| < \eta} \|\Delta_r y\|^2\, \mu_{nj}(dy). \end{aligned}$$

As $n \to \infty$, the left-hand side tends to

$$2 \int_{l^2} [1 - \exp(-\|\Delta_r y\|^2)]\, \nu(dy)$$

with a probability measure $\nu$. This integral tends to 0 as $r \to \infty$. Thus for symmetrical probability measures we have proved the necessity of conditions (ii)–(iv) in Theorem 2.1. □

*Remark:* For convolution powers the condition (ii) in Theorem VI.2.1 may be written

$$\lim_{r \to \infty} \limsup_{n \to \infty} \int_{\|\Delta_r y\| \geq \varepsilon} \mu_n^{*k_n}(dy) = 0 \tag{5}$$

for any $\varepsilon > 0$. Now the mapping $T_r$ defined by $T_r y = \|\Delta_r y\|$ is continuous and hence (5) may be written

$$\lim_{r\to\infty} \limsup_{n\to\infty} \int_{|t|\geq\varepsilon} \mu_n^{*k_n}(T_r^{-1}\,dy) = 0 \tag{6}$$

for any $\varepsilon > 0$. In this way the condition (ii) is transformed into a corresponding condition on $R$. It holds if and only if the sequence $\{\mu_n^{*k_n}(T_r^{-1}\cdot)\}$ of measures on $R$ converges weakly to the unit measure.

Using this fact we get a new simple proof of the theorem above and the theorem below. It easily follows from Theorem IV.9.1 then that $\{\mu_n^{*k_n}(T_r^{-1}\cdot)\}$, $n \to \infty$, $r \to \infty$, converges weakly to the unit measure if and only if (i)–(iv) in Theorem VI.2.1 hold.

## 4. NECESSARY AND SUFFICIENT CONDITIONS FOR THE WEAK CONVERGENCE OF CONVOLUTION POWERS OF PROBABILITY MEASURES

**Theorem 1.**† *In order that the sequence $\{\mu_n^{*k_n}\}$ of $\sigma$-smooth measures on $l^2$, $k_n \uparrow \infty$ as $n \uparrow \infty$, to converge weakly to a measure different from the zero measure, it is necessary and sufficient that*

(i) $$\lim_{r\to\infty} \limsup_{n\to\infty} k_n \int_{\|\Delta_r x\|\geq\varepsilon} \mu_n(dx) = 0 \qquad \textit{for} \quad \varepsilon > 0,$$

(ii) $$\lim_{r\to\infty} \limsup_{n\to\infty} k_n \left\| \int_{\|\Delta_r x\|<\varepsilon} \Delta_r x\, \mu_n(dx) \right\| = 0 \qquad \textit{for} \quad \varepsilon > 0,$$

(iii) $$\lim_{r\to\infty} \limsup_{n\to\infty} k_n \int_{\|\Delta_r x\|<\varepsilon} \|\Delta_r x\|^2\, \mu_n(dx) = 0 \qquad \textit{for} \quad \varepsilon > 0,$$

*(iv) $\{[\mu_n(\pi_r^{-1}\cdot)]^{*k_n}\}$ converges weakly on $\pi_r l^2$ for $r = 1, 2, \ldots$, where $\pi_r$ is the projection defined in Section 1. When $\{\mu_n^{*k_n}\}$ converges weakly, its weak limit is uniquely infinitely divisible. For $l_n(r) = [k_n/r]$, $r$ positive integer, $\{\mu_n^{*l_n(r)}\}$ then converges weakly to a $\sigma$-smooth measure $\nu_r$ for $r = 1, 2, \ldots$, where $\nu = \nu_1 = \nu_r^{*r}$. If the $\mu_n$ are probability measures, then $\nu$ and the $\nu_r$ are probability measures.*

*Proof:* We first consider probability measures $\mu_n$. The sufficiency of the conditions follows by Theorem 2.1. As we have pointed out in Section 2, conditions (iv) and (2.3) are necessary. We show that (2.3) implies (i)–(iii) when $\{\nu_n\}$ converges weakly to $\nu$. By Theorem 3.1, this is true for symmetrical measures $\mu_n$. We use this fact in order to show that (i)–(iii) are also necessary

† Added in proof: A simplified proof of the necessity of (i)–(iii) is given in a remark at the end of this section.

in the general case. Put

$$\nu_n = \mu_n^{*k_n}. \tag{1}$$

If $\{\nu_n\}$ converges weakly to $\nu$ and $\bar{\mu}_n$ is defined by $\bar{\mu}_n(E) = \mu_{ni}(-E)$ for any measurable set $E$, $\bar{\nu}_n = (\bar{\mu}_n)^{*k_n}$ converges weakly to $\bar{\nu}$ and $\nu_n * \bar{\nu}_n = (\mu_n * \bar{\mu}_n)^{*k_n}$ converges weakly to $\nu * \bar{\nu}$. Here $\mu_n * \bar{\mu}_n$ is symmetrical. By Theorem 3.1, we then find that

$$\lim_{r\to\infty} \limsup_{n\to\infty} k_n \int_{||\Delta_r x|| \geq \varepsilon} \mu_n * \bar{\mu}_n(dx) = 0 \qquad \text{for} \quad \varepsilon > 0, \tag{2}$$

$$\lim_{r\to\infty} \limsup_{n\to\infty} k_n \left\| \int_{||\Delta_r x|| < \varepsilon} \Delta_r x \, \mu_n * \bar{\mu}_n(dx) \right\| = 0 \qquad \text{for} \quad \varepsilon > 0, \tag{3}$$

$$\lim_{r\to\infty} \limsup_{n\to\infty} k_n \left\| \int_{||\Delta_r x|| < \varepsilon} ||\Delta_r x||^2 \, \mu_n * \bar{\mu}_n(dx) \right\| = 0 \qquad \text{for} \quad \varepsilon > 0, \tag{4}$$

[(3) is trivial according to the symmetry.] Put $k_n \int_{||\Delta_r x|| \geq \varepsilon} \mu_n * \mu_n(dx) = \varepsilon_n^{(r)}$. By Lemma 1.1, there exist elements $x_n(r)$ *in* $l^2$ such that

$$\int_{||\Delta_r x|| \geq \varepsilon} \mu_n[x_n(r) + dx] \leq \varepsilon_n^{(r)}/k_n. \tag{5}$$

We choose $x_n(r)$ such that $\Delta_r x_n(r) = x_n(r)$, which is possible since (5) only depends on $\Delta_r x_n(r)$.

For some $\varepsilon_0 \geq 3\varepsilon$, $\varepsilon_0 < 1$, put $m_n^{(r)} = \int_{||\Delta_r x|| < \varepsilon_0} \Delta_r x \, \mu_n[x_n(r) + dx]$, where by this notation we understand the element $m_n(r)$ with the coordinates

$$\begin{aligned} m_n^{(i)}(r) &= 0 && \text{for} \quad i \leq r, \\ &= \int_{||\Delta_r x|| < \varepsilon_0} x^{(i)} \, \mu_n[x_n(r) + dx] && \text{for} \quad i > r. \end{aligned}$$

(Compare the remarks at the end of Section 1.) Then $m_n(r) \in l^2$ since for $i > r$

$$[m_n^{(i)}(r)]^2 \leq \int_{||\Delta_r x|| < \varepsilon_0} [x^{(i)}]^2 \, \mu_n(x_n^{(r)} + dx)$$

and, hence,

$$||m_n(r)||^2 \leq \int_{||\Delta_r x|| < \varepsilon_0} ||\Delta_r x||^2 \, \mu_n[x_n(r) + dx]. \tag{6}$$

Regarding (2), we find that we may choose $r_0$ sufficiently large and then $n_0(r)$ for $r > r_0$ such that $\varepsilon_n^{(r)} < \varepsilon$ for $r > r_0$, $n > n_0(r)$. By (5) and (6), we then obtain

$$||m_n(r)||^2 \leq \varepsilon^2 + \varepsilon_0^2 \int_{||\Delta_r x|| \geq \varepsilon} \mu_n[x_n(r) + dx] \leq \varepsilon^2 + \varepsilon_0^2 \varepsilon / k_n < 4\varepsilon^2$$

for sufficiently large $n$, and thus $\|m_n(r)\| < 2\varepsilon$ $(3\varepsilon < \varepsilon_0)$. Put $\mu_n^{(r)} = \mu_n[x_n(r) + m_n(r) + \cdot\,]$ and observe that the set $\{x : \|\Delta_r x\| \geq \varepsilon\}$ contains the set $\{x : \|\Delta_r[x - m_n(r)]\| \geq \varepsilon_0\}$ for $3\varepsilon < \varepsilon_0$, $\|m_n(r)\| < 2\varepsilon$. Then we get by (5) for $\varepsilon_n^{(r)} \leq \varepsilon$,

$$\int_{\|\Delta_r x\| \geq \varepsilon_0} \mu_n^{(r)}(dx) = \int_{\|\Delta_r(x - m_n(r))\| \geq \varepsilon_0} \mu_n[x_n(r) + dx]$$
$$\leq \int_{\|\Delta_r x\| \geq \varepsilon} \mu_n[x_n(r) + dx] \leq \varepsilon_n^{(r)}/k_n. \tag{7}$$

Furthermore, using Lemma 1.2 and observing that the set $\{x : \|\Delta_r x\| < \varepsilon\}$ is contained in the set $\{x : \|\Delta_r[x - m_n(r)]\| < \varepsilon_0\}$, we get by the transformation $x + m_n(r) \to y$ (which transforms $x^{(i)} + m_n^{(i)}(r)$ into $y^{(i)}$)

$$\left\|\int_{\|\Delta_r x\| < \varepsilon_0} \Delta_r x\, \mu_n^{(r)}(dx)\right\| = \left\|\int_{\|\Delta_r(y - m_n(r))\| < \varepsilon_0} \Delta_r[y - m_n(r)]\mu_n[x_n(r) + dy]\right\|$$
$$\leq \left\|\int_{\|\Delta_r y\| < \varepsilon} \Delta_r[y - m_n(r)]\mu_n[x_n(r) + dy]\right\|$$
$$+ \varepsilon_0 \int_{\|\Delta_r y\| \geq \varepsilon} \mu_n[x_n(r) + dy]$$
$$\leq \left\|\int_{\|\Delta_r y\| < \varepsilon_0} \Delta_r y\, \mu_n[x_n(r) + dy] - m_n(r)\right\|$$
$$+ 2(\varepsilon_0 + \varepsilon) \int_{\|\Delta_r y\| \geq \varepsilon} \mu_n[x_n(r) + dy]. \tag{8}$$

The first term on the right-hand side of the last inequality is equal to 0 by the definition of $m_n(r)$, and the second term is at most equal to $2(\varepsilon_0 + \varepsilon)\varepsilon_n^{(r)}/k_n$ according to (5). Hence we obtain from (7) and (8)

$$\lim_{r\to\infty} \limsup_{n\to\infty} k_n \int_{\|\Delta_r y\| \geq \varepsilon_0} \mu_n^{(r)}(dy) = 0, \tag{9}$$

$$\lim_{r\to\infty} \limsup_{n\to\infty} k_n \left\|\int_{\|\Delta_r x\| < \varepsilon_0} \Delta_r x\, \mu_n^{(r)}(dx)\right\| = 0. \tag{10}$$

We claim that also

$$\lim_{r\to\infty} \limsup_{n\to\infty} k_n \int_{\|\Delta_r x\| < \varepsilon} \|\Delta_r x\|^2\, \mu_n^{(r)}(dx) = 0. \tag{11}$$

Indeed, by Theorem 3.1, we have

$$\lim_{r\to\infty} \limsup_{n\to\infty} k_n \int_{\|\Delta_r x\| < 2\varepsilon_0} \|\Delta_r x\|^2 \mu_n^{(r)} * \bar{\mu}_n^{(r)}(dx) = 0, \tag{12}$$

since $\mu_n^{(r)} * \bar{\mu}_n^{(r)} = \mu_n * \bar{\mu}_n$ is symmetrical. Now

$$\begin{aligned}\int_{\|\Delta_r x\| < 2\varepsilon_0} &\|\Delta_r x\|^2 \mu_n^{(r)} * \bar{\mu}_n^{(r)}(dx) \\ &\geq \int_{\|\Delta_r y\| < \varepsilon_0,\, \|\Delta_r z\| < \varepsilon_0} \|\Delta_r y - \Delta_r z\|^2 \mu_n^{(r)}(dy)\, \mu_n^{(r)}(dz) \\ &\geq 2\int_{\|\Delta_r y\| < \varepsilon_0} \|\Delta_r y\|^2 \mu_n^{(r)}(dy) - 2\left\|\int_{\|\Delta_r y\| < \varepsilon_0} \Delta_r y\, \mu_n^{(r)}(dy)\right\|^2 \\ &\quad - 2\varepsilon_0^2 \int_{\|\Delta_r y\| \geq \varepsilon_0} \mu_n^{(r)}(dy).\end{aligned}$$

This inequality and (9), (12), and (10) imply (11). Applying Lemma 2.2, we find that the relations (9)–(11) imply

$$\lim_{r\to\infty} \limsup_{n\to\infty} \int_{l^2} [1 - \exp(-\|\Delta_r x\|^2)][\mu_n^{(r)}]^{*k_n}(dx) = 0. \tag{13}$$

By the definition of $\mu_n^{(r)}$, we have $\mu_n^{(r)} = \mu_n[x_n(r) + m_n(r) + \cdot\,]$. Hence, putting $\mu_n^{*k_n} = v_n$, $b_n(r) = k_n[x_n(r) + m_n(r)]$, we get $[\mu_n^{(r)}]^{*k_n} = v_n[b_n(r) + \cdot\,]$. Then by the transformation $b_n(r) + x = y$ and the fact that $\Delta_r b_n(r) = b_n(r)$, we can write (13)

$$\lim_{r\to\infty} \limsup_{n\to\infty} \int_{l^2} (1 - \exp[-\|\Delta_r[y - b_n(r)]\|^2]\, v_n(dy) = 0. \tag{14}$$

By assumption, $v_n$ converges weakly to a probability measure $v$ as $n \to \infty$. Since $\{x : \|\Delta_r x\| \geq \varepsilon\}$ is a closed set for any $\varepsilon > 0$, we obtain by Theorem III.1.1

$$\limsup_{n\to\infty} \int_{\|\Delta_r x\| \geq \varepsilon} v_n(dx) \leq \int_{\|\Delta_r x\| \geq \varepsilon} v(dx) \to 0 \qquad (r \to \infty).$$

Hence, (14) must hold if we integrate over $\{x : \|\Delta_r x\| < \varepsilon\}$ instead of $l^2$. It then easily follows that

$$\lim_{r\to\infty} \limsup_{n\to\infty} |b_n(r)| = 0. \tag{15}$$

Indeed, if the left-hand side is equal to $\beta > 0$, we may choose a sequence $N_r$ of positive integers for any $r = 1, 2, \ldots$ such that

$$\lim_{n\to\infty,\, n\in N_r} |b_n(r)| = \limsup_{n\to\infty} b_n(r)$$

and then a sequence $r_j$ of positive integers such that

$$\lim_{r_j\to\infty} \lim_{n\to\infty,\, n\in N_{r_j}} \|b_n(r)\| = \beta,$$

but if we let $n$ and $r$ tend to $+\infty$ in the same way in (14), we find that the limit is $\int_{l^2} [1 - \exp(-\beta^2)]\, \nu(dy) > 0$. Thus (15) holds. It implies (9) with $\mu_n$ instead of $\mu_n^{(r)}$ and thus also (5) with $x_n(r)$ equal to the zero element. Proceeding then as (15) above, we get the following relations (10), . . . , (15) with $x_n(r) = 0$. In particular, (15) reduces to

$$\lim_{r\to\infty} \limsup_{n\to\infty} k_n m_n(r) = 0,$$

and consequently (10) holds with $\mu_n$ instead of $\mu_n^{(r)} = \mu_n[m_n(r) + \cdot)]$. We then get (11) with $\mu_n$ instead of $\mu_n^{(r)}$. Thus (i)–(iii) in the theorem hold. We have so far considered probability measures $\mu_n$. The weak convergence of $\{\mu_n^{*k_n}\}$ to $\nu$ then contains the convergence of $\mu_n^{k_n}(l^2)$ to $\nu(l^2)$, and since $\mu_n(S) = 1$, also $\nu(l^2) = 1$, i.e., $\nu$ is a probability measure. If $\mu_n$ is a $\sigma$-smooth measure different from the zero measure, we have $\mu_n = \alpha_n \mu_n'$ with a probability measure $\mu_n'$ and constants $\alpha_n > 0$. If $\{\mu_n^{*k_n}\}$ converges weakly to a measure $\nu$ different from the zero measure, then necessarily $\mu_n^{k_n}(l^2)$ converges to a positive number $\nu(l^2)$, which is the case if and only if $k_n\mu_n(l^2)$ converges to a positive number $\geq 0$. If this condition is satisfied, then $\{\mu_n^{*k_n}\}$ converges weakly if and only if $\{[\mu_n']^{*k_n}\}$ converges weakly. Note that (i) implies the convergence of $k_n\mu_n(l^2)$ since $\mu_n(\pi_r^{-1}\pi_r l^2) = \mu_n(l^2)$.

It remains to prove that the weak limit $\nu$ of $\{\mu_n^{*k_n}\}$ is infinitely divisible when it exists and that we have the relation $\nu = \nu_r^{*r}$ stated in the theorem. Now for $k_n(s) = [k_n/s]$, $s$ positive integer, we easily find that the conditions for the weak convergence of $\{\mu_n^{*k_n}\}$ imply these conditions for the weak convergence of $\{\mu_n^{*k_n(s)}\}$ to a $\sigma$-smooth measure $\nu_s$. For any positive integer $r$, $\{\mu_n^{*k_n(s)}(\pi_r^{-1}\cdot)\}$ then converges weakly to $\nu_s(\pi_r^{-1}\cdot)$, since $\pi_r$ is continuous. However, $\nu(\pi_r^{-1}\cdot) = \nu_1(\pi_r^{-1}\cdot) = [\nu_s(\pi_r^{-1}\cdot)]^{*s}$ according to Theorem IV.9.1. Then by Theorem 1.2 we have $\nu = \nu_s^{*s}$ for all $s$. Since $\nu(\pi_r^{-1}\cdot)$ is uniquely infinitely divisible $\nu$ is also uniquely infinitely divisible. □

*Remark:* In order to prove that the weak convergence of $\mu_n^{*k_n}$ to a probability measure $\nu$ implies the conditions (i)–(iii) in Theorem 1, we observe that the mapping $T_r$ from $l^2$ into $[0, \infty)$, defined by $T_r x = \|\Delta_r x\|$, is continuous. Hence the weak convergence of $\mu_n^{*k_n}$ to $\nu$ implies the weak convergence of $\mu_n^{*k_n}(T_r^{-1}\cdot)$ to $\nu(T_r^{-1}\cdot)$ on $[0, \infty)$. By Theorem IV.9.1 and Remark 2 on this theorem, $k_n\mu_n(T_r^{-1}\cdot)$ then converges weakly to a Lévy measure $\lambda_r$ on $[0, \infty)$ and

$$\lim_{n\to\infty} k_n \int_{0<t<\varepsilon} t\, \mu_n(T_r^{-1}\, dt) = m_r(\varepsilon) < \infty \tag{16}$$

for any $\varepsilon$ such that $\varepsilon$ is a continuity point of $\lambda_r$, and

$$\lim_{n\to\infty} k_n \int_{0<t<\varepsilon} t^2\, \mu_n(T_r^{-1}\, dt) \downarrow \sigma^2 \geq 0 \qquad \text{as} \quad \varepsilon \downarrow 0. \tag{17}$$

However, it follows by (5) that $\sigma = 0$.

According to the dominated convergence theorem for integrals and the fact that $\|\Delta_r x\| \to 0$ $(r \to \infty)$ we get

$$\lim_{r\to\infty} \int_{t \geq \varepsilon} \nu(T_r^{-1}\, dt) = \lim_{r\to\infty} \int_{\|\Delta_r x\| \geq \varepsilon} \nu(dx) = 0.$$

Hence $\mu_n^{*k_n}(T_r^{-1}\cdot)$ necessarily converges weakly to the unit probability measure, as in order $n \to \infty$, $r \to \infty$. Applying again Theorem IV.9.1 we conclude that

$$\lim_{r\to\infty} \limsup_{n\to\infty} k_n \int_{t \geq \varepsilon} \mu_n(T_r^{-1}\, dt) = 0,$$

$$\lim_{r\to\infty} \limsup_{n\to\infty} k_n \int_{0<t<\varepsilon} t\, \mu_n(T_r^{-1}\, dt) = 0.$$

By the transformation $t = T_r x$ we find that these relations and (6) imply (i)–(iii) in Theorem 1. The proof given here can be used to establish corresponding relations in general Banach spaces with Schauder basis.

## 5. DIFFERENT FORMS OF NECESSARY AND SUFFICIENT CONDITIONS FOR THE WEAK CONVERGENCE OF CONVOLUTION POWERS OF PROBABILITY MEASURES ON $l^2$

As we remarked at the end of the preceeding section, the weak convergence of convolution products of $\sigma$-smooth measures is essentially a problem of weak convergence of convolution powers of probability measures. Hence, we deal here only with convolutions of such measures. A $\sigma$-smooth, $\sigma$-finite measure $\lambda$ on $l^2\backslash\{0\}$ is called a Lévy measure if

$$\int_{\|y\| \geq \beta} \lambda(dy) \begin{cases} < \infty & \text{for } \beta > 0, \\ \to 0 & \text{as } \beta \to \infty, \end{cases} \tag{1}$$

$$\int_{\|y\| < \beta} \|y\|^2\, \lambda(dy) < \infty \qquad \text{for some } \beta > 0. \tag{2}$$

**Theorem 1.** *In order that the sequence* $\{\mu_n^{*k_n}\}$ *of convolution powers of probability measures* $\mu_n$ *on* $l^2$, $k_n \uparrow \infty$ *as* $n \uparrow \infty$ *converges weakly to a probability measure* $\nu$, *it is necessary and sufficient that*

(i) $k_n \mu_n$ *converges weakly to a Lévy measure on* $l^2\backslash\{0\}$,

(ii) $$\lim_{n\to\infty} \left\| k_n \int_{l^2} \frac{x}{1 + \|x\|^2}\, \mu_n(dx) - \bar{m} \right\| = 0$$

*with an element* $\bar{m}$ *in* $l^2$,

(iii) $$\lim_{\varepsilon \downarrow 0} \overline{\lim_{n\to\infty}}\, k_n \int_{\|x\| < \beta} (t \cdot x)^2\, \mu_n(dx) = q(t) = \sum_{i=1}^{\infty} \sum_{j=1}^{\infty} q_{ij} t^{(i)} t^{(j)}$$

*for $t \in l^2$ uniformly with respect to $t \in l^2$ for $\|t\| \le \beta_0$, where $q(t)$ is a positive definite symmetrical quadratic form and*

$$\lim_{\varepsilon \downarrow 0} \overline{\lim_{n\to\infty}}\, k_n \int_{\|x\|<\varepsilon} \|x\|^2 \, \mu_n(dx) = \sum_{i=1}^{\infty} q_{ii} < \infty$$

*[i.e., the trace of the infinite matrix $(q_{ij})$ is finite].*

(iv) *The limits $\lambda$, $\bar{m}$, and $q(t)$ determine the Gaussian functional*

$$\begin{aligned}\bar{\Gamma}(\alpha, v, x \cdot t) &= \lim_{n\to\infty} k_n \int_{l^2} \left[\Phi\left(\frac{x\cdot t - y \cdot t}{\alpha}\right) - \Phi\left(\frac{x \cdot t}{\alpha}\right)\right] \mu_n(dy) \\ &= -\frac{1}{\alpha}\bar{m}\cdot t\Phi'\left(\frac{x\cdot t}{\alpha}\right) + \frac{1}{2\alpha^2} q(t)\Phi''\left(\frac{x\cdot t}{\alpha}\right) \\ &\quad + \int_{l^2\setminus\{0\}} \left\{\Phi\left(\frac{x\cdot t - y\cdot t}{\alpha}\right) - \Phi\left(\frac{x\cdot t}{\alpha}\right)\right. \\ &\quad \left. + \frac{y\cdot t}{\alpha(1 + \|y\|^2)}\Phi'\left(\frac{x\cdot t}{\alpha}\right)\right\} \lambda(dy),\end{aligned}$$

*where the convergence as $n \to \infty$ is uniform and the integral converges uniformly with respect to $x \in l^2$ on $\|t\| \le \beta_0$.*

(v) *The limit $q(t)$ determines the $q_{ij}$ and*

$$q_{ij} = \lim_{\varepsilon\downarrow 0} \overline{\lim_{n\to\infty}}\, k_n \int_{\|x\|<\varepsilon} [x^{(i)}]^2 \, \mu_n(dx).$$

*For any continuity set $\{x : \|x\| \ge \beta\}$ for $\lambda$ we have*

$$\lim_{n\to\infty} k_n \int_{\|x\|<\beta} (x^{(i)})^2 \, \mu_n(dx) = q_{ij} + \int_{0<\|x\|<\beta} [x^{(i)}]^2 \, \lambda(dy).$$

(vi) *The limit in* (ii) *determines the limits*

$$\bar{m}^{(i)} = \lim_{n\to\infty} k_n \int_{l^2} \frac{x^{(i)}}{1 + \|x\|^2} \, \mu_n(dy), \qquad i = 1, 2, \ldots.$$

(vii) *The weak limit of $\{\mu_n^{*k_n}\}$ is infinitely divisible.*

*Proof:* We shall prove that the conditions (i)–(iii) in Theorem 1 above and the conditions in Theorem 4.1 imply each other. The latter conditions will here be denoted by (i′)–(iv′).

(i′) $$\lim_{r\to\infty} \limsup_{n\to\infty} k_n \int_{\|\Delta_r x\| \ge \beta} \mu_n(dx) = 0 \qquad \text{for any} \quad \beta > 0,$$

(ii′) $$\lim_{r\to\infty} \limsup_{n\to\infty} k_n \left\| \int_{\|\Delta_r x\| < \beta} \Delta_r x \, \mu_n(dx) \right\| = 0 \qquad \text{for some} \quad \beta > 0,$$

(iii′) $$\lim_{r\to\infty} \limsup_{n\to\infty} k_n \int_{\|\Delta_r x\| < \beta} \|\Delta_r x\|^2 \, \mu_n(dx) = 0 \qquad \text{for some} \quad \beta > 0,$$

(iv′) $\{[\mu_n(\pi_r^{-1}\cdot)]^{*k_n}\}$ converges weakly for any $\pi_r$ as $n \to \infty$. The mappings $\pi_r$ and $V_r$ have been introduced in Section 1. They are continuous. The stated implications will follow from the lemmas below.

**Lemma 1.** *Condition* (iv′) *implies the following relations:*

(1°) $k_n\mu_n(\pi_r^{-1}\cdot)$ *converges weakly to a Lévy measure* $\tilde{\lambda}^{(r)}$ *on* $\pi_r l^2$,

$$(2°)\qquad \limsup k_n \int_{||V_r\pi_r x|| \geq \beta} \mu_n(dx) \begin{cases} < \infty & \textit{for} \quad \beta > 0, \\ \to 0 & \textit{for} \quad \beta \to \infty, \end{cases}$$

$$(3°)\qquad \lim_n \left\| k_n \int_{l^2} \frac{V_r\pi_r x}{1 + ||V_r\pi_r x||^2} \mu_n(dx) - \bar{m}_r \right\| = 0$$

*with* $\bar{m}_r \in l^2$, $\bar{m}_r^{(i)} = 0$ *for* $i > r$,

$$(4°)\qquad \lim_{n\to\infty} \left| k_n \int_{||V_r\pi_r x|| < \beta} (t \cdot V_r\pi_r x)^2 \mu_n(dx) - q_r(\beta, t) \right| = 0$$

*if* $\{\tilde{x}^{(r)} : ||\tilde{x}^{(r)}|| < \beta\}$ *is a continuity set for* $\beta$ *(hence, for all* $\beta > 0$ *except countably many). Here* $q_r(\beta, t)$ *is a positive definite symmetrical quadratic form for* $t \in l^2$.

$$(5°)\qquad \lim_{n\to\infty} \left| k_n \int_{||V_r\pi_r x|| < \beta} ||V_r\pi_r x||^2 \mu_n(dx) - \sigma_r^2(\beta) \right| = 0$$

*with finite* $\sigma_r^2(\beta)$ *for* $\beta > 0$.

*Proof:* From (iv′) and Theorem IV.9.1 we get relations of the form (2°)–(5°) with $\tilde{x}^{(r)} \in \pi_r l^2$ instead of $V_r\pi_r x$ corresponding to the measures $\mu_n(\pi_r^{-1}\cdot)$. By the transformation $\tilde{x}^{(r)} = \pi_r x$, we then get (2°)–(5°) with $\pi_r x$ instead $V_r\pi_r x$. However, then (2°)–(5°) also hold in the given form, except that then, instead of a finite-dimensional vector $\bar{m}_r$, we define $\bar{m}_r$ as an element in $l^2$ by putting $\bar{m}_r^{(i)} = 0$ for $i > r$. □

**Lemma 2.** (i′)–(iv′) ⇒ (i) and (i) ⇒ (i′)

*Proof that* (i)′–(iv)′ ⇒ (i): Condition (iv′) implies (1°)–(5°) in Lemma 1. Regarding 2° and (i′), we get

$$\limsup_{n\to\infty} k_n \int_{||x|| \geq \beta} \mu_n(dx)$$
$$\leq \limsup_{n\to\infty} k_n \int_{||V_r\pi_r x|| \geq \beta/2} \mu_n(dx) + \limsup_{n\to\infty} k_n \int_{||\Delta_r x|| \geq \beta/2} \mu_n(dx),$$

and thus

$$\limsup_{n\to\infty} k_n \int_{||x|| \geq \beta} \mu_n(dx) \begin{cases} < \infty & \text{for} \quad \beta > 0, \\ \to 0 & \text{for} \quad \beta \to \infty. \end{cases} \tag{3}$$

By (5°) and (iii′), we obtain

$$\limsup_{n\to\infty} k_n \int_{\|x\|<\beta} \|x\|^2 \mu_n(dx) \le \limsup_{n\to\infty} k_n \int_{\|V_r\pi_r x\|<\beta} \|V_r\pi_r x\|^2 \mu_n(dx)$$
$$+ \limsup_{n\to\infty} k_n \int_{\|\Delta_r x\|<\beta} \|\Delta_r x\|^2 \mu_n(dx) < \infty. \quad (4)$$

Put $\tilde{F}_\beta^{(r)} = \{\tilde{x}^{(r)}: \|\tilde{x}^{(r)}\| \ge \beta\}$. Then

$$F_\beta^{(r)} = \pi_r^{-1} F_\beta^{(r)} = \{x: \|\pi_r x\| \ge \beta\} = \{x: \|V_r\pi_r x\| \ge \beta\}$$

is a closed set. Clearly, $\beta \to \tilde{\lambda}^{(r)}(\tilde{F}_\beta^{(r)})$ is a nonincreasing function of $\beta$ and it is bounded for $\beta \ge \beta_0 > 0$, since $\tilde{\lambda}^{(r)}$ is a Lévy measure. Hence, we may choose $\beta$ such that $\tilde{F}_\beta^{(r)}$ is a continuity set for $\tilde{\lambda}^{(r)}$, $r = 1, 2, \ldots$, since a bounded monotone function has at most countably many discontinuity points. Then

$$\lim_{n\to\infty} k_n \mu_n(\pi_r^{-1} \tilde{F}_\beta^{(r)}) = \tilde{\lambda}^{(r)}(\tilde{F}_\beta^{(r)})$$

for $r = 1, 2, \ldots$. The weak convergence of $k_n\mu_n(\pi_r^{-1}\cdot)$ on $\pi_r l^2$ then implies this convergence on the spaces $\tilde{F}_\beta^{(r)}$ (by Theorem III.2.1). Put $F_\beta = x: \|x\| \ge \beta$. If $\tilde{F}^{(r)}$ is any closed set on $\pi_r F_\beta$, then by the weak convergence of $k_n\mu_n(\pi_r^{-1}\cdot)$ to $\tilde{\lambda}^{(r)}$ we get, by applying Theorem III.1.1,

$$\limsup_{n\to\infty} k_n \mu_n(\pi_r^{-1} \tilde{F}^{(r)}) \le \tilde{\lambda}^{(r)}(\tilde{F}^{(r)}).$$

Hence, $k_n\mu_n(\pi_r^{-1}\cdot)$ converges weakly for all projections of the measures $k_n\mu_n$ on $F_\beta$. Furthermore,

$$\limsup_{n\to\infty} k_n \int_{\|\Delta_r x\| \ge \varepsilon,\, x\in F_\beta} \mu_n(dx) \le \limsup_{n\to\infty} k_n \int_{\|\Delta_r x\| \ge \varepsilon} \mu_n(dx)$$
$$\to 0 \qquad (r \to \infty)$$

according to (i′). Applying Theorem 1.1, we conclude that $k_n\mu_n$ converges weakly to a $\sigma$-smooth measure $\lambda_\beta$ on $F_\beta$. We may now choose a decreasing sequence $\{\beta_i\}$ of positive numbers, tending to 0 as $i \to \infty$, such that $k_n\mu_n$ converges weakly to a $\sigma$-smooth measure $\lambda_{\beta_i}$ on $F_{\beta_i}$ and also such that $F_{\beta_{i-1}}$ is a continuity set for $\lambda_{\beta_i}$, so that $\lambda_{\beta_i}$ agrees with $\lambda_{\beta_{i-1}}$ on $F_{\beta_{i-1}}$. These measures $\lambda_{\beta_i}$ determine a $\sigma$-smooth $\sigma$-finite measure $\lambda$ on $l^2\backslash\{0\}$. It follows by (3) and (4) that

$$\int_{\|y\|\ge\beta} \lambda(dy) \begin{cases} < \infty & \text{for} \quad \beta > 0, \\ \to 0 & \text{for} \quad \beta \to \infty, \end{cases}$$

$$\int_{\|y\|<\beta} \|y\|^2 \lambda(dy) < \beta \qquad \text{for} \quad \beta > 0.$$

Thus $\lambda$ is a Lévy measure on $l^2\backslash\{0\}$.

*Proof that* (*i*) $\Rightarrow$ (*i*′): By Theorem III.1.1 and the $\sigma$-smoothness of $\lambda$:

$$\limsup_{n\to\infty} k_n \int_{\|\Delta_r x\|\geq\varepsilon} \mu_n(dx) \leq \int_{\|\Delta_r x\|\geq\varepsilon} \lambda(dx) \to 0 \qquad (r\to\infty). \quad \square$$

**Lemma 3.** (i)–(iii) $\Rightarrow$ (iv), (v), (vi), (iv′), (iii′), (ii′):

*Proof:* (i)–(iii) $\Rightarrow$ (iv), (v), (vi): Note that (ii) requires

$$k_n \int_{l^2} \frac{x}{1+\|x\|^2}\,\mu_n(dy) \in l^2.$$

Furthermore, for a sequence $\{\bar{m}_n\}$ in $l^2$, the relation $\|\bar{m}_n - \bar{m}\| \Rightarrow 0$ implies $\bar{m}_n \cdot t \to \bar{m}\cdot t$ as $n\to\infty$ (by Lemma I.11.2; besides, it is obvious). Hence, (ii) implies

$$\lim_{n\to\infty} k_n \int \frac{x\cdot t}{1+\|x\|^2}\,\mu_n(dx) - \bar{m}\cdot t = 0, \qquad t\in l^2.$$

This relation implies (vi). We may now write

$$\lim_{n\to\infty} k_n \int_{l^2}\left[\Phi\left(\frac{x\cdot t - y\cdot t}{\alpha}\right) - \Phi\left(\frac{x\cdot t}{\alpha}\right)\right]\mu_n(dx) = -\frac{1}{\alpha}\,\bar{m}\cdot t\Phi'\left(\frac{x\cdot t}{\alpha}\right)$$
$$+\lim_{n\to\infty}\int_{l^2}\left[\Phi\left(\frac{x\cdot t - y\cdot t}{\alpha}\right) - \Phi\left(\frac{x\cdot t}{\alpha}\right) + \frac{1}{\alpha}\frac{y\cdot t}{1+\|y\|^2}\,\Phi'\left(\frac{x\cdot t}{\alpha}\right)\right]\mu_n(dy),$$

provided that this limit exists. Consider first the contribution to this limit when we integrate over the set $F_\varepsilon = \{x:\|x\|\geq\varepsilon\}$ instead of $l^2$, where $F_\varepsilon$ is a continuity set for the Lévy measure $\lambda$, which exists according to (i). Then $k_n\mu_n$ converges weakly to $\lambda$ on $F_\varepsilon$, and the contribution to the limit of the part belonging to $F_\varepsilon$ is

$$\int_{F_\varepsilon}\left\{\Phi\,\frac{x\cdot t - y\cdot t}{\alpha} - \Phi\left(\frac{x\cdot t}{\alpha}\right) + \frac{1}{\alpha}\frac{y\cdot t}{1+\|y\|^2}\,\Phi'\left(\frac{x\cdot t}{\alpha}\right)\right\}\lambda(dy). \tag{5}$$

It follows as on $R$ that this convergence is uniform with respect to $x\in l^2$. Consider now that part of the limit, which belongs to $\{x:\|x\|<\varepsilon\}$. Writing

$$y\cdot t/(1+\|y\|^2) = y\cdot t - (y\cdot t)\|y\|^2/(1+\|y\|^2)$$

and using the Taylor expansion, we get

$$\left|\Phi\left(\frac{x\cdot t - y\cdot t}{\alpha}\right) - \Phi\left(\frac{x\cdot t}{\alpha}\right) + \frac{1}{\alpha}\frac{y\cdot t}{1+\|y\|^2}\,\Phi'\left(\frac{x\cdot t}{\alpha}\right) - \frac{1}{2\alpha^2}(y\cdot t)^2\Phi''\left(\frac{x\cdot t}{\alpha}\right)\right|$$
$$\leq \frac{\|y\cdot t\|^3}{6\alpha^3}\sup_\tau|\Phi'''(\tau)| + \frac{\|t\|\cdot\|y\|^3}{\alpha[1+\|y\|^2]}\sup_\tau|\Phi'(\tau)|.$$

Multiplying this inequality by $k_n$ and then integrating over the set $x:\|x\| < \varepsilon$, we get 0 uniformly with respect to $x \in l^2$ on the left-hand side, when in order $n \to \infty, \varepsilon \downarrow 0$, by the second relation in (iii). Furthermore, regarding properties (1) and (2) of the Lévy measure, we find that integral (5) converges uniformly with respect to $x \in l^2$ and $\|t\| \le \beta_0$ as $\varepsilon \downarrow 0$. Thus (i)–(iii) ⇒ (iv). If $t^{(i)} = 1$, $t^{(j)} = 0$ for $i \neq j$, then $q(t)$ reduces to

$$q_{ij} = \lim_{\varepsilon \downarrow 0} \overline{\lim_{n\to\infty}}\, k_n \int_{\|x\|<\varepsilon} [x^{(i)}]^2 \mu_n(dy).$$

If $\{x:\varepsilon \le \|x\| \le \beta\}$ is a continuity set for $\lambda$, we obtain

$$\lim_{n\to\infty} k_n \int_{\varepsilon \le \|x\| \le \beta} [x^{(i)}]^2 \mu_n(dx) = \int_{\varepsilon \le \|x\| \le \beta} [x^{(i)}]^2 \lambda(dx)$$
$$\to \int_{0<\|x\| \le \beta} [x^{(i)}]^2 \lambda(dx) \qquad (\varepsilon \downarrow 0).$$

Thus the relations in (v) hold.

*Proof:* *(i)–(iii) ⇒ (iv′), (iii′), (ii′):* Any $t \in l^2$ determines a continuous mapping $\pi(t):\pi(t)x = t \cdot x = \tilde{x}$, $\tilde{x} \in R$. Then if we put $\tilde{x} = t \cdot x$, we get by transformation of the integral

$$\lim_{n\to\infty} k_n \int_R \left[\Phi\left(\frac{\tilde{x} - \tilde{y}}{\alpha}\right) - \Phi\left(\frac{\tilde{x}}{\alpha}\right)\right] \mu_n\{[\pi(t)]^{-1}\, d\tilde{y}\}$$
$$= \lim_{n\to\infty} k_n \int_{l^2} \left[\Phi\left(\frac{x\cdot t - y\cdot t}{\alpha}\right) - \Phi\left(\frac{x\cdot t}{\alpha}\right)\right] \mu_n(dy).$$

The last limit exists uniformly with respect to $x \in l^2$ and $\|t\| \le \beta_0$, according to (iv). In particular, this is true for any $t$ with only finitely many coordinates different from 0. Then by Theorem IV.9.1, $\{\mu_n[\pi(t)^{-1}\cdot]\}^{*k_n}$ converges weakly. Thus (iv′) holds. If $\{x:\|x\| \ge \beta\}$ is a continuity set for $\lambda$, we obtain by (v),

$$\lim_{n\to\infty} k_n \int_{\|x\|<\beta} \|V_r \pi_r x\|^2 \mu_n(dx) = \sum_{i=1}^{r} q_{ii} + \int_{0<\|x\|<\beta} \|V_r \pi_r x\|^2 \lambda(dy).$$

Furthermore, in the same way as we have proved (v), we obtain

$$\lim_{n\to\infty} k_n \int_{\|x\|<\beta} \|x\|^2 \mu_n(dx) = \lim_{\varepsilon\downarrow 0} \overline{\lim}\, k_n \int_{\|x\|<\varepsilon} \|x\|^2 \mu_n(dx) + \int_{0<x<\beta} \|x\|^2 \lambda(dx)$$
$$= \sum_{i=1}^{\infty} q_{ii} + \int_{0<\|x\|<\beta} \|x\|^2 \lambda(dy).$$

As $r \to \infty$, $\|V_r \pi_r x\| \to \|x\|^2$. Hence, by these relations we obtain

$$\lim_{r\to\infty} \limsup_{n\to\infty} k_n \int_{\|x\|<\beta} \|\Delta_r x\|^2 \, \mu_n(dx) = \lim_{r\to\infty} \sum_{i=r+1}^{\infty} q_{ii} = 0,$$

since the series $\sum q_{ii}$ is convergent. Then by (i),

$$\lim_{r\to\infty} \limsup_{n\to\infty} k_n \int_{\|\Delta_r x\|<\beta} \|\Delta_r x\|^2 \, \mu_n(dx)$$

$$\leq \lim_{r\to\infty} \limsup_{n\to\infty} \left\{ k_n \int_{\|x\|<\beta} \|\Delta_r x\|^2 \, \mu_n(dx) + k_n \int_{\|\Delta_r x\|<\beta,\ \|x\|\geq\beta} \|\Delta_r x\|^2 \, \mu_n(dx) \right\}$$

$$= \lim_{r\to\infty} \limsup_{n\to\infty} k_n \int_{\|\Delta_r x\|<\beta,\ \|x\|\geq\beta} \|\Delta_r x\|^2 \, \mu_n(dx)$$

$$= \lim_{r\to\infty} \int_{\|\Delta_r x\|<\beta,\ \|x\|\geq\beta} \|\Delta_r x\|^2 \, \lambda(dx) = 0. \tag{6}$$

Hence, (iii′) holds. In order to show that (ii′) is also satisfied, we observe that according to (vi),

$$\lim_{n\to\infty} \left\| k_n \int_{l^2} \frac{V_r \pi_r x}{1+\|x\|^2} \, \mu_n(dy) - \bar{m}_r \right\| = 0,$$

where $\bar{m}_r^{(i)} = \bar{m}^{(i)}$ for $i \leq r$, $\bar{m}_r^{(i)} = 0$ for $i > r$, $\|\bar{m}_r - \bar{m}\| \to 0$ $(r \to \infty)$. Hence, by (ii),

$$\lim_{r\to\infty} \limsup_{n\to\infty} \left\| k_n \int_{l^2} \frac{\Delta_r x}{1+\|x\|^2} \, \mu_n(dx) \right\| = 0. \tag{7}$$

Here, for any $\beta > 0$

$$k_n \int_{l^2} \frac{\Delta_r x}{1+\|x\|^2} \, \mu_n(dx) = k_n \int_{\|\Delta_r x\|<\beta} \Delta_r x \, \mu_n(dx) - k_n \int_{\|\Delta_r x\|<\beta} \frac{\|x\|^2}{1+\|x\|^2} \Delta_r x \, \mu_n(dx) + k_n \int_{\|\Delta_r x\|\geq\beta} \frac{\Delta_r x}{1+\|x\|^2} \, \mu_n(dx). \tag{8}$$

Applying Lemma 1.2, we obtain

$$\limsup_{n\to\infty}\left\|k_n\int_{\|\Delta_r x\|<\beta}\frac{\|x\|^2}{1+\|x\|^2}\Delta_r x\,\mu_n(dx)\right\|^2$$
$$\leq \limsup_{n\to\infty}\left[k_n\int_{\|\Delta_r x\|<\beta}\left(\frac{\|x\|^2}{1+\|x\|^2}\right)^2\mu_n(dx)\right]$$
$$\times\left[k_n\int_{\|\Delta_r x\|<\beta}\|\Delta_r x\|^2\,\mu_n(dx)\right]=0. \tag{9}$$

Indeed, on the right-hand side, the first factor remains bounded according to (3) and (4) as $(n\to\infty)$, and the second factor tends to 0 according to (iii′) as first $n\to\infty$ and then $r\to\infty$. By Lemma 1.2 and (i′), we get

$$\limsup_{n\to\infty}\left\|k_n\int_{\|\Delta_r x\|\geq\beta}\frac{\Delta_r x}{1+\|x\|^2}\,\mu_n(dx)\right\|$$
$$\leq \limsup_{n\to\infty}k_n\int_{\|\Delta_r x\|\geq\beta}\mu_n(dx)\to 0 \qquad (r\to\infty). \tag{10}$$

Combining (7)–(10), we obtain (ii′). □

**Lemma 4.** (i′)–(iv′) ⇒ (iii).

*Proof:* (i′)–(iv′) ⇒ (i) according to Lemma 2. Choose $\beta$ such that $\{x:\|V_r\pi_r x\|\leq\beta,\ \|x\|\geq\beta\}$ is a continuity set for $\lambda$, $r=1,2,\ldots$ and $\lambda\{x:\|x\|=\beta\}=0$. Then

$$\limsup_{n\to\infty}\left|k_n\int_{\|x\|<\beta}\|x\|^2\,\mu_n(dx)-k_n\int_{\|V_r\pi_r x\|<\beta}\|V_r\pi_r x\|^2\,\mu_n(dx)\right|$$
$$\leq\beta^2\limsup_{n\to\infty}k_n\int_{\|V_r\pi_r x\|\leq\beta,\ \|x\|\geq\beta}\mu_n(dx)$$
$$+\limsup_{n\to\infty}k_n\int_{\|x\|<\beta}\|\Delta_r x\|^2\,\mu_n(dx). \tag{11}$$

On the right-hand side, the first term is equal to

$$\beta^2\int_{\|V_r\pi_r x\|\leq\beta,\ \|x\|\geq\beta}\lambda(dx),$$

and it tends to 0, as $r\to\infty$ since $V_r\pi_r x\uparrow x$ as $r\uparrow\infty$. The second term on the right-hand side tends to 0 as $r\to\infty$ according to (iii′) and (6), which only depends on (i) and (iii′). Thus we have proved that the left-hand side of (11) tends to 0 as $r\to\infty$. Combining this relation with (5°) in Lemma 1, which holds true because of (iv′), we obtain

$$\lim_{r\to\infty}\limsup_{n\to\infty}\left|k_n\int_{||x||<\beta}||x||^2\,\mu_n(dx)-\sigma_r^2(\beta)\right|=0. \tag{12}$$

This relation implies the Cauchy convergence and, hence, the convergence of

$$k_n\int_{||x||<\beta}||x||^2\,\mu_n(dx)$$

to a finite limit $\sigma^2(\beta)$. Then it follows by (12) that

$$\lim_{r\to\infty}\sigma_r^2(\beta)=\sigma^2(\beta).$$

Clearly

$$\sigma^2(\beta)\downarrow\sigma^2(0+),\qquad \beta\downarrow 0. \tag{13}$$

It is easily seen that (11) remains true if we change $||x||^2$ to $(x\cdot t)^2$ and $||V_r\pi_r x||^2$ to $(V_r\pi_r x\cdot t)^2$ with $||t||\le 1$ (but do not change the sets for integration). Then it follows by 4° in Lemma 1 that

$$\lim_{r\to\infty}\limsup_{n\to\infty}\left|k_n\int_{||x||<\beta}(x\cdot t)^2-q_r(\beta,t)\right|=0, \tag{14}$$

and, proceeding as above, we conclude that

$$k_n\int_{||x||<\beta}(x\cdot t)^2\,\mu_n(dx)\to q(\beta,t)\qquad(n\to\infty),$$

$$q(\beta,t)\downarrow q(t)\qquad(\beta\downarrow 0),\qquad q_r(\beta,t)\to q(\beta,t)\qquad(r\to\infty).\quad\square$$

**Lemma 5.** (i′)–(iv′) ⇒ (ii).

*Proof:* By Lemmas 2 and 4, (i) and (iii) are satisfied when (i′)–(iv′) hold. Applying Lemma 1.2, we obtain

$$\begin{aligned}
&\limsup_{r\to\infty}\limsup_{n\to\infty}k_n\left\|\int_{l^2}\left[\frac{V_r\pi_r x}{1+||V_r\pi_r x||^2}-\frac{V_r\pi_r x}{1+||x||^2}\right]\mu_n(dx)\right\|\\
&=\limsup_{r\to\infty}\limsup_{n\to\infty}k_n\left\|\int_{l^2}\frac{||\Delta_r x||^2 V_r\pi_r x}{[1+||x||^2][1+||V_r\pi_r x||^2]}\,\mu_n(dx)\right\|\\
&\le\limsup_{r\to\infty}\limsup_{n\to\infty}k_n\int_{l^2}\frac{||\Delta_r x||^2}{1+||x||^2}\,\mu_n(dx)\\
&\le\limsup_{r\to\infty}\limsup_{n\to\infty}k_n\int_{||\Delta_r x||<\beta}||\Delta_r x||^2\,\mu_n(dx)\\
&\quad+\limsup_{r\to\infty}\limsup_{n\to\infty}k_n\int_{||\Delta_r x||\ge\beta}\mu_n(dx)=0,
\end{aligned} \tag{15}$$

according to (iii′) and (i′). Furthermore,

$$\left\| k_n \int_{l^2} \frac{x}{1+\|x\|^2}\, \mu_n(dx) - k_n \int_{l^2} \frac{V_r \pi_r x}{1+\|x\|^2}\, \mu_n(dx) \right\|$$
$$= \left\| k_n \int_{l^2} \frac{\Delta_r x}{1+\|x\|^2}\, \mu_n(dx) \right\|. \tag{16}$$

The right-hand side tends to 0 as first $n \to \infty$ and then $r \to \infty$. This follows by (8)–(10) since (i′), (ii′), and (iii′) hold. Combining (3°) in Lemma 1 with (14) and (15), we obtain

$$\lim_{r\to\infty} \limsup_{n\to\infty} \left\| k_n \int_{l^2} \frac{x}{1+\|x\|^2}\, \mu_n(dx) - \bar{m}_r \right\| = 0. \tag{17}$$

By this relation we find that

$$\left\{ k_n \int_{l^2} \frac{x}{1+\|x\|^2}\, \mu_n(dx) \right\}$$

is Cauchy convergent and thus converges in the norm to an element $\bar{m}$, and then by (16) we get $\|\bar{m} - \bar{m}_r\| \to 0$ $(r \to \infty)$. □

By these lemmas we have proved the implications (i)–(iii) ⇔ (i′)–(iv′) and (i)–(iii) ⇒ (iv), (v), (vi). Furthermore, (vii) follows by Theorem 4.1. □

## 6. INVARIANTS OF INFINITELY DIVISIBLE $\sigma$-SMOOTH MEASURES ON $l^2$. GAUSSIAN FUNCTIONALS

**Theorem 1.** (1°) *An infinitely divisible probability measure $\mu$ on $l^2$ determines uniquely a Lévy measure $\lambda$ on $l^2$, a mean-value vector $\bar{m}$ in $l^2$, and a positive definite symmetrical quadratic form $q(t)$ such that for $\mu = \mu_n^{*n}$*

(i) *$n\mu_n$ converges weakly to $\lambda$,*

(ii) $$\lim_{n\to\infty} \left\| n \int_{l^2} \frac{y}{1+\|y\|^2}\, \mu_n(dy) - \bar{m} \right\| = 0,$$

(iii) $$\lim_{\varepsilon\downarrow 0} \overline{\lim_{n\to\infty}}\, n \int_{\|y\|<\varepsilon} (t\cdot y)^2\, \mu_n(dy) = q(t)$$

*for any $t \in l^2$, where*

(iii′) $$q(t) = \sum_{i=1}^{\infty} \sum_{j=1}^{\infty} q_{ij} t^{(i)} t^{(j)}, \qquad q_{ij} = q_{ji}, \qquad \sum_{i=1}^{\infty} q_{ij} < \infty.$$

(2°) *Conversely, a given Lévy measure* $\lambda$, *a given element* $\bar{m}$ *in* $l^2$, *and a positive definite symmetrical form given by* (iii′) *determine uniquely an infinitely divisible probability measure* $\mu$ *such that* (i)–(iii) *hold.*

(3°) $\lambda$, $\bar{m}$ *and* $q(t)$ *are invariants of* $\mu$. *They determine uniquely the projected Gaussian functional*

$$\bar{\Gamma}(\alpha, \mu, x \cdot t) = -\frac{1}{\alpha}\bar{m} \cdot t\Phi'\left(\frac{x \cdot t}{\alpha}\right) + \frac{1}{2\alpha^2} q(t)\Phi''\left(\frac{x \cdot t}{\alpha}\right)$$

$$+ \int_{l^2}\left[\Phi\left(\frac{x \cdot t - y \cdot t}{\alpha}\right) - \Phi\left(\frac{x \cdot t}{\alpha}\right) + \frac{y \cdot t}{\alpha(1 + \|y\|^2}\Phi'\left(\frac{x \cdot t}{\alpha}\right)\right]\lambda(dy)$$

*for* $t \in l^2$, *and* $\mu$ *is uniquely determined by* $\bar{\Gamma}(\alpha, \mu, x \cdot t)$ (*for different* $\alpha$, $x$, *and* $t$).

(4°) *The infinitely divisible probability measures* $\nu$ *form a commutative semigroup under convolution, and it is* 1.1 *isomorphic to the additive group of the projected Gaussian functionals*

$$\nu \leftrightarrow \bar{\Gamma}(\alpha, \nu, x \cdot t),$$
$$\nu_1 * \nu_2 \leftrightarrow \bar{\Gamma}(\alpha, \nu_1, x \cdot t) + \bar{\Gamma}(\alpha, \nu_2, x \cdot t).$$

*Remark:* An infinitely divisible $\sigma$-smooth measure, different from the zero measure, has the form $\beta\mu$ with a probability measure $\mu$ and a positive number $\beta$. □

*Proof:* We start by proving (4°). By Theorem 1.2, a $\sigma$-smooth measure $\nu$ on $l^2$ is uniquely determined by its projections $\nu\{[\pi(t)]^{-1}\cdot\}$, $\pi(t)x = x \cdot t$. If $\nu$ is infinitely divisible, then $\nu[\pi(t)^{-1}\cdot]$ is infinitely divisible and it determines the projected Gaussian functional uniquely, which in turn determines $\nu[\pi(t)^{-1}\cdot]$ uniquely. On $R$, the semigroup under convolution of infinitely divisible probability measures is 1.1 isomorphic to the additive group of the Gaussian functionals. Hence, this is also true on $l^2$. Note, however, that the projected Gaussian functional of $\nu \in l^2$ is determined by $\lambda$, $\bar{m}$, and $q(t)$, which, according to (i)–(iii), must have certain properties in order to determine an infinitely divisible probability measure.

If $\nu$ is infinitely divisible, then $\nu = \nu_n^{*n}$ for all $n$ and thus is the limit of a sequence of convolution powers. Applying Theorem 5.1, we conclude that $\nu$ determines a Lévy measure $\lambda$, a vector $\bar{m}$, and a positive definite symmetrical quadratic form, according to (i)–(iii).

*Proof of* (3°): We get $\bar{\Gamma}(\alpha, \nu, x \cdot t)$ by the transformation $\tilde{x} = x \cdot t$ of the Gaussian functional belonging to $\nu[\pi(t)^{-1}\cdot]$.

We have to show that $\lambda$, $\bar{m}$, and $q(t)$ determine an infinitely divisible probability measure such that (i)–(iii) hold. Then, according to (4°), it is sufficient to consider the following three cases:

(a) $\bar{m}$ zero element, $q(t)$ *identically zero,*
(b) $\lambda$ identically zero, $\bar{m}$ zero element,
(c) $\lambda$ identically zero, $q(t)$ identically 0.

*Case (a):* To given $n = 1, 2, \ldots$ choose $\beta_n > 0$ such that

$$\lim_{n\to\infty} \frac{1}{\sqrt{n}} \int_{\|x\| \geq \beta_n} \lambda(dx) = 0, \qquad \beta_n \downarrow 0 \quad \text{as} \quad n \uparrow \infty, \tag{1}$$

and also such that $F_{\beta_n} = \{x : \|x\| \geq \beta_n\}$, $n = 1, 2, \ldots$, is a continuity set for $\lambda$. Then define the (finite) measure $\lambda_n$ by $\lambda_n(E) = \lambda[E \cap F_{\beta_n}]$ for any Borel set $E \in l^2$, and put

$$\bar{m}_n = \int_{l^2} \frac{y}{1 + \|y\|^2} \lambda_n(dy) = \int_{F_{\beta_n}} \frac{y}{1 + \|y\|^2} \lambda(dy), \tag{2}$$

$$\begin{aligned} \mu_n &= \left[1 + \frac{1}{n}\lambda(F_{\beta_n})\right]^{-1} \left(e + \frac{\lambda_n}{n}\right) * e\left(\cdot + \frac{\bar{m}_n}{n}\right) \\ &= \left[1 + \frac{1}{n}\lambda(F_{\beta_n})\right]^{-1} \left[e\left(\cdot + \frac{\bar{m}_n}{n}\right) + \frac{1}{n}\lambda_n\left(\cdot + \frac{\bar{m}_n}{n}\right)\right]. \end{aligned} \tag{3}$$

Note that $\bar{m}_n \in l^2$ and that $\mu_n$ is a probability measure since

$$\mu_n(l^2) = \left[1 + \frac{1}{n}\lambda(F_{\beta_n})\right]^{-1} \left[1 + \frac{1}{n}\lambda(F_{\beta_n})\right] = 1.$$

Applying Theorem 5.1, we show that $\mu_n^{*n}$ converges weakly to an infinitely divisible probability measure. By Lemma 1.2, we get

$$\|\bar{m}_n\|^2 \leq \int_{l^2} \frac{\|y\|^2}{1 + \|y\|^2} \lambda(dy) \int_{F_{\beta_n}} \lambda(dy),$$

and, hence, by (1) and the property of $\lambda$,

$$\lim_{n\to\infty} \frac{1}{\sqrt{n}} \|\bar{m}_n\|^2 = 0. \tag{4}$$

Then we get by (3) for any closed set $F$ belonging to $\{x : \|x\| \geq \beta\}$ for some $\beta$

$$\limsup_{n\to\infty} n\mu_n(F) = \limsup_{n\to\infty} \lambda(F + \bar{m}_n/n) = \lambda(F).$$

Thus $n\mu_n$ converges weakly to $\lambda$. Furthermore,

$$\begin{aligned} \left\| n \int_{l^2} \frac{y}{1 + \|y\|^2} \mu_n(dy) \right\| = \left[1 + \frac{1}{n}\lambda(F_{\beta_n})\right]^{-1} \Bigg\| &-\bar{m}_n \frac{1}{1 + \|\bar{m}_n/n\|^2} \\ &+ \int_{l^2} \frac{y}{1 + \|y\|^2} \lambda_n\left(dy + \frac{\bar{m}_n}{n}\right)\Bigg\|. \end{aligned} \tag{5}$$

Observing the definition of $m_n$ and transforming the integral on the right-hand side of (5), we find that this side is at most equal to

$$\frac{1}{n^2}\|\bar{m}_n\|^3 + \left\| \int_{l^2} \left[ \frac{y - \bar{m}_n/n}{1 + \|y - \bar{m}_n/n\|^2} - \frac{y}{1 + \|y\|^2} \right] \lambda_n(dy) \right\|$$

$$\leq \frac{1}{n^2}\|\bar{m}_n\|^3 + \frac{|\bar{m}_n|}{n}\lambda(F_{\beta_n}) + \left\| \int_{l^2} \left[ \frac{1}{1 + \|y - \bar{m}_n/n\|^2} - \frac{1}{1 + \|y\|^2} \right] y\,\lambda_n(dy) \right\|.$$

On the right-hand side, the first two terms tend to 0 as $n \to \infty$ by (1) and (4). We estimate the third term by the help of Lemma 1.2, observing that

$$\left| \frac{1}{1 + \|y - \bar{m}_n/n\|^2} - \frac{1}{1 + \|y\|^2} \right| \leq \frac{2(\|\bar{m}_n\|/n)\|y\| + \|\bar{m}_n\|^2/n^2}{1 + \|y\|^2},$$

and so find that this integral is at most equal to

$$\left\{ \lambda(F_{\beta_n}) \int_{l^2} \left[ \frac{2(\|\bar{m}_n\|/n)\|y\| + \|\bar{m}_n\|^2/n^2}{1 + \|y\|^2} \right]^2 \|y\|^2\,\lambda(dy) \right\}^{1/2}.$$

It follows by (1) and (4) and the properties of $\lambda$ that this term tends to 0 as $n \to \infty$. Thus by (5)

$$\lim_{n\to\infty} \left\| n \int_{l^2} \frac{y}{1 + \|y\|^2}\,\mu_n(dy) \right\| = 0. \tag{6}$$

For the second-order moments, we obtain

$$\limsup_{\varepsilon\downarrow 0}\,\limsup_{n\to\infty}\, n \int_{\|y\|<\varepsilon} (y \cdot t)^2\,\mu_n(dy)$$

$$\leq \|t\|^2 \limsup_{\varepsilon\downarrow 0}\,\limsup_{n\to\infty}\, n \int_{\|y\|<\varepsilon} \|y\|^2\,\mu_n(dy) = 0. \tag{7}$$

Indeed, we have

$$n \int_{\|y\|<\varepsilon} \|y\|^2\,\mu_n(dy) \leq n\,\frac{\|\bar{m}_n\|^2}{n^2} + \int_{\|y\|<\varepsilon} \|y\|^2\,\lambda_n\!\left(dy + \frac{\bar{m}_n}{n}\right).$$

On the right-hand side, the first term tends to 0 as $n \to \infty$, according to (4). For the integral, we get

$$\limsup_{n\to\infty} \int_{\|y\|<\varepsilon} \|y\|^2\,\lambda_n\!\left(dy + \frac{\bar{m}_n}{n}\right) \leq 2\limsup_{n\to\infty} \int_{\|y - \bar{m}_n/n\|<\varepsilon} \|y\|^2\,\lambda(dy)$$

$$+ 2\limsup_{n\to\infty} \frac{\|\bar{m}_n\|^2}{n^2}\lambda(F_{\beta_n}) \to 0,$$

according to (1), (4) and the property of $\lambda$. Applying Theorem 5.1 we conclude that $\{\mu_n^{*n}\}$ converges to an infinitely divisible probability measure $\nu$ with the projected Gaussian functional

$$\bar{\Gamma}(\alpha, \nu, x \cdot t) = \int_{l^2} \left[ \Phi\left(\frac{x \cdot t - y \cdot t}{\alpha}\right) - \Phi\left(\frac{x \cdot t}{\alpha}\right) + \frac{y \cdot t}{\alpha(1 + \|y\|^2} \Phi'\left(\frac{x \cdot t}{\alpha}\right) \right] \lambda(dy). \quad (8)$$

*Case (b)*: Let the infinite positive definite symmetrical quadratic form

$$q(t) = \sum_{i=1}^{\infty} \sum_{j=1}^{\infty} q_{ij} t^{(i)} t^{(j)}, \qquad \sum_{i=1}^{\infty} q_{ij} < \infty, \qquad t \in l^2 \quad (9)$$

be given, and put

$$q_r(t) = \sum_{i=1}^{r} \sum_{j=1}^{r} q_{ij} t^{(i)} t^{(j)}.$$

To $q_r(t)$ there belongs a symmetrical Gaussian measure $\tilde{\nu}^{(r)}$ in $\pi_r l^2$ such that

$$\lim_{n \to \infty} n \int_{\|\tilde{y}^{(r)}\| \geq \varepsilon} \tilde{\nu}^{(r)}(\sqrt{n}\, d\tilde{y}^{(r)}) = 0, \qquad \text{for any} \quad \varepsilon > 0, \quad (10)$$

$$\lim_{\varepsilon \downarrow 0} \overline{\lim_{n \to \infty}}\, n \int_{\|\tilde{y}^{(r)}\| < \varepsilon} (\tilde{y}^{(r)} \cdot t) \tilde{\nu}^{(r)}(\sqrt{n}\, d\tilde{y}^{(r)}) = q_r(t). \quad (11)$$

Then $\tilde{\nu}^{(r)}$ determines the Gaussian measure $\tilde{\nu}^{(r)}(V_r^{-1} \cdot)$ on $l^2$. To given $r$, $r = 1, 2, \ldots$, we may choose $n_r$ such that

$$n_r \int_{\|y\| \geq \varepsilon} \tilde{\nu}^{(r)}(\sqrt{n_r}\, V_r^{-1}\, dy) = n_r \int_{\|\tilde{y}^{(r)}\| \geq \varepsilon} \tilde{\nu}^{(r)}(\sqrt{n_r}\, d\tilde{y}^{(r)}) < \frac{1}{r},$$

$$\left\| n_r \int_{\|y\| < \varepsilon_r} (y \cdot t)^2\, \tilde{\nu}^{(r)}(\sqrt{n_r}\, V_r^{-1}\, dy) - q_r(t) \right\|$$
$$= \left\| n_r \int_{\|\tilde{y}^{(r)}\| < \varepsilon_r} (\tilde{y}^{(r)} \cdot t)^2 \tilde{\nu}^{(r)}(\sqrt{n_r}\, d\tilde{y}^{(r)}) - q_r(t) \right\| < \frac{1}{r}, \quad (12)$$

where $\varepsilon_r \downarrow 0$ as $r \to \infty$. Then the left-hand sides of (11) and (12) tend to 0 as $r \to \infty$, and we find by Theorem 5.1 that

$$[\tilde{\nu}^{(r)}(\sqrt{n_r}\, V_r^{-1} \cdot)]^{*n_r}$$

converges weakly to a Gaussian measure $\nu$ with the Gaussian projected functional

$$\bar{\Gamma}(\alpha, \nu, x \cdot t) = (1/2\alpha^2) q(t) \Phi'[(x \cdot t)/\alpha]. \quad (13)$$

*Case (c):* To a given element $\bar{m} \in l^2$ there belongs the infinitely divisible probability measure $e(\cdot - \bar{m}/n)$, with

$$[e(\cdot - \bar{m}/n)]^{*n} = e(\cdot - \bar{m}),$$

$$\lim_{n\to\infty} n \int_{l^2} \frac{y}{1+\|y\|^2}\, e\left(dy - \frac{\bar{m}}{n}\right) = \lim_{n\to\infty} \frac{1}{1+\|\bar{m}/n\|^2}\,\bar{m} = \bar{m},$$

and the projected Gaussian functional

$$\lim_{n\to\infty} n\left[\Phi\left(\frac{x\cdot t - y\cdot t}{\alpha}\right) - \Phi\left(\frac{x\cdot t}{\alpha}\right)\right] e\left(dy - \frac{\bar{m}}{n}\right)$$

$$= \lim_{n\to\infty} n\left[\Phi\left(\frac{x\cdot t - \bar{m}\cdot t/n}{\alpha}\right) - \Phi\left(\frac{x\cdot t}{\alpha}\right)\right] = \frac{1}{\alpha}\,\bar{m}\Phi'\left(\frac{x\cdot t}{\alpha}\right). \tag{14}$$

If follows from (8), (13), (14), and the stated semigroup isomorphism that there exists an infinitely divisible probability measure with the projected Gaussian functional as in (3°) of the theorem. □

## 7. THE CHARACTERISTIC FUNCTION OF PROBABILITY MEASURES ON $l^2$

By Theorem 1.2, a probability measure $\mu$ on $l^2$ is uniquely determined by the projected measures $\mu(\pi_r^{-1}\cdot)$, and by Theorem IV,4 these are uniquely determined by their characteristic functions

$$\int_{\pi_r l^2} \exp[i(t\cdot\tilde{x}^r)]\,\mu(\pi_r^{-1}\,d\tilde{x}^{(r)}), \tag{1}$$

where $t \in \pi_r l^2$. By the transformation $\tilde{x}^{(r)} = \pi_r x$, this integral may be written

$$\tilde{\mu}(t) = \int_{l^2} \exp(it\cdot x)\,\mu(dx) \tag{2}$$

if we let $t$ be the element $t = t^{(1)}, \ldots, t^{(r)}, 0, 0 \ldots)$ in $l^2$. However, the integral in (2) has meaning for any $t \in l^2$. For $t = (t^{(1)} \ldots t^{(r)}, 0 \ldots)$ it is equal to (1). We call $\mu$ the characteristic function of $\mu$. We have just proved

**Theorem 1.** *A probability measure $\mu$ on $l^2$ determines uniquely its characteristic function $\tilde{\mu}$, which conversely determines $\mu$.*

*Remark:* Note that $\tilde{\mu}(t)$ can be written

$$\tilde{\mu}(\tau) = \int_R \exp(i\tau\cdot\tilde{x})\,\mu[\pi(t)^{-1}\,d\tilde{x}],$$

where $\pi(t)$ is the projection

$$\pi(t)x = t\cdot x = \tau\cdot\tilde{x},$$

and thus $\tilde{\mu}(\tau)$ is the characteristic function of the measure $\mu[\pi(t)^{-1}\cdot]$ on the line.

We now claim

**Theorem 2.** *The characteristic function $\tilde{\nu}$ of an infinitely divisible probability measure $\nu$ has the representation*

$$\text{(i)}\quad \tilde{\nu} = \exp\left\{ i\bar{m}\cdot t - \tfrac{1}{2}q(t) + \int_{l^2\setminus\{0\}} \left[ \exp(iy\cdot t) - 1 - \frac{iy\cdot t}{1+\|y\|^2} \right] \lambda(dy) \right\},$$

*where $\lambda$ is a Lévy measure, $\bar{m}$ an element in $l^2$, and $q(t)$ a positive definite symmetrical quadratic form determined as in Theorem 6.1.*

**Corollary:** *A sequence $\{\mu_n^{*k_n}\}$, $k_n \uparrow \infty$ as $n \uparrow \infty$, of convolution powers of probability measures $\mu_n$ converges weakly to a probability measure $\nu$ if and only if the characteristic function of $\{\mu_n^{*k_n}\}$, i.e., $(\tilde{\mu}_n)^{k_n}$ converges to a function with the representation* (i), *$\bar{m}$, $q(t)$, and $\lambda$ determined in Theorem 5.1. Then $\tilde{\nu}$ in* (i) *is the characteristic function of this weak limit $\nu$.*

*Proof:* If $\nu$ is an infinitely divisible probability measure, then $\nu = \nu_n^{*n}$ for all $n$, and thus the weak limit of $\nu_n^{*n}$. Let us then all at once consider the case when $\{\mu_n^{*k_n}\}$ converges to $\nu$, $\mu_n$ is the probability measure, $k_n \uparrow \infty$ as $n \uparrow \infty$. Then for the projection $\pi(t)x = t\cdot x$, the sequence $\{\mu_n[\pi(t)^{-1}\cdot]\}^{*k_n}$ converges weakly, and its characteristic function is equal to

$$\tilde{\nu}(t,\tau) = \exp \lim_{n\to\infty} k_n \int_R [\exp(i\tau\tilde{y} - 1)]\, \mu_n[\pi(t)^{-1}\, d\tilde{y}].$$

(Compare IV.11.5.) By transformation of the integral, we obtain for $\tilde{y} = \pi(t)y = y\cdot t$,

$$\tilde{\nu}(t,\tau) = \exp \lim_{n\to\infty} k_n \int_{l^2} [\exp(i\tau y\cdot t - 1)]\, \mu_n(dy). \tag{3}$$

We may write this relation for $\tau = 1$

$$\ln \tilde{\nu}(t) = \lim_{n\to\infty} k_n \int_{l^2} \left[ \exp\left( iy\cdot t - 1 - \frac{iy\cdot t}{1+\|y\|^2} \right) \right] \mu_n(dy) + i\bar{m}\cdot t \tag{4}$$

with

$$\lim_{n\to\infty} \left\| \bar{m} - k_n \int_{l^2} \frac{y}{1+\|y\|^2}\, \mu_n(dy) \right\| = 0.$$

The limit $\bar{m}$ exists and is finite, according to Theorem 5.1. We estimate the limit in (4) exactly in the same way as in the proof of Lemma 5.1 and so get the right-hand side of (i) in the theorem above. Thus we have proved the theorem and the "if" part of the corollary. The "only if" part follows by Theorem 5.1. □

APPENDIX

# A PRODUCT–SUM IDENTITY†

Consider polynomials of a special form in variables $x_1, \ldots, x_n, y_1, \ldots, y_n$ with rational coefficients. We use the notation

$$\sum! \, y_1 y_2 \cdots y_k x_{k+1} \cdots x_n$$

for the sum of all monomials

$$y_1 y_2 \cdots y_k x_{k+1} \cdots x_n$$

corresponding to the different permutations of the numbers 1, 2, . . . , *n*, the first *k* factors being always *y*s, the other factors *x*s. The number of such monomials is obviously $n!$. Now we define the polynomial

$$f^{(n)} = \frac{1}{(n+1)!} \{\sum! \, x_1 x_2 \cdots x_n + \sum! \, y_1 x_2 \cdots x_n + \sum! \, y_1 y_2 x_3 \cdots x_n + \cdots + \sum! \, y_1 y_2 \cdots y_n\} \tag{1}$$

and the polynomials $f_k^{(n)}$, $\bar{u}_{kj}$, $\bar{v}_{kj}$, and

$$\bar{f}^{(n)} = \frac{1}{n} \sum_{k=1}^{n} f_k^{(n)}, \tag{2}$$

where $f_k^{(n)}$ is obtained from $f^{(n)}$ when $x_k$ and $y_k$ are specialized to 1, $\frac{1}{2}\bar{u}_{kj}^{(n)}$ is obtained from $f_k^{(n)}$ when $x_j$ is specialized to 1 and $y_j$ to 0, and $\frac{1}{2}\bar{v}_{kj}^{(n)}$ is obtained from $f_k^{(k)}$ when $x_j$ is specialized to 0 and $y_j$ to 0; that is,

$$\begin{aligned} f_k^{(n)} &= (f^{(n)})_{x_k = y_k = 1}, \\ \tfrac{1}{2}\bar{u}_{kj}^{(n)} &= (f_k^{(n)})_{x_j = 1,\, y_j = 0}, \\ \tfrac{1}{2}\bar{v}_{kj}^{(n)} &= (f_k^{(n)})_{x_j = 0,\, y_j = 1}. \end{aligned} \tag{3}$$

† Quoted from Harald Bergström, Some remarks on an algebraic identity, *Math. Scand.* **8**, 39–42 (1960).

The identity to be proved then has the form

$$\prod_{k=1}^{n} x_k - \prod_{k=1}^{n} y_k = \bar{f}^{(n)} \sum_{k=1}^{n} (x_k - y_k) + \frac{1}{2n} \sum_{k=1}^{n} \sum_{j=1}^{n} \{\bar{u}_{kj}^{(n)}(x_k - y_k)(x_j - x_k) + \bar{v}_{kj}^{(n)}(x_k - y_k)(y_j - y_k)\}. \tag{4}$$

It may be observed that $\bar{f}^{(n)}$, $\bar{u}_{kj}^{(n)}$, and $\bar{v}_{kj}^{(n)}$ are arithmetic means of monomials in the variables. This is of importance for applications. If, for instance, all $x_k$ and $y_k$ belong to a convex commutative semigroup, then $\bar{f}^{(n)}$, $\bar{u}_{kj}^{(n)}$, and $\bar{v}_{kj}^{(n)}$ belong to the same semigroup.

When we specialize all $x_k$ to $x$ and all $y_k$ to $y$, the identity (4) reduces to

$$x^n - y^n = (x - y)(x^{n-1} + x^{n-2}y + \cdots + y^{n-1}), \tag{5}$$

of which (4) is a generalization.

Before proving the identity, we observe that $f^{(n)}$ is symmetrical in the variables $x_1, x_2, \ldots, x_n$ as well as in the variables $y_1, y_2, \ldots, y_n$ and invariant under the transformation $x_k \to y_k$, $y_k \to x_k$, $k = 1, 2, \ldots, n$. Moreover, $f^{(n)}$ is linear in each variable, and every monomial contains either $y_k$ or $x_k$ as a factor, but never both of them.

Since $f^{(n)}$ does not change if $j$ changes into $k$ and $k$ into $j$, we obviously have

$$\bar{u}_{kj}^{(n)} = \bar{u}_{jk}^{(n)}, \qquad \bar{v}_{kj}^{(n)} = \bar{v}_{jk}^{(n)}. \tag{6}$$

We now prove the identity (4). For this purpose we use the following notations:

$$\prod_{\lambda,\mu} x_k = \begin{cases} 1 & \text{for } \lambda \geqq \mu, \\ x_{\lambda+1} \cdots x_\mu & \text{for } \mu > \lambda, \end{cases}$$

$$\sum_{\lambda,\mu} x_k = \begin{cases} 0 & \text{for } \lambda \geqq \mu, \\ x_{\lambda+1} + x_{\lambda+2} + \cdots + x_\mu & \text{for } \mu > \lambda. \end{cases}$$

Writing

$$\prod_{0,n} x_k - \prod_{0,n} y_k = x_1 \prod_{1,n} x_v - y_1 \prod_{1,n} x_v + \left(\prod_{0,1} y_v\right) x_2 \prod_{2,n} x_v - \left(\prod_{0,1} y_v\right) y_2 \prod_{2,n} x_v + \cdots + \left(\prod_{0,n-1} y_v\right) x_n - \left(\prod_{0,n-1} y_v\right) y_n,$$

we get the identity (which still holds true when the letters denote elements in a noncommutative ring)

$$\prod_{0,n} x_k - \prod_{0,n} y_k = \sum_{k=1}^{n} \left(\prod_{0,k-1} y_v\right)(x_k - y_k)\left(\prod_{k,n} x_v\right). \tag{7}$$

Since $x_1, \ldots, x_n, y_1, \ldots, y_n$ are arbitrary, (7) still holds if we permute $x_1, \ldots, x_n$ and correspondingly $y_1, \ldots, y_n$ in all possible ways; the permutation, of course, leaves the left-hand side unaltered. Adding the identities ($n!$ in number) which correspond to the different permutations, and dividing the obtained equality by $n!$, we get an identity of the form

$$\prod_{0,n} x_k - \prod_{0,n} y_k = \sum_{0,n} f_k^{(n)}(x_k - y_k), \tag{8}$$

where obviously the coefficients $f_k^{(n)}$ are the polynomials defined above. In fact, the coefficient of $x_n - y_n$ is

$$\frac{1}{n!}\{\textstyle\sum! x_1x_2 \cdots x_{n-1} + \sum! y_1x_2 \cdots x_{n-1} + \cdots + \sum! y_1y_2 \cdots y_{n-1}\},$$

and the coefficient of $x_k - y_k$ is obtained from the last polynomial by interchanging $k$ and $n$. Introducing the mean value (2), we may write

$$f_k^{(n)} = \bar{f}^{(n)} + f_k^{(n)} - \bar{f}^{(n)}, \tag{9}$$

where

$$f_k^{(n)} - \bar{f}^{(n)} = \frac{1}{n}\sum_{j=1}^{n} \{f_k^{(n)} - f_j^{(n)}\}. \tag{10}$$

Now we observe that

$$f_k^{(n)} = Ax_j + By_j,$$

where $A$ and $B$ are polynomials independent of $x_j$ and $y_j$. Hence we recognize $A$ and $B$ as the polynomials $\frac{1}{2}\bar{u}_{kj}^{(n)}$ and $\frac{1}{2}\bar{v}_{kj}^{(n)}$, respectively, and get

$$f_k^{(n)} = \tfrac{1}{2}\{\bar{u}_{kj}^{(n)}x_j + \bar{v}_{kj}^{(n)}y_j\}.$$

According to (6), we may also write

$$f_j^{(n)} = \tfrac{1}{2}\{\bar{u}_{kj}^{(n)}x_k + \bar{v}_{kj}^{(n)}y_k\}.$$

Thus we obtain

$$f_k^{(n)} - f_j^{(n)} = \tfrac{1}{2}\{\bar{u}_{kj}^{(n)}(x_j - x_k) + \bar{v}_{kj}^{(n)}(y_j - y_k)\}. \tag{11}$$

Combining (8), (9), (10) and (11), we obtain identity (4). □

## NOTES AND COMMENTS

A main tool in the presentation of weak convergence of sequences of measures in this book is functional analysis. Weak convergence of a sequence of measures means weak convergence of bounded, linear functionals on certain Stone vector lattices of bounded continuous functions. A main source for my presentation is Alexandrov's fundamental papers [2]. His work is founded on the classical theory on Riez–Radon representations of linear functionals by Radon measures, but Alexandrov generalizes this theory by the introduction of $\sigma$-topological spaces. Furthermore, he deals with additive set functions, not necessarily nonnegative or $\sigma$-smooth, which he calls charges, here called signed measures. In this book only nonnegative additive set functions, called measures, are considered, and so Alexandrov's theory could be simplified and some of his conditions weakened.

Essential for the procedures in my presentation are three theorems, here called Alexandrov's first, second, and third theorem, which deal with the connections between functionals and measures. In the second theorem, we may consider functionals on a smaller Stone vector lattice than in the first one when we deal with the weak convergence of a sequence of measures. This means that we require a smaller lattice when the weak limit is known. For instance, the Stone vector lattice of bounded, uniformly continuous functions is then sufficient for a metric space provided that the measures in the sequence are $\sigma$-smooth. This lattice is also always sufficient for the first theorem when the metric space is pseudocompact (a concept introduced here), particularly if it is compact. This fact explains the importance of the tightness condition, which to a certain extent reduces the weak convergence problem in a metric space to a corresponding problem in a compact subspace of this space, but then we must have a metric which admits certain compactness properties of the given space. The tightness condition is essential for Prokorov's weak compactness theorem [32, 33]. In the applications in this book, the tightness condition is used only in the finite-dimensional vector spaces $R^{(k)}$, where it is easy to verify. The tightness condition has been replaced by a weaker condition, which we may call a reduction condition since it permits a reduction of the weak limit problem in a given space to other spaces, where it may be easier to solve. We have this situation for the $C$- and $D$-spaces and the Hilbert space $l^2$, where the weak limit problem can be reduced to limit problems in finite-dimensional subspaces of $R^{(\infty)}$. The reduction principle, which we here apply, depends on the consistence theorem of Kolmogorov [24] about the construction of a $\sigma$-smooth measure on $R^{(\infty)}$ by a consistent family of $\sigma$-smooth measures on the finite-dimensional subspaces of $R^{(\infty)}$. This reduction principle also depends on suitable approximations of elements in the given space by Schauder sequences (Schauder series). Special Schauder series are analyzed in [22] and used in connection with the Wiener measure by Ciesielski [10] and McKeen Jr. [26]. The reduction condition was introduced in [5], but the sufficiency of this condition for metric spaces does not follow so directly from Alexandrov.s theorems as was stated there. A complete presentation is given in [6] and here in Section III.6.

For these applications, the weak convergence of sequences of $\sigma$-smooth measures on $k$-dimensional vector spaces $R^{(k)}$ is essential. Of particular interest here is the weak convergence of convolution products. By the use of seminorms, the weak convergence of these products is a special form of convergence of quite general products. I have applied Gaussian transforms for the examination of weak convergence on $R^{(k)}$ in several earlier papers, and I use the seminorm method in [7].

In the applications in Chapter V and Chapter VI, the weak convergence in the $C$- and $D$-spaces and the Hilbert space $l^2$ is reduced to weak limit problems in $R^{(k)}$. The reduction condition requires an analysis of fluctuations of sums analogous to but simpler than the verification of the tightness condition. Thus, to a certain extent, I use the same lemmas about fluctuations as in [9]. These applications are close to the applications in [8], [9], [20], [21], [32], and [39].

The weak convergence of sequences of measures goes back to elementary limit theorems in probability theory by Laplace, Moivre, and others, and central limit theorems by Liapounov [29] and Lindeberg [30] for distributions on the line, further developed by several mathematicians, particularly by Gnedenko, Kolmogorov, and Lévy (compare [18] and [27] and the references in these books). The weak convergence was considered on finite-dimensional vector spaces by Bernstein [8], de Finneti [15], and Lévy [27].

The starting point for the theory of measures on function spaces was Wiener's fundamental paper [40], where he introduced that measure on the $C$-space, which is now called the Wiener measure. More general measures on this and other function spaces were then constructed by methods developed by Doob [14], Kolmogorov [24, 25], Prokorov [33, 34], Parthasarathy [32], and Varadan [37].

The study of weak convergence on function spaces was initiated by Doob [13] and began with Donsker's paper [11]. Then, in the short period 1951–1956, there appeared a series of important works in this field by Gihman [16], Skorokhod [36], Prokorov [33, 34], and Le Cam [28]. Presentations of weak convergence in metric spaces are given in [9] and [32].

Almost standing by itself is Alexandrov's early framework [2], which is not directly concerned with probability theory, but which nevertheless must be considered as a fundamental work on weak convergence. As a matter of fact, it gives together with Kolmogorov's consistent theorem, as shown in Section III.6, an essential part of the whole theory.

It should be noted that my book is concerned exclusively with weak convergence of sequences of measures, and deals with random functions (stochastic processes) only so far as they are directly connected with measures. For the extensive theory of stochastic processes, I refer to literature in this field (see for instance, [17, Vols. I–III] and the references there). It should also be noted that the rate of convergence is not considered in my book and, hence, no special references are given for the various estimations of remainder terms.

# BIBLIOGRAPHY


1. Achieser, N. J., and Glassmann, J. M. "Theorie der Linearen Operatoren in Hilbert Raum." Akademie-Verlag, Berlin, 1958.
2. Alexandrov, A. D. Additive set functions in abstract spaces, *Mat. Sb.* **8**, 307–348 (1940); **9**, 536–628 (1941); **13**, 169–238 (1943).
3. Bauer, H. "Probability Theory and Elements of Measure Theorie." Holt, New York, 1972.
4. Bergström, H. "Limit Theorems for Convolutions." Almqvist & Wiksell, Stockholm, 1963.
5. Bergström, H. On weak convergence in normal spaces, *Symposia Mathematica, 21st* pp. 73–89. Instituto Nazionale di Alta Matematica, Rome, 1977.
6. Bergström, H. Reduction of weak limit problems by transformations, lecture notes on the conference *Analytische Methoden in der Wahrscheinlichkeitsrechnung, Oberwolfach, 8–14 June, 1980.* Springer-Verlag, Berlin and New York, 1980.
7. Bergström, H. General limit theorems for products with applications to convolution products of measure, *in* "Contributions to Probability" (J. Gani and V. K. Rohatgi, eds.). Academic Press, New York, 1981.
8. Bernstein, S. N.: Sur l'extension du théoreme limit du calcul des probabilités aux sommes de quantités dépendentes, *Math. Ann.* **97**, 1–59 (1927).
9. Billingsley, P. "Convergence of Probability Measures." Wiley, New York, 1968.
10. Ciesielaki, Z. Hölder condition for realizations of Gaussian processes, *TAMS* **99**, 403–413, (1961).
11. Donsker, M. An invariance principle for certain probability limit theorems, *Mem. Amer. Math. Soc.* **6**, 1–12 (1951).
12. Donsker, M. Kolmogorov–Smirnov theorems, *Ann. Math. Statist.* **23**, 277–281 (1952).
13. Doob, J. L. Heuretic approch to the Kolmogorov–Smirnov theorems, *Ann. Math. Statist.* **20**, 393–403 (1949).
14. Doob, J. L. "Stochastic Processes." Wiley, New York, 1953.
15. de Finneti, B. Sulle funzioni a incremento aleatorio, *Rend. Acaad. Naz. Lincei, Cl. Sci. Fiz. Math. Nat.* **10** (6), 163–168 (1929).
16. Gihman, I. I. On a theorem of Kolmogorov, *Nauch. Zap. Kiev, Un-ta, Mat. Sb.* **7**, 76–94 (1953).
17. Gihman, I. I., and Skorohod, A. V. "The Theory of Stochastic Processes," Vols. I–III. Springer-Verlag, Berlin and New York, 1974–1979.

18. Gnedenko, B. V., and Kolmogorov, A. N. "Limit Distributions for Sums of Independent Random Variables." Addison-Wesley, Reading, Massachusetts, 1954.
19. Halmos, P. R. "Measure Theorie." Van Nostrand-Reinhold, Princeton, New Jersey, 1962.
20. Ibragimov, I. A. Some limit theorems for stationary processes, *Teor. Prob. Appl.* **7**, 349–382 (1962).
21. Ibragimov, I. A. A central limit theorem for a class of dependent random variables, *Theor. Prob. Appl.* **8**, 83–89 (1963).
22. Kazmars, S., and Steinhaus, H. "Theorie der Ortogonalreihen." Chelsea, Bronx, New York, 1951.
23. Kelley, J. L. "General Topology." Van Nostrand-Reinhold, New York, 1961.
24. Kolmogorov, A. N. Grundbegriffe der wahrscheinlichkeitsrechnung, *Erg. Math.* **2** (3), (1933). Springer Verlag, Berlin and New York.
25. Kolmogorov, A. N. On Skorokod convergence, *Theor. Prob. Appl.* **1**, 215–222 (1956).
26. Mc Kean, Jr., H. P. "Stochastic Integrals." Academic Press. New York, 1969.
27. Lévy, P. "Théorie de l'Addition des Variables Aléatoires." Gauthier-Villars, Paris, 1937.
28. Le Cam, L. Convergence in distribution of stochastic processes, *Univ. California Publ. Statist.* **2** (11), 207–236.
29. Liapounov, A. Nouvelle forme du théorème sur la limit de probabilité, *Mem. Acad. Sc. St. Pétersbourg* **12** (5), 1–24 (1901).
30. Lindeberg, J. W. Eine neue herleitung des exponalgesetzes in der wahrscheinlichkeitsrechnung, *Math. Z.* **15**, 211–225 (1922).
31. Loève, M. "Probability Theory." Van Nostrand-Reinhold, Princeton, New Jersey, 1963.
32. Parthasarathy, K. R. "Probability Measures on Metric Spaces." Academic Press, New York, 1967.
33. Prokorov, Yu. V.: Probability distributions in functional spaces, *Uspehi, Mat. Nauk (N.S.)* **55**, 167 (1953).
34. Prokorov, Yu. V. Convergence of random processes and limit theorems in probability theory, *Theor. Prob. Appl.* **1**, 157–214.
35. Sazonov, V. V. On characteristic functionals, *Theor. Prob. Appl.* **3**, 201–205 (1958).
36. Skorokhod, A. V. Limit theorems for stochastic processes with independent increments, *Theor. Prob. Appl.* **2**, 138–171 (1957).
37. Varadan, S. R. S. Limit theorems for sums of random variables with values in a Hilbert space, *Sankhyā Ser A*, **24**, 213–238 (1962).
38. Varadarajan, V. S. A useful convergence theorem, *Sankhyā* **20**, Ser A 221–222 (1958).
39. Varadarajan, V. S. Measures on topological spaces, *Mat. Sb.* **55**, 35–100 (1961).
40. Wiener, N. Differential space, *J. Phys. Mass. Inst. Tech.* **2**, 131–174 (1923).
41. Yosida, K. "Functional Analysis." Springer-Verlag, Göttingen, 1965.

# INDEX

# Probability and Mathematical Statistics

*A Series of Monographs and Textbooks*



---

Thomas Ferguson. Mathematical Statistics: A Decision Theoretic Approach. 1967

Howard Tucker. A Graduate Course in Probability. 1967

K. R. Parthasarathy. Probability Measures on Metric Spaces. 1967

P. Révész. The Laws of Large Numbers. 1968

H. P. McKean, Jr. Stochastic Integrals. 1969

B. V. Gnedenko, Yu. K. Belyayev, and A. D. Solovyev. Mathematical Methods of Reliability Theory. 1969

Demetrios A. Kappos. Probability Algebras and Stochastic Spaces. 1969

Ivan N. Pesin. Classical and Modern Integration Theories. 1970

S. Vajda. Probabilistic Programming. 1972

Sheldon M. Ross. Introduction to Probability Models. 1972

Robert B. Ash. Real Analysis and Probability. 1972

V. V. Fedorov. Theory of Optimal Experiments. 1972

K. V. Mardia. Statistics of Directional Data. 1972

H. Dym and H. P. McKean. Fourier Series and Integrals. 1972

Tatsuo Kawata. Fourier Analysis in Probability Theory. 1972

Fritz Oberhettinger. Fourier Transforms of Distributions and Their Inverses: A Collection of Tables. 1973

Paul Erdös and Joel Spencer. Probabilistic Methods in Combinatorics. 1973

K. Sarkadi and I. Vincze. Mathematical Methods of Statistical Quality Control. 1973

Michael R. Anderberg. Cluster Analysis for Applications. 1973

W. Hengartner and R. Theodorescu. Concentration Functions. 1973

Kai Lai Chung. A Course in Probability Theory, Second Edition. 1974

L. H. Koopmans. The Spectral Analysis of Time Series. 1974

L. E. Maistrov. Probability Theory: A Historical Sketch. 1974

William F. Stout. Almost Sure Convergence. 1974

E. J. McShane. Stochastic Calculus and Stochastic Models. 1974

Robert B. Ash and Melvin F. Gardner. Topics in Stochastic Processes. 1975

Avner Friedman, Stochastic Differential Equations and Applications, Volume 1, 1975; Volume 2. 1975

Roger Cuppens. Decomposition of Multivariate Probabilities. 1975

Eugene Lukacs. Stochastic Convergence, Second Edition. 1975

H. Dym and H. P. McKean. Gaussian Processes, Function Theory, and the Inverse Spectral Problem. 1976

N. C. Giri. Multivariate Statistical Inference. 1977

Lloyd Fisher and John McDonald. Fixed Effects Analysis of Variance. 1978

Sidney C. Port and Charles J. Stone. Brownian Motion and Classical Potential Theory. 1978

Konrad Jacobs. Measure and Integral. 1978

K. V. Mardia, J. T. Kent, and J. M. Biddy. Multivariate Analysis. 1979

Sri Gopal Mohanty. Lattice Path Counting and Applications. 1979

Y. L. Tong. Probability Inequalities in Multivariate Distributions. 1980

Michel Metivier and J. Pellaumail. Stochastic Integration. 1980

M. B. Priestly, Spectral Analysis and Time Series. 1980

Ishwar V. Basawa and B. L. S. Prakasa Rao, Statistical Inference for Stochastic Processes. 1980

M. Csörgö and P. Révész. Strong Approximations in Probability and Statistics. 1980

Sheldon Ross. Introduction to Probability Models, Second Edition. 1980

P. Hall and C. C. Heyde. Martingale Limit Theory and Its Application. 1980

Imre Csiszár and János Körner, Information Theory: Coding Theorems for Discrete Memoryless Systems. 1981

A. Hald. Statistical Theory of Sampling Inspection by Attributes. 1981

H. Bauer. Probability Theory and Elements of Measure Theory. 1981

M. M. Rao. Foundations of Stochastic Analysis. 1981

Jean-Rene Barra. Mathematical Basis of Statistics. Translation and Edited by L. Herbach. 1981

Harald Bergström. Weak Convergence of Measures. 1981